Structure-Activity Relationship Studies in Drug Development by NMR Spectroscopy
(Volume 1)

Editor

Atta-ur-Rahman, FRS

Kings College
University of Cambridge
Cambridge
UK

&

Iqbal Choudhary

H.E.J. Research Institute of Chemistry
University of Karachi, Karachi 75270
Pakistan

Structure-Activity Relationship Studies in Drug Development by NMR Spectroscopy

Volume 1

Contents

PREFACE

Drug discovery is still a process which involves screening of a large number of chemical compounds against a defined biological target. However, this approach is now being gradually replaced by more rational and knowledge-based approaches, such as *in silico* screening and target based drug development. Increase in the cost of production of NME (New Molecular Entities) and productivity crises in pharmaceutical industry is catalyzing this change.

While NMR is generally valued as a tool for structure determination and more recently as a non-invasive diagnostic technique, its potential in drug discovery is based on its capacity to map molecular interactions at the atomic level. Chemical shifts, cross relaxation, and exchange of protons are among the NMR parameters which are highly sensitive to exact environment of the atom, and therefore yield information about whether a small molecule (candidate compound) binds to a target protein (receptor) or to other macromolecules. These NMR parameters are also used to exactly map the part of the macromolecular target to which the ligand is bound. As a result, NMR is now extensively used as an efficient tool in the drug (ligand) discovery and optimization process or for the assessment of target drug ability.

Spectacular advances in the field of use of NMR spectroscopy in drug discovery and development have been triggered by a greater understanding of the disease process at the molecular level. Recent advances in NMR hardware and methodology have provided a fresh impetus in the area of pharmaceutical innovation and productivity.

The book entitled, "*Structure – Activity Relationship Studies in Drug Development by NMR Spectroscopy*" is aimed to present recent cutting edge knowledge, practices and experiences in this important field. This book will hopefully fulfill an important need of the scientific community by providing comprehensive reviews on various topics written by leading experts in the field.

The book comprises seven scholarly written articles on this multifaceted technique. Mailavaram and Devarakonda skillfully summarize the broad approaches of NMR-based drug discovery in their introductory review. Park and Kim in their review focus on the use of NMR spectroscopic techniques in the study of drug delivery, release and other pharmaceutic parameters. Roberto and Marina provide a chronological account of fragment-based drug discovery by using NMR-based screening methods. Specific utilization of NMR-based techniques in the discovery and development of anticancer drugs is reviewed by Solomon and Lee. Ramalho *et al.* review another fascinating application of NMR in understanding the behavior and radio sensitizing properties of bioreductive drugs. Zoppetti comprehensively reviews the application of NMR spectroscopy in the study of host-guest cyclodextrin complexes, particularly their use as drug delivery vehicles. A concise review is contributed by Nhat and Hong on recent advances in the field of application of NMR in structure-activity relationship study of nanostructure drugs. These contributions of outstanding group of experts make this a very useful treatise of highly readable articles. Our felicitations and gratitude goes to them.

We extend our warmest thanks to staff of Bentham Science Publisher, particularly, to Mr. Mahmood Alam (Director), Ms. Taqdees Malik (Editorial Assistant) and Ms. Sadaf Idrees Khan (Composer) for undertaking the important task of closely coordinating with the contributors and preparing the content lists, index, etc.

Prof. Dr. Atta-ur-Rahman, *FRS*
Honorary Life Fellow
Kings College
University of Cambridge
Cambridge
UK

Prof. M. Iqbal Choudhary
International Center for Chemical and Biological Sciences
H.E.J. Research Institute of Chemistry
University of Karachi
Karachi
Pakistan

Contributors

Raghu Prasad Mailavaram	Department of Pharmaceutical Chemistry, Shri Vishnu College of Pharmacy [Affiliated to Andhra University, Visakhapatnam], Padmabhushan Dr. B. V Raju Foundation and Sri Vishnu Educational Society, Vishnupur, Bhimavaram, West Godavari District, Andhra Pradesh, 534202, India
Murty Devarakonda	Department of Pharmaceutical Chemistry, Shri Vishnu College of Pharmacy [Affiliated to Andhra University, Visakhapatnam], Padmabhushan Dr. B. V Raju Foundation and Sri Vishnu Educational Society, Vishnupur, Bhimavaram, West Godavari District, Andhra Pradesh, 534202, India
Soo-Jin Park	Department of Chemistry, Inha University, 253 Yonghyun-dong, Nam-gu Incheon 402-751, Korea
Ki-Seok Kim	Department of Chemistry, Inha University, 253 Yonghyun-dong, Nam-gu Incheon 402-751, Korea
Consonni Roberto	Institute for the Study of Macromolecules (ISMAC), NMR Laboratory, National Council of Research, Milan, Italy
Veronesi Marina	Italian Institute of Technology, Genoa, Italy
V. Raja Solomon	Tumour Biology Group, Northeastern Ontario Regional Cancer Program at the Sudbury Regional Hospital, 41 Ramsey Lake Road, Sudbury, Ontario P3E 5J1, Canada Department of Biology, Laurentian University, 935 Ramsey Lake Road, Sudbury, Ontario P3E 2C6, Canada
Hoyun Lee	Tumour Biology Group, Northeastern Ontario Regional Cancer Program at the Sudbury Regional Hospital, 41 Ramsey Lake Road, Sudbury, Ontario P3E 5J1, Canada Department of Biology, Laurentian University, 935 Ramsey Lake Road, Sudbury, Ontario P3E 2C6, Canada Department of Medical Sciences, the Northern Ontario School of Medicine, 935 Ramsey Lake Road, Sudbury, Ontario P3E 2C6, Canada
Teodorico C. Ramalho	Universidade Federal de Lavras, Campus Universitário – UFLA, Dept. de Química, 37200-000, Lavras-MG - Brazil
Elaine F. F. da Cunha	Universidade Federal de Lavras, Campus Universitário – UFLA, Dept. de Química, 37200-000, Lavras-MG - Brazil
Marcus V. J. Rocha	Universidade Federal de Lavras, Campus Universitário – UFLA, Dept. de Química, 37200-000, Lavras-MG - Brazil

Giorgio Zoppetti   R&D Pharma Department, IBSA Institut Biochimique SA –Via al Ponte 13-
                            6903, Lugano 3 (CH), Switzerland

Hoang Nam Nhat   Department of Technical Physics and Nanotechnology, UET, Vietnam
                            National University Hanoi, 144 Xuan Thuy, Cau Giay, Hanoi, Vietnam

Tran Thi Hong   Department of Technical Physics and Nanotechnology, UET, Vietnam
                            National University Hanoi, 144 Xuan Thuy, Cau Giay, Hanoi, Vietnam

NMR-A Gate Way to Drug Discovery

Raghu Prasad Mailavaram[*] and Murty Devarakonda

Department of Pharmaceutical Chemistry, Shri Vishnu College of Pharmacy [Affiliated to Andhra University, Visakhapatnam], Padmabhushan Dr. B. V Raju Foundation and Sri Vishnu Educational Society, Vishnupur, Bhimavaram, West Godavari District, Andhra Pradesh, 534202-India

Abstract: The aim of pharmaceutical research in the present scenario is to modulate the targets with the intention of evoking therapeutic benefits. The advent of DNA recombinant technology has facilitated the preparation of almost any type of target in the purest form and in required amounts. In parallel, combinatorial chemistry and high throughput organic synthetic approaches have dramatically expanded the number of compounds that can be evaluated for biological activity. Both these advances have paved the way to develop high throughput screening systems for evaluation of millions of compounds against these targets in a rapid and efficient manner. However, overall return on investment in these technologies has been meager and disappointing. Thus an alternate approach which has the ability to potently and specifically modulate protein targets with small organic molecules is a need of the hour. All these facts have contributed to the emergence of multidimensional NMR as a gate way to drug design and development.

Fragment based drug design is the concept, where ligand binding site is dissected into different regions and the binding of diverse compounds in individual sub sites are analyzed separately. Fragments with fairly good affinity at all sub sites are collected and are interlinked in various possible manners, intuitively to generate a new set of compounds. The affinity of these compounds will be the sum of the affinity of individual fragments with additional gains due to entropy. In this approach, ligand-target interaction is measured and characterized with at most accuracy and precision as it caters valuable information in the areas like quantitative determination of binding affinities of potential ligands, ligand binding site determination on protein, conformation information of ligands and proteins and local flexibilities in the structure for complex and free components. In addition NMR helps to determine ligand specificity and selectivity towards different targets and also provide valuable hints to predict possible side effects and toxicological outcomes at various levels. Another advantage of these techniques includes attaining capability to eliminate the false positives. They are also useful in further optimization of the potential leads.

Drug receptor interaction is quantified based on various techniques through NOE, relaxation and diffusion editing, isotope editing/ filtering, shapes strategy, HSQC strategy, etc. Fragment linking and elaboration strategies are employed for linking the fragments together and design the new molecules. Enormous number of compounds with high affinity towards several targets from FKBP to Bcl-2 has been reported in the literature.

Based on the above facts, NMR can play a decisive role from in various stages of drug discovery i.e., from structural elucidation to identifying medicinally valid drug candidates. This article is an effort to throw a light on the basic principles involved in each and every technique of NMR, their advantages and limitations, along with its applicability.

Keywords: NMR spectroscopy, drug discovery by NMR, multidimensional NMR, NMR techniques.

INTRODUCTION

The search for drugs to cure the diseases and disorders affecting the performance and efficiency of a person is as old as mankind. Earlier, the drug discovery process involved the irrational *in vivo* screening of a relatively small number of organic compounds. In this *Systems-Centered or Holistic Approach* (Fig. **1**) there was a rigid priori assumption that any molecular targets involved in the observed phenotype were in fact both disease/disorder modifying and druggable. It has paved a way to number of serendipitous discovery of most successful drugs, including blockbuster, valproic acid for bipolar disorder. However, this strategy has severe limitations not only on the number of compounds that could be evaluated (limiting the

* Corresponding author: Tel: +91-9490720070; E-mail: raghumrp@mailcity.com

Atta-ur-Rahman / M. Iqbal Choudhary (Eds.)

chemical diversity that could be sampled), but also on properties of the compounds (examples: Bioavailability, solubility, potency, adverse effects etc.) that can otherwise be addressed with medicinal chemistry. It was often the case that the molecular target responsible for the observed phenotype remained unknown, hampering efforts to understand the biological mechanism of action of therapeutic agent under study. In fact the enzymes targeted by valproic acid, histone deacetylases have only been recently identified [1, 2].

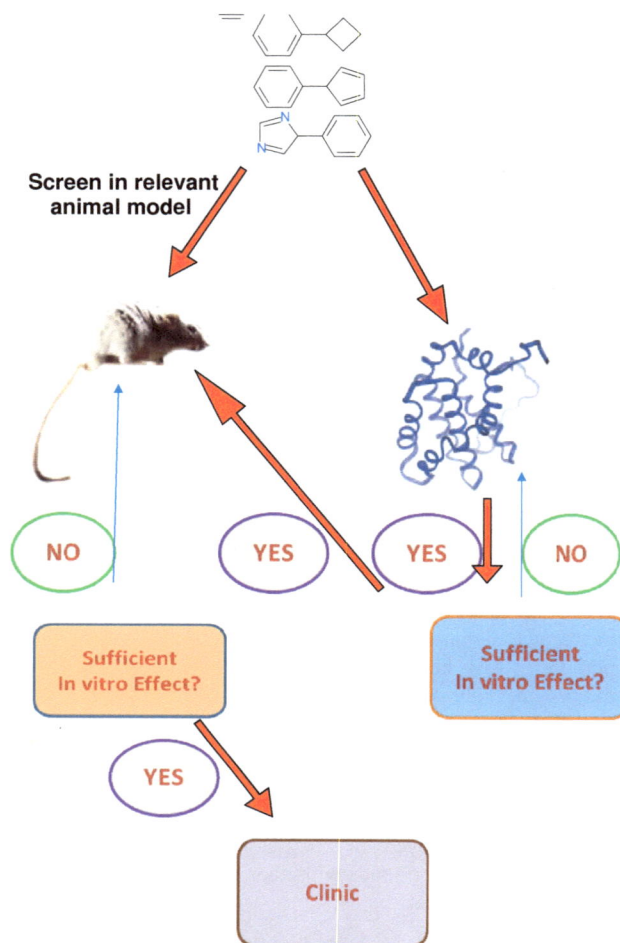

Fig. (1). System centric approach.

With the advent of rational approach in drug discovery, the efforts are put towards identifying the targets that if sufficiently modulated, will alter either the onset or progression of human disease and also to know its druggability [3] which is known as *Target-Centric or Reductionist Approach*. (Fig. **2**) The improvement in the DNA recombinant technology has facilitated the preparation of almost any type of target in the purest form and in required quantities [4]. In parallel, combinatorial chemistry and high throughput organic synthetic techniques have dramatically expanded the number of compounds that can be evaluated for biological activity [5]. Both these advances has paved the way for the development and standardization of high throughput screen systems for testing of millions of compounds prepared on these targets in a rapid and efficient manner [6]. Though these techniques has led to the discovery of several novel drugs including secenovir, a potent HIV protease inhibitor but overall returns on the investments in these technologies is totally discouraging [7,8].

In this reductionist drug discovery paradigm, after identification of a distinct molecular target, an HTS of an approximately 10^6 compound serves as an entry point to lead identification and optimization. In spite of that it is failing to produce high-quality clinical candidates. Two reasons were identified for this failure. First, the leads obtained lacked physicochemical properties such as solubility, lipophilicity, stability, oral bioavailability etc. for clinical use [9-11]. These drawbacks were over come to certain extent by introduction of high throughput ADME into drug discovery process which increased the chances of lead compound to emerge as a drug candidate [12, 13]. The second reason for failure is that it results in leads which are not worthy for further discovery which is rather difficult to overcome [14]. This may be due to the fact that chemical compound required for inhibition of particular target did not existed in the compound library itself perhaps this situation was acknowledged by many pharmaceutical companies and has resulted in expansion of both depth and breadth of chemical diversity contained in their corporate repositories [15, 16] or may be due to that protein cannot be targeted with these small molecules.

The reason for not getting targeted may be due to inadequate binding of the small molecules at distinct points of the active site of the protein (receptor or enzyme or hormone). It was taught of interest by many medicinal chemists to fragment the ligand-binding site into number of pockets and to identify the individual building blocks that bind to distinct sub sites within the pocket. Intuitive alignment of these building blocks can result in the molecules with good binding affinity towards the target. This was the basis of fragment based drug design which helped to increase the success rate of converting a lead molecule to an active drug candidate [17, 18]. For primary fragment screening biophysical techniques like surface phenomena

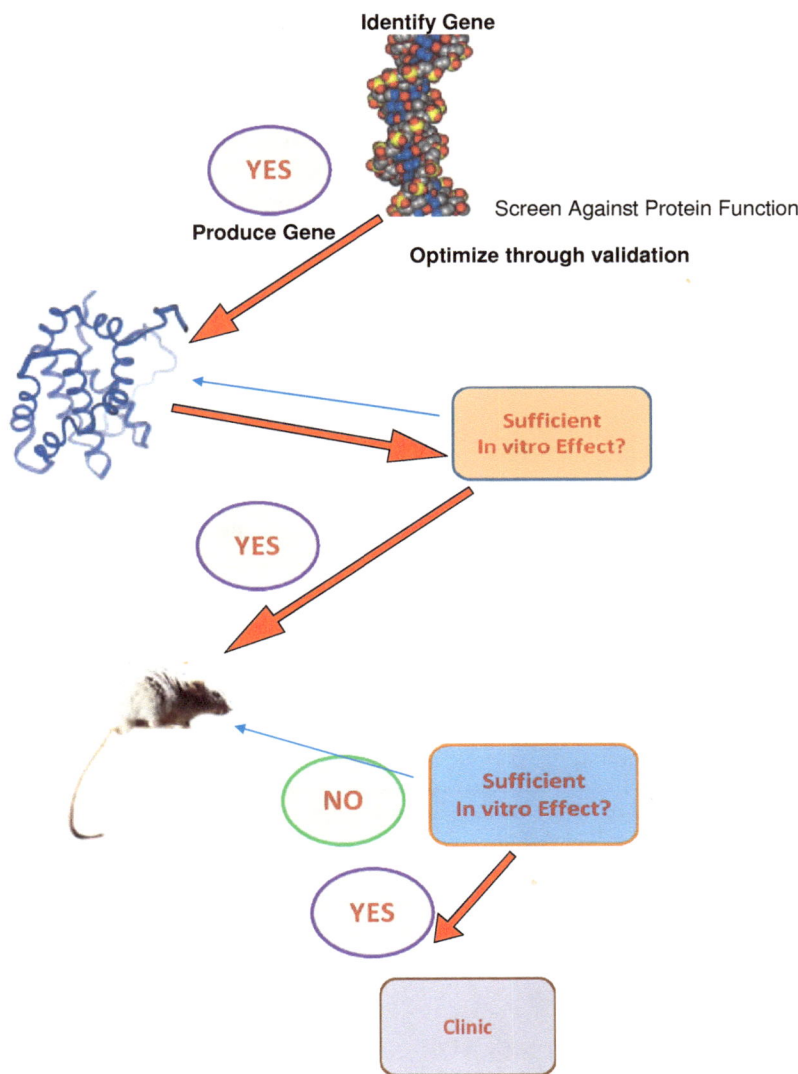

Fig. (2). Target centric approach.

resonance (SPR) [19, 20], X ray crystallography [21], and NMR have been utilized as effective tools. In X ray crystallography large amount of protein is required and results in numerous false positives whereas in SPR, immobilization of protein on ligand is a major drawback. NMR has emerged as an effective tool over others not only for characterizing the structure of the prepared compounds but also extended to know and quantify the interaction between building blocks and distinctive sites of fragmented pockets and alignment of the blocks. These facts contributed for the emergence of multidimensional NMR as a gateway to drug design and discovery.

TECHNIQUES IN NMR-BASED DRUG DESIGN

NMR is basically a technique used to identify and characterize the molecules based on chemical shifts of nuclei present in the compound. NMR methods have been introduced for quantitative analysis in order to determine the impurity profile of a drug, to characterize the composition of drug products, and to investigate metabolites of drugs in body fluids. For pharmaceutical technologists, solid state measurements can provide information about polymorphism of drug powders, conformation of drugs in tablets etc. Micro-imaging can be used to study the dissolution of tablets, and whole-body imaging (Magnetic Resonance Imaging-MRI) is a powerful tool in clinical diagnostics.

Advancements in NMR in the last decade lead to the advent of multidimensional NMR which also opened up gates for solving the structure of proteins, peptides, and drug molecules of smaller molecular weight. In recent years it has been developed as an alternate tool to X-ray crystallography for the structural determination of small proteins which are impossible or difficult to crystallize. Additionally it offers many possibilities to study the physical flexibilities of proteins in its free form and when any endogenous substances or drug molecules interacts with them, which would in turn helps to understand the structure activity relationship, function, mechanism of action and binding specificities in biological system. Due to this uniqueness, information on the dynamic behavior of the systems under study can be gained. These aspects are important in leading us from the often merely static definition of "structure" towards a much more realistic view of biologically active molecules and their interactions. In the following section different approaches to study the intermolecular interactions by NMR have been presented in brief.

The available techniques may be divided into two different classes, ligand-detected and protein-detected methods (Table 1). The protein-detected methods yield more information on the ligand-protein interaction, but the ligand-detected methods are more widely used in the pharmaceutical and biotech industries since they can be applied to a much wider range of target proteins and require less upfront resources.

Table 1. Examples of NMR Techniques for Monitoring Protein - Ligand Interactions

NMR technique or observable	Nonbinding ligand	Protein	Bound ligand	Representative references
Saturation transfer difference	Unsaturated	Saturated	Saturated	[29]
Water-LOGSY	Positive NOE	Negative NOE	Negative NOE	[43, 44]
Line width	Narrow	Broad	Broadened	[198]
Transverse relaxation/selective longitudinal relaxation	Slow	Rapid	Increased	[51, 54]
Translational diffusion	Rapid	Slow	Slowed	[51, 199]
Transferred NOE	Weak positive NOE Slow relaxation	Strong negative NOE	Weak negative NOE	[24, 26]
Paramagnetic spin label	Slow relaxation	-	Rapid relaxation	[55, 56]
Protein Chemical shift perturbation	-	Unshifted	Shifted	[32]

NUCLEAR OVERHAUSER ENHANCEMENT - NOE

In ^{13}C NMR spectra, decoupling of the spins of the nuclei (^{13}C, ^{1}H) is accomplished by simultaneously irradiating the protons attached to carbons using a second tunable radiofrequency generator, the decoupler. Irradiation causes the proton to become saturated and they undergo rapid upward and downward transitions, among all their possible spin states. These rapid transitions decouple any spin-spin interactions between the hydrogens and the ^{13}C nuclei. In effect, all spin interactions are averaged to zero by the rapid changes. The carbon nucleus senses only one average spin state for the attached hydrogens rather than two or more distinct spin states. This not only simplifies the spectra by removing the coupling but also increases the intensities of the carbon peaks. This effect is called as Heteronuclear overhauser effect. The same effect also occurs in ^{1}H - ^{1}H spin decoupling but the effect is small and is called as Homonuclear overhauser effect.

The through-space nuclear overhauser effect has been utilized to determine the ^{1}H-^{1}H proximity within or between molecules. A proton that is close in space to irradiated proton is affected by the NOE whether or not it is coupled to the irradiated proton. Polarization - that is a change in population of the energy levels - by the weak irradiation results, through space, in an increase in the population of the higher energy level by the nearby non-irradiated proton. This excess population undergoes T_1 relaxation to a lower energy level, thereby increasing the signal intensity of the nearby protons The NOE is seen operating through the space when the nuclei are at the distance of less than 4.0 Å. This has been used as a powerful tool to know the interaction between the ligands and protein [22-27].

SATURATION TRANSFER DIFFERENCE [STD]

The origins of the STD experiment [28] can be traced to the spin-saturation transfer experiment or Forsén-Hoffman experiment from the 1960s [29]. STD experiments are based on NOE driven polarization transfer from the receptor spins to the ligand spins in the transient complex. The resonance of the target proteins are targeted selectively by an irradiation frequency of < -1 to > 8 ppm, where normally no proton resonances of the small ligands are bound. The polarization is distributed via spin diffusion to all spins of the protein and then to the bound ligand. In large complexes, a negative NOE is observed i.e., the signal intensity decreased. When bound ligand exchanges rapidly with free ligand the intensity change is transferred to the free ligand and can be observed as a reduction in its signal. It can also be incorporated in 2D-NMR like 2D-STD-TOCSY [30].

In the STD experiment, a subset of the protein ^{1}H resonances are saturated by means of a train of frequency-selective radiofrequency pulses applied to a narrow spectral region devoid of ligand resonances. The saturation is transferred by spin diffusion (^{1}H-^{1}H cross relaxation pathways) to the rest of the protein, a process that becomes more efficient with increasing rotational correlation time of the protein. As a consequence, the experiment will become more sensitive with increasing molecular weight of the target protein. The saturation of the protein will then be transferred, via intermolecular ^{1}H-^{1}H cross-relaxation, to weakly binding ligands at the ligand-protein interface. After the ligand has dissociated, the long spin-lattice relaxation times (T_1) of the free ligand protons ensure that it is possible to detect the transferred saturation as an attenuation of the ligand signals. During the time of the saturation, more unsaturated ligand molecules will bind and dissociate, leading to an increasing population of saturated ligands. The saturated spectrum is then subtracted from a spectrum obtained under nonsaturating conditions to obtain an STD spectrum showing only the signals from compounds binding reversibly to the target protein. This simplicity of the resulting spectrum is a very attractive feature of the STD experiment. In practice, the pulse sequence is written in such a way that the subtraction is performed automatically in every other scan, i.e., the individual spectra are never observed. The STD experiment allows for very high ligand:protein ratios to be used, often as high as 50 or 100. Thus, the protein concentration can be kept very low, in fragment screening it is possible to use less than 1 μM target protein in favorable cases. The protein signals are not visible in the STD spectrum due to the low protein concentration and/or the presence of a spinlock in the pulse sequence. The sensitivity of the method depends critically on how efficiently the protein resonances have been saturated [31]. The duration of the saturating selective pulse train is typically 1-4 s, where the longer saturation times in this interval can be used for smaller proteins. The upper limit of the saturation time for an efficient saturation transfer is determined by the T_1 of the ligand in the bound state. A spectral region suitable for efficient saturation is the protein methyl groups which usually has a maximum intensity close to 0.7 ppm. In order to avoid direct saturation of ligands, the saturation is usually applied further up field where most target proteins have up field shifted methyl resonances. The STD experiment is preferably performed in samples containing 100 % D_2O. This makes optimal water suppression less crucial and minimizes exchange-mediated saturation leakage via dipole-dipole interactions between saturated protein protons and hydration water molecules [32, 33]. The intensity of the STD signals contains information on the ligand affinity and it should in principle be possible to rank fragment hits directly with respect to affinity. There are, however, several caveats to be aware of and caution should be exercised when interpreting data. First, the individual STD peak heights should be normalized to the corresponding peak heights in a reference spectrum. This could either be the spectrum obtained under nonsaturating conditions (collected as part of the STD experiment) or a T_2-filtered ^{1}H 1D spectrum of the same sample, using the same length of the spinlock and interscan delay as in the STD experiment. The parameters in the STD experiments, such as saturation frequency, saturation time and relaxation delay, should be identical for all samples. For the normalized STD response to be directly related to the relative affinity of different ligands to the target protein, there are a number of assumptions that should be fulfilled: the contribution to the STD signals should come from binding to a single site only, the exchange must be fast (should be fulfilled for weakly binding fragments) and the T_1 values for the proton signals to be compared must be similar. The last condition is important since the STD responses are highly dependent on the T_1 value for the observed proton [33, 34] so that, for example, a smaller T_1 value would result in a smaller STD response.

The variation of the normalized STD responses for different protons within a ligand also contains information on how the ligand binds to the target protein [35, 36]. The magnitude of a normalized STD response is related to the corresponding ligand proton proximity to target protons, a consequence of the distance dependence of intermolecular ^1H-^1H cross relaxation. Thus, a stronger STD response implies closer intermolecular contact, which can be interpreted as information on which part of the ligand is responsible for most of the binding energy. To compare directly the normalized STD responses, the same assumptions as described above should be fulfilled. This 'ligand epitope mapping' approach is best suited for weakly binding ligands and/or the use of sufficiently short saturation times, in order to ascertain that the magnetizations of the ligand protons are not equalized due to spin diffusion during the lifetime of the ligand-target complex. This approach has been substantially extended to determine indirectly the conformation of a ligand when bound to a target protein with a known structure. In this method, called SOS-NMR [37], STD is applied to a ligand complexed to a series of perpetuated target protein samples, each with different specific amino acid types protonated. The relative STD intensities of the ligand peaks from the different samples contain quantitative information on which protons on the ligand are in the vicinity of which type of amino acid in the target protein. With a known target protein structure, the structure of the ligand-target complex can be deduced from this information provided that experiments have been performed on a sufficient number of differently labeled protein samples to define the binding site uniquely. An analysis of 272 unique crystal structures of ligand-protein complexes showed that 3-9 differently labeled protein samples would be enough to identify unambiguously the ligand binding site of more than 90 % of the ligand-protein complexes. Obviously, SOS-NMR is a very resource-intensive method, but could prove highly valuable in cases where it has not been possible to obtain any structural information on the ligand-target complex either by X-ray crystallography or by 'traditional' NMR methods.

Finally, STD has been demonstrated to be applicable for the detection of ligand binding in very demanding systems such as a virus [38] and an integral membrane protein either reconstituted in liposomes [39] or on living cells [40].

WATER LIGAND OBSERVE VIA GRADIENT SPECTROSCOPY [WATER-LOGSY]

The Water-LOGSY experiment [41, 42] relies on water-mediated magnetization transfer to compounds that bind to the target protein. The most favored version of this experiment [44] is essentially an NOE experiment starting with selective inversion of the bulk water magnetization followed by a long mixing time (up to several seconds). The advantage of this method is that in a single experiment many different compounds can be tested and binding compounds can be directly identified if corresponding resonances are known. It is useful in identifying novel scaffolds of micromolar affinity which can be further optimized by computational methods. Titration Water-LOGSY in combination with competition binding is used for the evaluation of the dissociation of binding constant after correction. The high sensitivity of the technique in combination with the easy deconvolution of the mixtures for the identification of the active components significantly reduces the amount of material needed for NMR screening process [43, 45]. The inverted bulk water magnetization is transferred to the target protein and binding compounds via several possible magnetization transfer pathways [46]: (i) direct ^1H-^1H cross-relaxation between compounds and tightly bound water molecules at the binding site, (ii) chemical exchange of inverted bulk water with protein hydroxyl and amine groups at the binding site, which in turn will transfer the inverted magnetization to the bound compound protons, and (iii) chemically exchanging hydroxyl and amine groups on the protein that are not situated in the binding site may contribute via spin diffusion through the protein. During the mixing time, the ^1H-^1H cross-relaxation will give rise to NOEs with opposite signs depending on whether the magnetization transfer takes place in the bound state (slow tumbling, long rotational correlation time) or in solution (fast tumbling, short rotational correlation time). Reversibly binding compounds will experience magnetization transfer in both the bound and free states, whereas nonbinding compounds will only experience magnetization transfer in the free state. For the binding compounds, the contribution from the bound state will dominate provided that the ligand excess is not extremely high and/or the interaction very weak. Consequently, the resulting Water-LOGSY spectrum will consist of positive signals from compounds binding to the target protein and negative signals from compounds not binding to the protein. To detect also very weak binders, Water-LOGSY spectra should be collected in both the presence and absence of target protein. The signals from a very weak binder may be negative, but less negative compared with when the target protein is absent. As with the STD experiment, efficient detection of reversibly binding compounds is based on the long T_1 values of the free ligand protons and the use of an excess of ligand leading to a build-up of the population of ligands that has experienced magnetization transfer during the mixing time. Both STD and Water-LOGSY are very popular fragment screening techniques and share many features. However, STD is an incoherent technique, i.e. it relies on transfer of saturation (incoherence), whereas Water-LOGSY takes advantage of the same physical processes but in a coherent way, i.e. polarization transfer. Both techniques use a relatively high ligand: protein ratio and the sensitivity with respect to both protein consumption and detectable dissociation constant range is comparable. However, for targets with a low proton density, e.g. RNA, the use of Water-LOGSY is clearly advantageous [47]. Targets with a low proton density suffer from inefficient spin diffusion, leading to low STD sensitivity. Water-LOGSY, on the other hand, is much less dependent on spin diffusion efficiency since the method mainly relies on magnetization transfer from the water molecules surrounding the target.

Nevertheless, the low spin diffusion efficiency in DNA can be exploited in STD experiments to obtain information on the binding site of ligands. By saturating at different frequencies (and therefore different DNA regions) and observing the relative STD responses, base-pair intercalators have been distinguished from minor groove binders [48]. For the best sensitivity, the Water-LOGSY experiment should be performed in H_2O with only small amounts of D_2O for field-frequency locking. The intense H_2O signal, however, raises the need for efficient water suppression schemes and is a source of potential spectral artifacts. Since signals from nonbinding compounds appear in the spectrum as negative peaks, the spectrum will be more complex than the corresponding STD spectrum.

LINE BROADENING AND RELAXATION FILTER EXPERIMENTS

The spin-spin relaxation times (T_2) of protons in a slowly tumbling entity such as a target protein are much shorter than those of protons from a faster tumbling small organic compound. Since the ^1H signal line width at half-height is proportional to $(\pi T_2)-1$, a simple approach is to measure the line widths of the ligand ^1H signals in the absence and presence of a target protein. The method requires that the magnet field homogeneity is identical for the samples. This is achieved by employing a gradient shimming step for each sample and by checking that the shimming quality is identical for all samples by observing the linewidth of an inert compound present in all samples, e.g. the calibration standard DSS (2, 2- Dimethyl-2-silapentane-5-sulfonic acid). Upon binding of the ligand to the target protein, the tumbling rate of the small molecule is decreased. This results in a decreased T_2, which is manifested by both a broadening and a concomitant decrease of ligand ^1H peak heights. Further, if fast exchange kinetics are assumed and possible exchange contributions to the linewidths are small or negligible, it is also straightforward to estimate the binding affinity [49]. However, since the primary NMR screening is often performed on cocktails of fragments, spectral overlap both from other compounds and from the target protein may render accurate line shape analysis difficult. In order to detect weakly binding ligands reliably, the molar excess of ligand over target protein must be kept fairly low. Rather than measuring the ligand line widths, a more common setup is to apply a relaxation filter, i.e. a spinlock (e.g. a CPMG sequence) [50] directly after the 90° detection pulse and before the acquisition of the signal [51]. The relaxation filter experiments are performed on two samples that are identical except that the target protein is present in one and absent in the reference sample. The spectra collected on these two samples are compared for changes in NMR signal intensities of the small molecules. A long enough relaxation filter will eliminate the target protein signals. With the assumptions that the exchange contributions to the line widths are negligible and that the ligands bind with a similar on-rate (e.g. a diffusion-controlled on-rate) and to one site only, the signals for a higher affinity ligand will decrease more than for a lower affinity ligand with a given time length of the spinlock. Therefore, in general, it is possible to use the outcome of the relaxation-filtered experiment to rank ligands with respect to affinity [52, 53]. A convenient way to affinity rank ligands is to collect relaxation filtered spectra with different durations of the spinlock. The duration of the spinlock for a given target determines the detection cut-off and, therefore, when longer spinlock times are used, weaker binding ligands are detected. For example, simulations have suggested that for a relatively small target protein of MW [52, 53]15 kDa, a spinlock of 400 ms will eliminate the signals from ligands binding with an affinity of 500 μM or tighter for equimolar amounts of protein and ligand [52]. Here, it can be noted that it is often erroneously claimed that fast exchange is always a requirement for direct detection of binding by ligand-detected methods. That is not the case for the transverse relaxation filter technique provided that no molar excess of ligand over target protein is used. In the extreme case of a covalent binder, for example, all ligand molecules would then be bound to the target protein during the spinlock time and no NMR signal from the ligand would pass the relaxation filter. It is also possible to utilize the differences in spin-lattice relaxation (T_1) between a large protein and small organic molecule to detect binding by applying an inversion-recovery pulse sequence [54]. However, detection of binding is only possible for *selective* T_1 measurements, i.e. the inversion of fragment ^{1}H signals must be performed by the use of frequency-selective pulses. As a consequence, it is not practical to use this technique for primary fragment screening if the screened fragment signals are to be observed, since a separate inversion pulse would have to be designed for every fragment in the library. Instead, a weakly binding 'spy' molecule should be employed in competition experiments where only selectively inversed signals from the 'spy' molecule are repeatedly monitored [54].

PARAMAGNETIC SPIN LABELS

A paramagnetic spin label is an atom with an unpaired electron, which will enhance the relaxation of nearby protons due to the strong electron-proton dipole-dipole interaction. By covalently attaching a paramagnetic spin label on selected target protein side-chains, the transverse ^{1}H relaxation times of any fragments binding in the vicinity of the spin label will decrease dramatically. For detection, a transverse relaxation filter experiment is usually applied. This technique has been dubbed SLAPSTIC (spin labels attached to protein side-chains as a tool to identify interacting compounds) [55, 56]. The strong practical aspects of using NMR dipole-dipole interaction between an unpaired electron and a proton will dominate over all other relaxation mechanisms and is effective over much longer through-space distances than is typical for ^{1}H-^{1}H relaxation (up to 15-20 Å). Further, the strong transverse relaxation effect makes it possible to use low protein concentrations, on the order of a factor 10 lower than in a normal transverse relaxation filter experiment [57]. The main disadvantage of the method is the need to introduce the spin label at a suitable amino acid side-chain near the binding site of interest. There is a risk that the introduction of the spin label might alter the structure and/or the binding properties of the target protein. An attractive variant of the method is the use of second-site screening. A spin-label is then attached to a known ligand and used to detect fragments binding to a site adjacent to the binding site of the spin-labeled ligand. The quenching effect of the spin label on compounds is observed if, and only if, both the spin-labeled compound and the other compound bind at the same time and in the vicinity of each other. Here, it is important that the attachment of the spin label does not alter either the first or the second binding site.

FLUORINE RELAXATION FILTER

The ^{19}F nucleus exhibits a number of features that make it attractive to use for detection in NMR binding experiments [58]: (i) the ^{19}F transverse relaxation rate and chemical shift are very sensitive to changes in the microenvironment (e.g. upon ligand-target complex formation), (ii) ^{19}F occurs at 100% natural abundance and has a gyromagnetic ratio nearly as high as that for ^{1}H and (iii) the absence of ^{19}F in biological macromolecules, most organic compounds and buffer components results in very clean ^{19}F spectra. However, direct observation of ^{19}F in a primary fragment screen would require that all fragments contain at least one fluorine atom. In practice, therefore, ^{19}F detection in screening campaigns is applied in competition experiments where a fluorine containing 'spy' molecule is employed [59, 60]. The observed parameter of the spy molecule is typically the ^{19}F signal intensity after a spin-echo filter.

PROTEIN-DETECTED TECHNIQUES

Protein-detected techniques rely on detecting changes in the NMR observables of a protein upon exposure to a ligand. Protein-detected techniques require isotopically enriched samples, but have the benefit (provided that sequence-specific resonance assignments have been obtained) of being able to identify directly the binding epitope on the target, which also makes it possible to distinguish between specific and nonspecific binding events. It is also possible to assess whether any significant conformational changes of the target protein occur upon binding. Further, protein-detected methods do not rely on fast exchange to retrieve information from the bound state, making it possible to detect both low- and high-affinity hits. The major drawbacks of protein-detected methods are the limited applicability to typical pharmaceutical target proteins and the high demands on resources. Most therapeutically important target proteins have molecular weights above 30 kDa, and obtaining sequential resonance assignments for such proteins is a major task in terms of both resources and time, despite recent advances in data acquisition schemes. An indirect and more limited approach would be to assign only resonances of active site residues by the use of known substrates or ligands. There are, in any case, substantial demands on the protein production. Milligram amounts of isotopically labeled target protein (e.g. ^{13}C, ^{15}N, ^{2}H) are to be produced in a suitable expression host. The purified protein must be soluble, stable for at least a couple of weeks at room temperature and remain monodisperse at the high concentrations needed. Even if it were possible to obtain the sequence-specific resonance assignments for a target protein of 30-40 kDa, the overlap in the 2D spectrum would be considerable due to both severe line-broadening effects and the large number of peaks in the spectrum. Consequently, it would be very difficult to reliably re-assign the peaks that have moved upon ligand binding. Therefore, efforts have been dedicated to develop protein-detected methods that are more generally applicable to larger proteins.

TWO DIMENSIONAL NMR

1D protein spectra of proteins are far too complex for interpretation as most of the signals overlap heavily. By the introduction of additional spectral dimensions these spectra are simplified and some extra information is obtained.

The construction of a 2D experiment is simple: In addition to preparation and detection which are already known from 1D experiments the 2D experiment has an indirect evolution time t_1 and a mixing sequence.

This scheme can be viewed as: Do something with the nuclei (preparation), Let them precess freely (evolution), Do something else (mixing), Detect the result (detection, of course) (Fig. **3**).

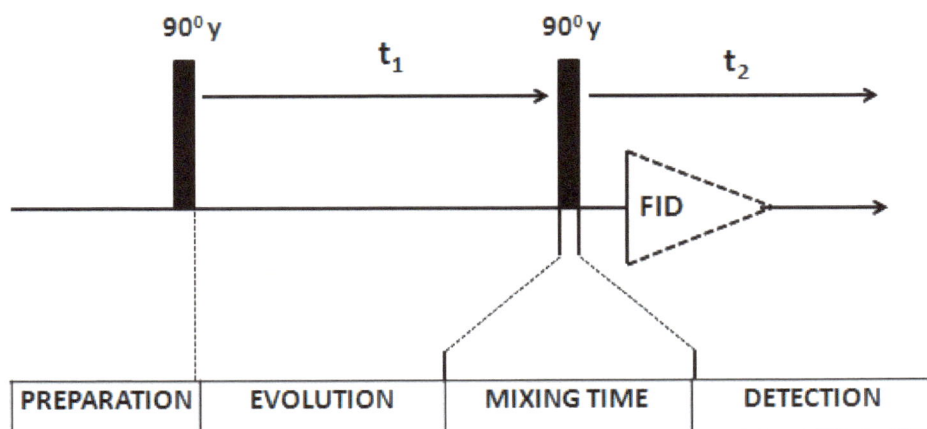

Fig. (3). Anatomy of 2D-NMR experiment.

After preparation the spins can precess freely for a given time t_1. During this time the magnetization is labelled with the chemical shift of the first nucleus. During the mixing time magnetization is then transferred from the first nucleus to a second one. Mixing sequences utilize two mechanisms for magnetization transfer: scalar coupling or dipolar interaction (NOE). Data are acquired at the end of the experiment (detection, often called direct evolution time); during this time the magnetization is labelled with the chemical shift of the second nucleus.

In one dimensional NMR, the signal is presented as a function of single parameter, usually the chemical shift whereas in two dimensional NMR there are two coordinate axes. Generally these axes also represent ranges of chemical shifts. The signals are presented as function of each of these chemical shift ranges. The data are plotted as a grid, one axis represents one chemical shift range, the second axis represents the second chemical shift range and the third dimension constitutes the intensity of observed signal (Fig. **4**).

A diagonal of signals (A and B) divides the spectrum in two equal halves. Symmetrical to this diagonal, there are more signals (X), called cross signals. The diagonal results from contributions of the magnetization that has not been changed by the mixing sequence (equal frequency in both dimensions) i.e. from contributions which remained on the same nucleus during both evolution times.

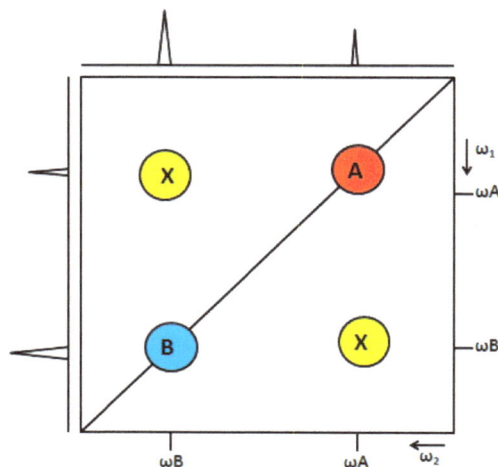

Fig. (4). 2D-NMR spectrum of ethanol.

The cross signals originate from nuclei that exchanged magnetization during the mixing time (frequencies of the first and second nucleus in each dimension, respectively). They indicate an interaction of these two nuclei. Therefore, the cross signals contain the really important information of 2D NMR spectra [61].

Most widely used two-dimensional spectroscopy are ^{1}H-^{1}H Correlation Spectroscopy also known as COSY and Heteronuclear Correlation spectroscopy also known as HETCOR. In a COSY experiment, the chemical shift range of the proton spectrum is plotted on both axes whereas in HETCOR experiment, the chemical shift range of the proton spectrum is plotted on one axis while the chemical shift range ^{13}C spectrum for the sample is plotted on the second axis. There are three 2D spectra which are widely used for the structure determination of proteins with a mass of up to 10 kD: 2D COSY, 2D TOCSY and 2D NOESY.

2D COSY: [CORRELATION SPECTROSCOPY]

In the COSY experiment, magnetization is transferred by scalar coupling. Protons that are more than three chemical bonds apart give no cross signal because the ^{4}J coupling constants are close to 0. Therefore, only signals of protons which are two or three bonds apart are visible in a COSY spectrum (red signals). The cross signals between H^N and H^{alpha} protons are of special importance because the *phi* torsion angle of the protein backbone can be derived from the ^{3}J coupling constant between them (Fig. **5**).

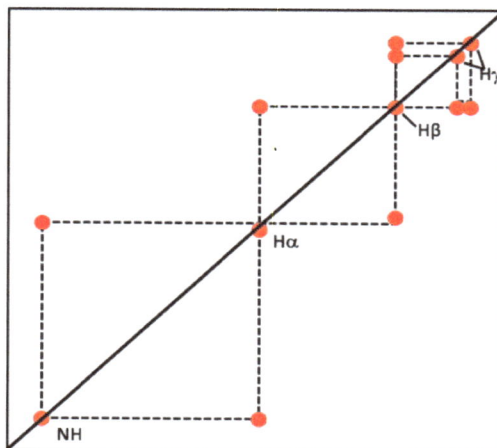

Fig. (5). 2D-Correlation spectroscopy.

2D TOCSY: (TOTAL CORRELATION SPECTROSCOPY)

In the TOCSY experiment, magnetization is dispersed over a complete spin system of an amino acid by successive scalar coupling. The TOCSY experiment correlates all protons of a spin system. Therefore, not only the red signals are visible (which also appear in a COSY spectrum) but also additional signals (green) which originate from the interaction of all protons of a spin system that are not directly connected via three chemical bonds (Fig. **6**).

Thus a characteristic pattern of signals results for each amino acid from which the amino acid can be identified. However, some amino acids have identical spin systems and therefore identical signal patterns. They are: cysteine, aspartic acid, phenylalanine, histidine, asparagine, tryptophane and tyrosine ('AMX systems') on the one hand and glutamic acid, glutamine and methionine ['AM(PT)X systems'] on the other hand.

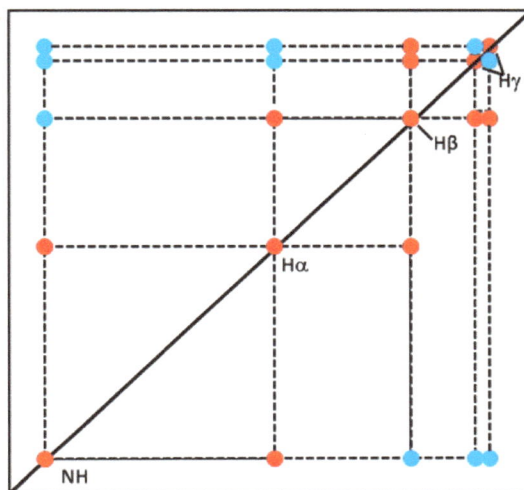

Fig. (6). 2D-Total correlation spectroscopy.

2D NOESY

The NOESY experiment is crucial for the determination of protein structure. It uses the dipolar interaction of spins (the nuclear Overhauser effect, NOE) for correlation of protons. The intensity of the NOE is in first approximation propotional to $1/r^6$, with r being the distance between the protons: The correlation between two protons depends on the distance between them, but normally a signal is only

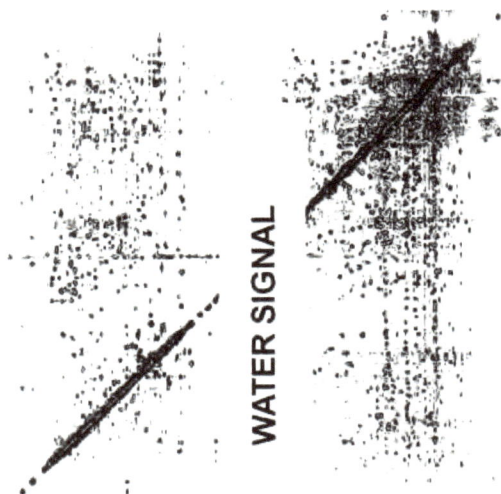

Fig. (7). 2D-NOESY.

observed if their distance is smaller than 5 Å. The NOESY experiment correlates all protons which are close enough. It also correlates protons which are distant in the amino acid sequence but close in space due to tertiary structure. This is the most important information for the determination of protein structures (Fig. 7).

THE HSQC EXPERIMENT

The most important inverse NMR experiment is the HSQC (heteronuclear single quantum correlation) the pulse sequence of which is shown below (Fig. **8**). It correlates the nitrogen atom of an NH_x group with the directly attached proton. Each signal in a HSQC spectrum represents a proton that is bound to a nitrogen atom.

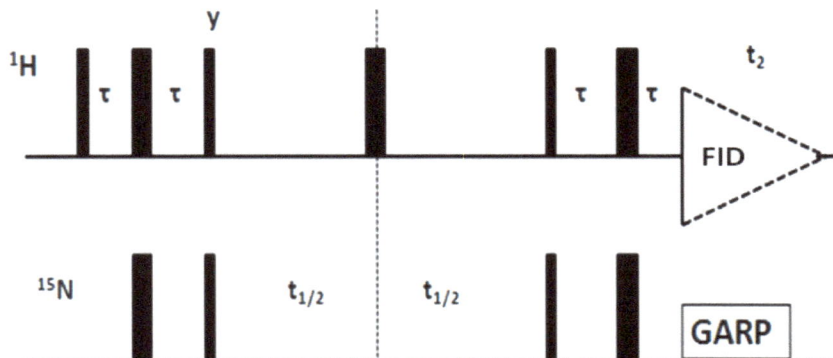

Fig. (8). Anatomy of HSQC experiment.

The spectrum contains the signals of the H^N protons in the protein backbone. Since there is only one backbone H^N per amino acid, each HSQC signal represents one single amino acid. The HSQC also contains signals from the NH_2 groups of the side chains of Asn and Gln and of the aromatic H^N protons of Trp and His. A HSQC has no diagonal like a homonuclear spectrum, because different nuclei are observed during t_1 and t_2. An analogous experiment (^{13}C-HSQC) can be performed for ^{13}C and ^1H (Fig. **9**).

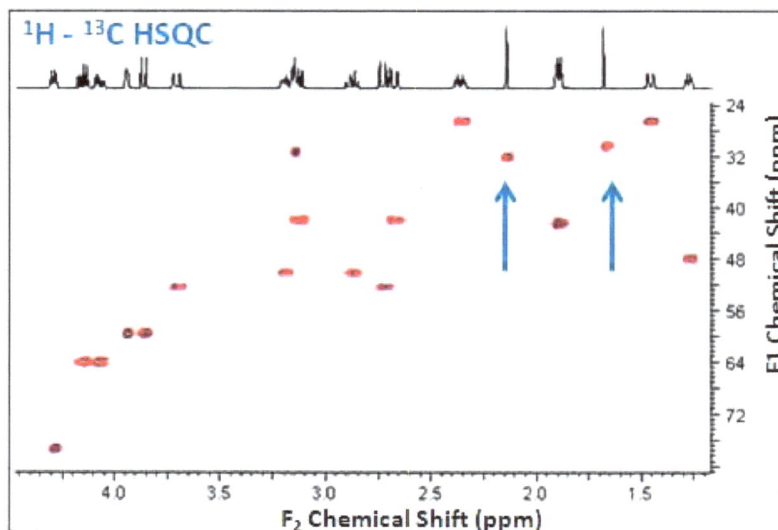

Fig. (9). HSQC spectrum.

3D-NMR SPECTROSCOPY

A three dimensional NMR experiment can easily be constructed from a two dimensional one by inserting an additional indirect evolution time and a second mixing period between the first mixing period and the direct data acquisition. Each of the different indirect time periods (t_1, t_2) is incremented separately (Fig. **10**).

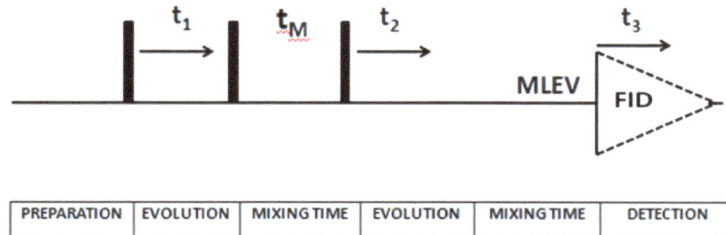

Fig. (10). Anatomy of a 3D NMR experiment.
   ∞   There are two principal classes of 3D experiments
   ∞   Experiments consisting of 'two 2D experiments after another'
   ∞   The triple resonance experiments.

NOESY-HSQC, TOCSY-HSQC

2D spectra (like NOESY or TOCSY) of larger proteins are often crowded with signals. Therefore, these spectra are spreading out in a third dimension (usually 15N or 13C), so that the signals are distributed in a cube instead of a plane. This spread out is achieved by combining HSQC and NOESY in a single 3D experiment: The NOESY experiment is extended by an HSQC step. Acquisition starts after this HSQC step instead of at the end of the NOESY mixing time. The resulting experiment is called 3D NOESY-HSQC. In a similar way, a TOCSY-HSQC can be constructed by combining the TOCSY and the HSQC experiment.

HCCH-TOCSY

The experiments HCCH-TOCSY and HCCH-COSY are alternatives to the ^{13}C-TOCSY-HSQC experiment which shows a markedly reduced sensitivity for larger proteins. In both experiments the magnetization is transferred via direct ^{1}J couplings between the atoms, allowing a much faster magnetization transfer as in the TOCSY-HSQC (Fig. **11**).

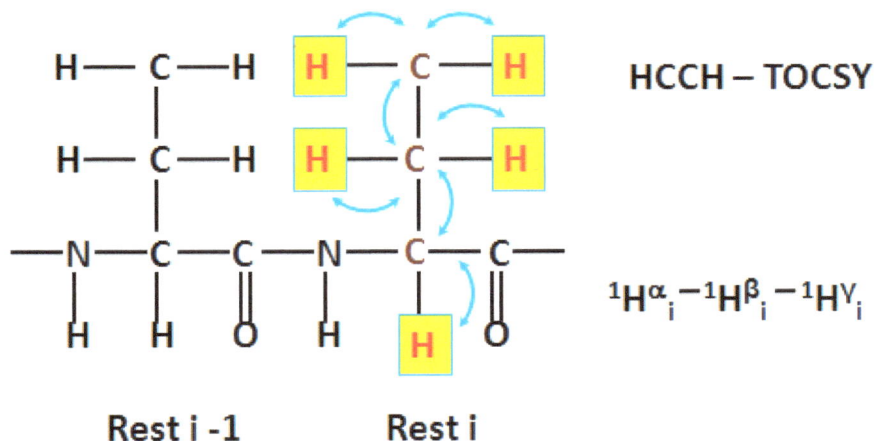

Fig. (11). HCCH-TOCSY.

Magnetization is transferred (blue arrows) from a side chain (or backbone) proton (red) to the direct attached carbon atom (yellow), by ^{1}J coupling to the neighboring carbon atoms (yellow) and finally to their attached protons (red). Therefore, a HCCH-TOCSY experiment looks in principle the same like a ^{13}C-TOCSY-HSQC. The type of an amino acid can be recognized from the peak pattern in the usual way, as can be seen from the picture below. It shows some regions from different planes of a HCCH-TOCSY which together contain the complete spin system of a leucine residue (Fig. **12**).

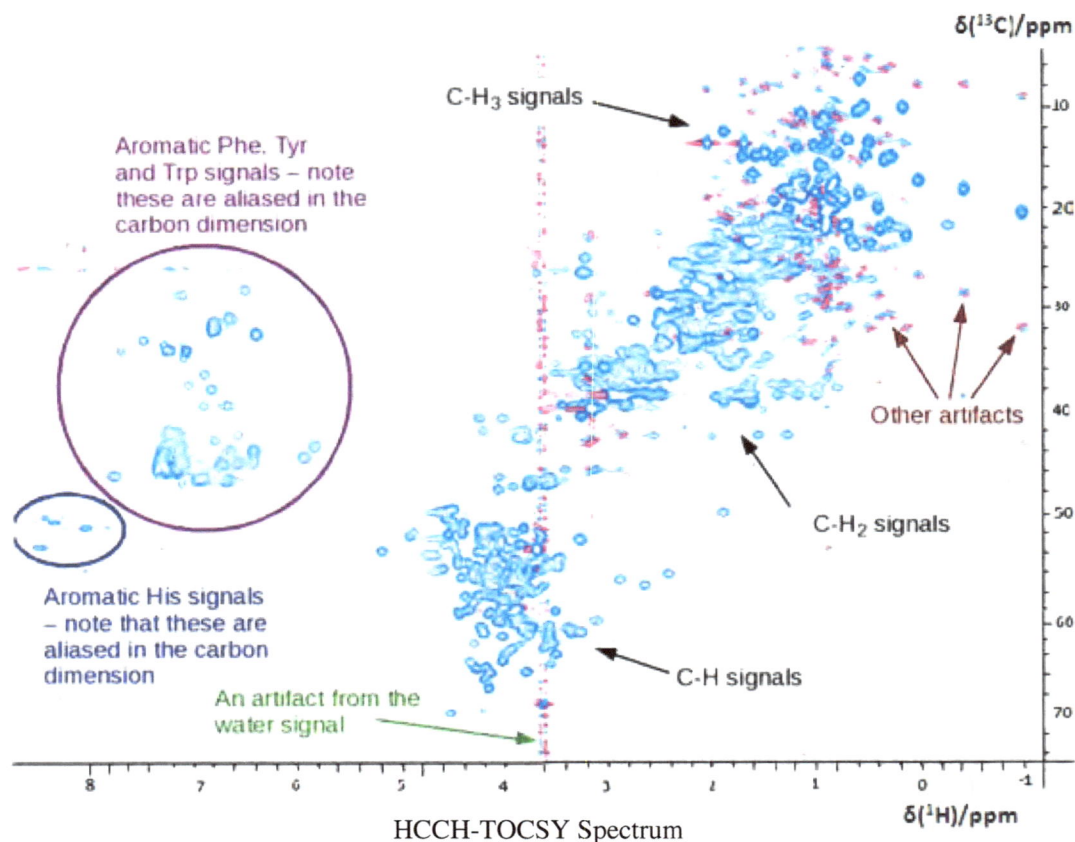

HCCH-TOCSY Spectrum

Fig. (12). HCCH-TOCSY spectrum.

TRIPLE RESONANCE EXPERIMENTS

Triple resonance experiments are the method of choice for the sequential assignment of larger proteins (> 150 AA). These experiments are called 'triple resonance' because three different nuclei (^{1}H, ^{13}C, ^{15}N) are correlated. The experiments are performed on doubly labelled (^{13}C, ^{15}N) proteins.

The most important advantage of the triple resonance spectra is their simplicity: They contain only a few signals on each frequency-often only one. The problem of spectral overlap is therefore markedly reduced (this is the main reason, why proteins of more than 20 kDa can be assigned with triple resonance experiments). However, the coordinates of clearly separated signals from different amino acids can accidentally be identical ('degeneration' of signals). The correct choice of connectivities between amino acids is the main problem in the assignment of triple resonance spectra. Another advantage of triple resonance spectra is their high sensitivity which is caused by an efficient transfer of magnetization. The magnetization is transferred via ^{1}J or ^{2}J couplings (i.e. directly via the covalent chemical bonds). Therefore, the transfer times are shorter and the losses due to relaxation are smaller than in homonuclear experiments. The disadvantage of all triple resonance experiments is the necessity of doubly labeled proteins, the preparation of which is often expensive.

NOMENCLATURE OF TRIPLE RESONANCE EXPERIMENTS

The names of triple resonance experiments sound very cryptic at first glance, but they are very descriptive: The names of all nuclei which are used for magnetization transfer during the experiment are listed in the order of their use, bracketing the names of nuclei which are used only for transfer and whose frequencies are not detected. Two experiments HNCO (Fig. **13**) and HN(CA)CO (Fig. **14**) have been used as examples to illustrate this. In the HNCO experiment the magnetization is transferred (blue arrows) from the H^N(i) proton via the N(i) atom to the directly attached CO(i-1) carbon atom and return the same way to the H^N(i) nucleus which is directly detected. The frequencies of all three nuclei (red) are detected.

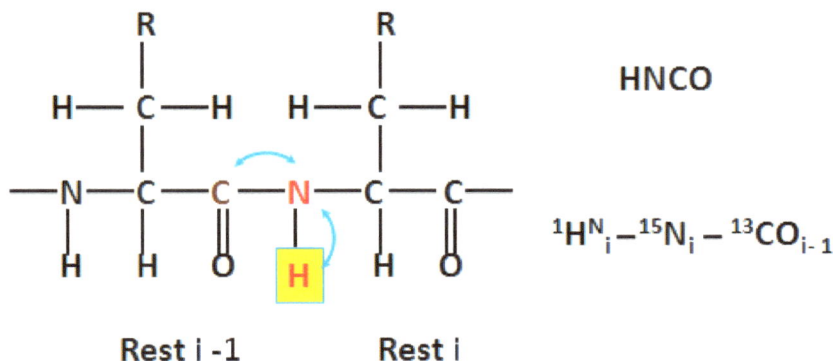

Fig. (13). HNCO coupling.

In the HN(CA)CO experiment the magnetization is transferred from the $H^N(i)$ proton via the N(i) atom and the CA nucleus [$C^{alpha}(i)$] to the CO(i) carbon atom and back the same way. The C^{alpha} atom (yellow) acts only as relay nucleus, its frequency are not detected. It is only the frequencies of H^N, N and CO (red) which are detected.

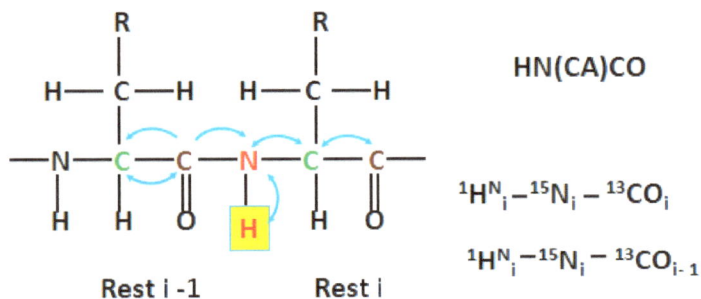

Fig. (14). H(CA)CO coupling.

This nomenclature has the advantage, that the spectra can be easily imagined by their names: In HNCO an amide proton is correlated with the CO atom of the preceding amino acid, whereas in HN(CA)CO the correlation to the intra residual CO atom is also visible.

THE HNCA EXPERIMENT

The HNCA experiment is the prototype of all triple resonance experiments. Starting at an amide proton (H) the magnetization is transferred to the directly attached nitrogen atom (N) which is measured as the first spectral dimension. Then the magnetization is transferred to the C^{alpha} nucleus (CA) which is measured as second dimension. Afterwards, the magnetization is transferred back the same way to the amide proton which is measured as the third (direct) dimension (Fig. **15**).

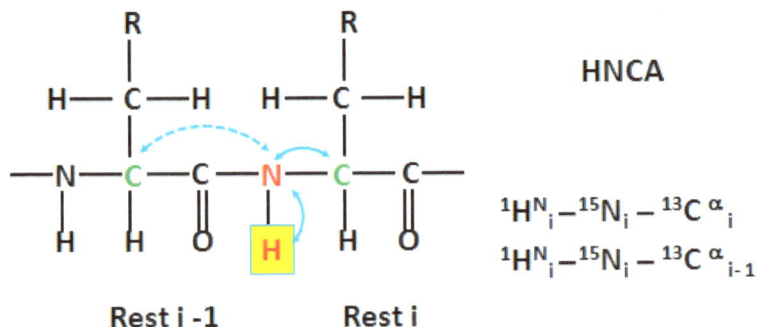

Fig. (15). HNCA coupling.

In each step magnetization is transferred via strong ^{1}J couplings between the nuclei. The coupling which connects the nitrogen atom with the C^{alpha} carbon of the preceeding amino acid (^{2}J= 7 Hz) is only marginally smaller than the coupling to the directly attached C^{alpha} atom (^{1}J = 11 Hz). Thus, the nitrogen atom of a given amino acid is correlated with both C^{alpha} - its own and the one of the preceding amino acid.

Therefore, it is possible to assign the protein backbone exclusively with an HNCA spectrum. But usually more triple resonance experiments are needed because the cross signal of the preceeding amino acid has to be identified and degenerate resonance frequencies have to be resolved [19].

NMR AND DOCKING OF COMPOUNDS

While NMR spectroscopy has become an accepted alternative to X-ray diffraction for elucidating high-resolution structures of biomacromolecules such as proteins, DNA, RNA and oligosaccharides, it can be applied to drug design in many more ways. Besides delivering the three-dimensional structures of the free proteins as raw material for modeling studies on ligand binding, NMR can directly yield valuable experimental data on biologically important protein-ligand complexes themselves [62].

1). *Qualitative and quantitative binding assay*: changes in NMR parameters allow us to detect and quantitatively determine binding affinities of potential ligands.

2). *Determination of binding site*: based on changes in the assigned NMR signals of a protein upon ligand addition, a ligand's binding site on the protein can be located.

3). *Conformational information*: the conformations of the protein and/or ligand in the complex can be determined and compared to their free states.

4). *Dynamic information*: in addition to the static structure, local flexibilities can be measured for the complex and the free components to yield a more accurate picture of the intermolecular interactions.

ASPECTS OF BINDING AFFINITY

The behavior of protein-ligand complexes in NMR measurements depends largely on their thermodyna-mic and kinetic properties (*i.e.*, dissociation constants and on/off rates). The characteristics of the system under study and the kind of information desired determine the adequate approach for an NMR investigation. Protein concentrations in solution are often limited by solubility and aggregation (and, of course, protein availability) to a maximum of 0.5-5 mM. Today most NMR structural studies are performed at 1-2 mM protein concentrations, although it can be expected that new technologies leading to increased sensitivity in NMR experiments (increased field strength, improved probe design, cryo-probes etc.) will lower the concentration requirements to the 100-∝M range within the next few years. For a typical millimolar protein concentration, three different regimes can be distinguished.

Thus we have:

1). Strong binding, with dissociation constants K_D in the sub-micromolar range

2). Moderate binding, with micro molar K_D values

3). Weak interactions with millimolar dissociation constants

(for KD > 1 M essentially no complex formation will occur at all and no observable effects be detected in NMR measurements). For strongly binding ligands, under normal protein NMR conditions, dissociation constants in the nanomolar range lead to negligible concentrations of free protein and ligand under stoichiometric conditions and we can assume that both the protein and ligand exist almost exclusively (>95 %) in the bound state. The complex behaves like a single stable molecule, and its spectra will not show any signs of the free species or of exchange between the free and bound species. Protein complexes with high-affinity ligands, e.g., enzyme inhibitors, can be treated as single molecules in NMR and X-ray studies and thus poses no special problems, except for their larger size and possible changes in solubility. Because of their high stability, such complexes usually crystallize readily and are therefore easily accessible to X-ray diffraction studies.

In the moderately binding case, under stoichiometric conditions both protein and ligand will exist as bound *and* free species in significant percentages. Depending on the kinetic stability of the complex, exchange between the free and bound species might also affect the NMR spectra. For a 1 mM dissociation constant, in a stoichiometric mixture of protein and ligand (1 mM each) only half of the molecules exist as protein-ligand complex, the other half being free ligand and free protein (in slow or fast exchange with the complex). It is obvious that under such conditions the NMR spectrum of the system will be far more complex and difficult to analyze than for the isolated complex alone. However, for micro molar KD values the protein can be driven into a completely complexed state by adding a moderate (10- to 100-fold) excess of the ligand. Signals of the excess component can be eliminated from the NMR spectra by isotope filtering techniques (see below). Of course, ligand solubility has to be sufficiently high, but this can often be accomplished by adding small amounts of organic solvents (e.g., DMSO) to the aqueous protein solution. Chemical and biological processes often include such moderately strong intermolecular interactions leading to a dynamic equilibrium between free and bound states of the components involved. Since the resulting mixtures of free and bound species in most cases cannot be crystallized, NMR is the method of choice for gaining structural information on these very interesting systems.

In the weakly bound case, it will not be possible to force a complex component into a completely complexed state, even with a large excess of the other compound. Under NMR conditions, only a small fraction of the molecules will exist in the complexed state, so that a conventional three-dimensional structure determination becomes impossible. Nevertheless, a lot of information about the complex can be gained from NMR measurements (usually involving isotope filtering), e.g., indications of conformational changes upon complexation and the molecular regions involved in binding. When screening for lead compounds from a library weakly binding ligands are the routine case that will still have to undergo extensive optimization to yield a strongly binding ligand. Again, X-ray diffraction methods are generally not applicable to these cases and NMR becomes the method of choice for structural studies. Hence it is extended to fragment based drug design [26].

NMR Screening Strategies

Objective of any screening strategy is directed towards the discovery of drug candidates to achieve the desired qualities like specificity, selectivity, potency, safety, duration of action. Different strategies of fragment based drug design are categorized based on, in which primary screening hits are followed up: linking the initial hits, searching for complex compounds using initial hits or by changing and combining hits. Based on the mentioned approaches Combination strategy, elaboration strategy and variation strategy are considered respectively.

Combination Strategy

In this approach, number of available smaller molecules is screened for its possible binding with the target protein and affinities are calculated. As smaller fragments have very limited potential for making favorable interactions, these smaller compounds are linked through chemical synthesis and screened again. Once these smaller units with less binding affinities are joined, there may be chance of getting compounds with higher affinity with nano molar concentrations due to change in entropy. Desired scaffold class of compounds can be prepared and screened by linking simple smaller compounds in this manner.

Various approaches have been suggested for combining molecular fragments [63]. Fragment fusion is the testing of compounds (usually purchased from commercial sources) that contain two or more substructures known to bind to the target. Combinatorial target-guided ligand assembly uses reactive groups built into synthetic fragments to make all possible combinations of binding fragments. Unfortunately, unless the proximity and relative orientations of the individual bound fragments are known, with the latter methods there is no way to know that the blindly linked compounds will have the correct conformation to bind to the protein. To address this problem, SAR (structure activity relationship) by NMR [64] guides the synthesis of linked compounds using NMR derived structural information: close contacts between bound fragments found using NOE experiments [65] and binding sites modeled using chemical shift perturbation data [66, 67]. Alternatively, it is possible to determine the structures of fragments soaked into protein crystals using X-ray crystallography (Fig. **16**).

Elaboration Strategy

In this approach primary screening hits are used as building blocks upon systematically incorporating several moieties and ring systems to prepare more complex molecules. The goal is to incorporate several functional groups and ring systems that will form new contacts with target and increase potency without

Library	SAR by NMR	Fragment Fusion	Combinatorial Library

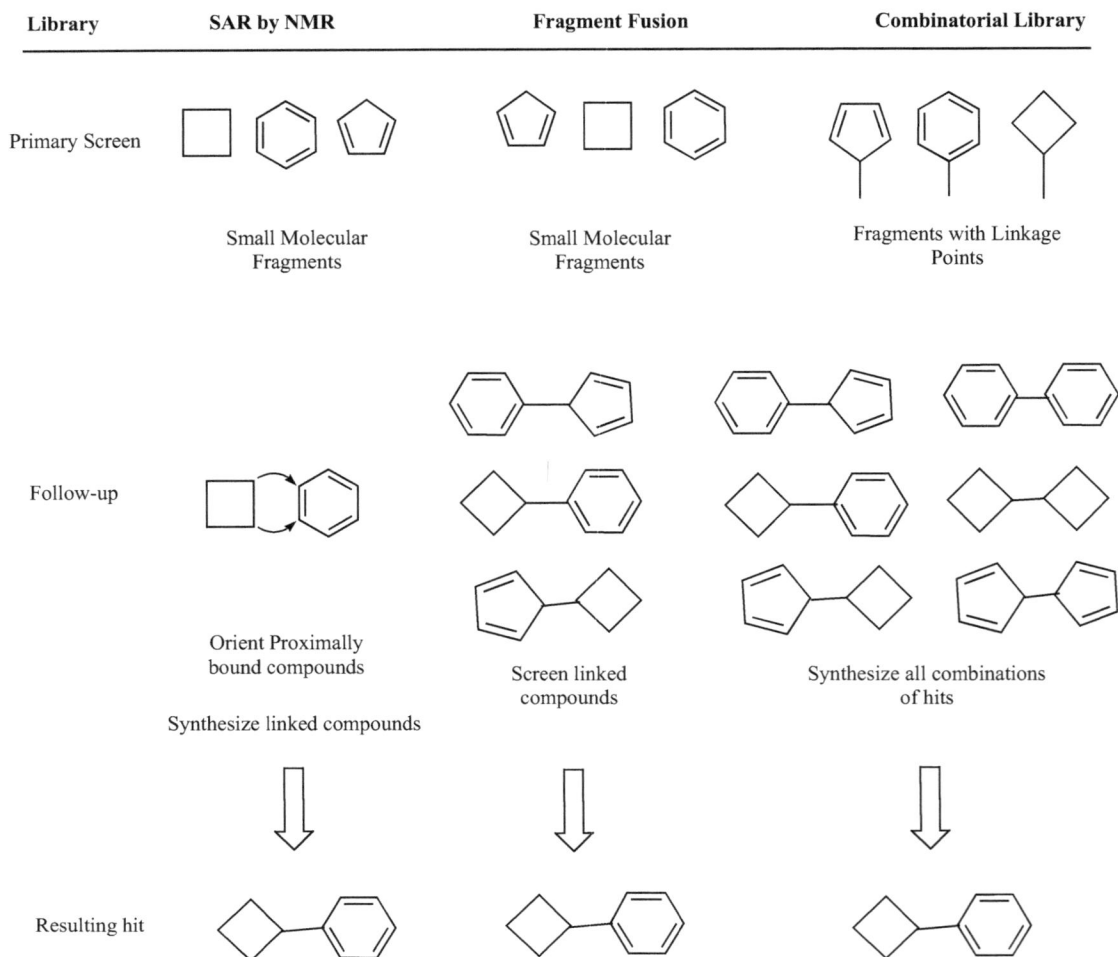

Fig. (16). Types of screening strategies.

disturbing binding. This is a sought of building essential features (pharmacophore) of the molecule by adding up several moieties, which might be responsible for biological activity. Series of compounds are prepared and screening is carried out using HTS assays.

In elaboration strategy the prerequisite is the compounds which are screened for binding affinities with proteins should be amenable to synthetic modification or for which many analogues are commercially available. Advantage with elaboration strategy is compounds are designed using already existing scaffolds rather than to assemble a completely new one from small fragments (Fig. **17**).

Variation Strategy

In variation strategy, compounds from shapes linking library and combinatorial scaffolds are selected and linked systematically that are synthetically amenable to being indepen-dently varied. All the possible combination of molecules are synthesized and subjected for HTS screening. Alternately, one may screen individual combinatorial scaffolds that contain multiple positions for substitution. This approach is much advantageous in finding drug candidates comparing others because the compounds are designed based on exploring already available screening hits.

Variation strategy involves building different new groups and ring systems over available scaffolds from shapes linking library and screened for its affinity and efficacy, so optimization of the drug candidate for safety, variations in kinetics, and stability aspects among synthesized molecules is attainable easily [63, 68] (Fig. **18**).

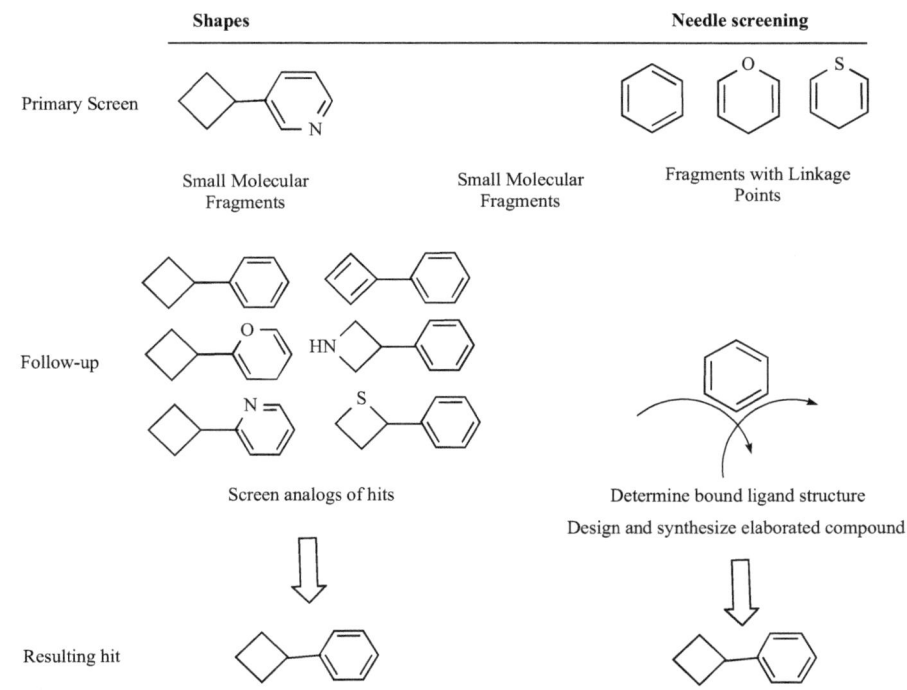

Fig. (17). Elaboration strategy.

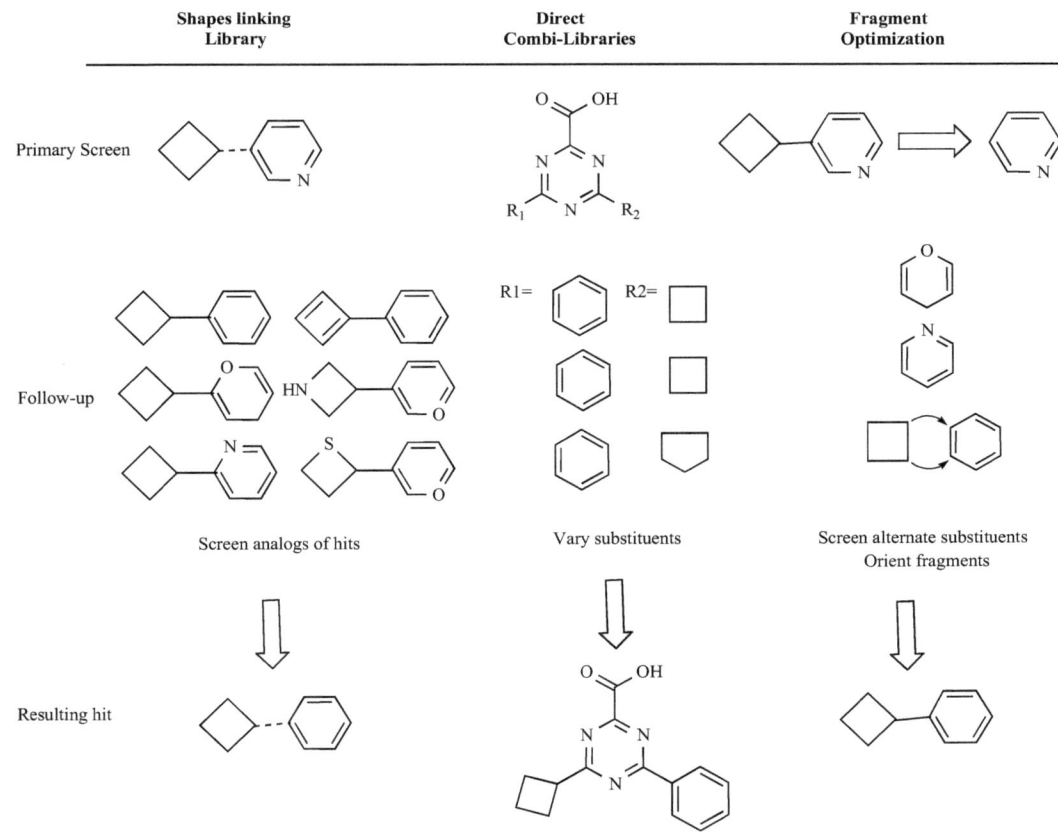

Fig. (18). Variation strategy.

Choosing an Experimental Method

The choice of experimental method used for the screen is based primarily on the strategy and the nature of the target. SAR by NMR [64] and fragment optimization [63] employ protein-detected heteronuclear methods that require isotopically labeled protein, but problems with spectral complexity and line broadening have in the past limited these methods to proteins of less than approximately 30 kDa. Recent methods propose to increase this limit to 100 kDa by using ^{13}C-labeled methyls in combination with deuterated protein [69], selective observation of solvent-exposed amides in a TROSY experiment using deuterated protein [70], or protein containing a unique pair of ^{13}C and ^{15}N labels in adjacent amino acids [71]. Despite these advances, protein-based methods are still not applicable to very large or membrane-bound proteins, and because of the current cost and difficulty of expressing isotopically labeled proteins in insect cells or yeast, they are generally limited to proteins that express well in bacteria.

Ligand-detected screening methods are preferred for targets that have high molecular weights, are membrane-bound, cannot be labeled, have low solubility, are expressed in nonbacterial systems, or cannot be obtained in large quantities. Although these methods do not identify the ligand binding site, this information can be obtained by a variety of other methods (competition experiments, enzyme assays, chemical shift perturbation, crystallography, etc.). The advantages of using unlabeled protein, low protein consumption, and unlimited molecular weight typically outweigh the disadvantage of having to determine the binding site in a separate experiment.

Saturation transfer difference [72, 73] is by far the most sensitive ligand-detected method, capable of screening at protein concentrations as low as 100 nM. This is 500- fold less protein than required by protein-detected methods and is even below the concentration used in some enzyme assays; 1000 compounds may be screened (in mixtures of five) against a 50 kDa target using 0.5 mg of protein. It is possible to screen relatively insoluble receptor proteins by immobilizing or stabilizing them in synthetic membranes [74]. For example, the *trans*-membrane integrin_IIb_3 has been successfully screened while embedded in liposomes [75].

Another advantage of the STD method is that the threshold of detection may be set at the desired level by adjusting the protein concentration. If a targeted library is being screened, then it may be advisable to screen using line-broadening experiments [22, 70] at low ligand: protein ratio in order to detect relatively tight binding (< 1μM) ligands.

The NMR screening program at Vertex used the SHAPES strategy, which was based on ligand-detected screening of drug-like scaffolds [63, 76]. At various times, they have used transferred NOE, line broadening, relaxation difference, diffusion and saturation transfer difference methods to detect binding. Currently they favor STD because of its high sensitivity [63]. Information about binding scaffolds was used to drive combination, elaboration, and variation strategies. NMR screens were carried out using the original drug-like SHAPES library as well as a combi-chem-friendly SHAPES linking library, gene-family specific libraries, and follow-up libraries based on previous hits.

The original SHAPES library consisted of commercially available compounds that represented scaffolds commonly found in known drugs. Two modeling studies using shape descriptors [77, 78] found that a surprisingly small number of molecular scaffolds and side chains were sufficient to describe a large percentage of known drugs covering a wide variety of indications. A library was constructed by Vertex from commercially available compounds that contained a selection of the most common of these structures. Compounds with undesirable physicochemical properties and functional groups were excluded by using REOS ("Rapid Elimination of Swill") software filter [79, 80]. This filter routinely rejects approximately 60% of candidate molecules, though care must be taken to avoid over-filtering. For example, with overly conservative settings the REOS filter rejects 73 % of known drugs from the Comprehensive Medicinal Chemistry database. Similar problems have been reported for other classification methods [81, 82]. A combination of filtering and diversity ranking is very effective for narrowing a list of candidate compounds. Applying an 85 % similarity threshold to a REOS-filtered set of commercially available compounds removes 42 % of the compounds; the combined pass rate for the two filters is only 23 %.

The SHAPES Linking Library was designed to facilitate the use of combinatorial chemistry to follow up screening hits [76]. This library consists primarily of commercially available compounds containing two drug-like scaffolds connected by a linkage that is synthetically accessible. To construct this library, a database of commercially available compounds was filtered to select for drug-likeness and the presence of the desired molecular features, then edited based on predicted solubility [76]. This reduced set was subjected to Jarvis-Patrick clustering [83], and compounds were manually selected from the centroids of the clusters. In addition to facilitating the purchase of follow-up compounds, centroid picking has the advantage of selecting less complex compounds; the latter is an unintended artifact of clustering based on molecular fingerprints [84]. Cascaded clustering [85] was used to reduce the number of small clusters and singletons. Selected compounds were prepared as stock solutions in DMSO using sonication to aid dissolution. In addition to LC-MS analysis to check compound purity and identity, NMR was used to test the solubility of individual compounds and mixtures of five compounds. By the end of this process, a database of 1.2 million compounds was narrowed to a final library of ca. 500 compounds [76].

Targeted libraries of Vertex comprised of compounds that inhibit members of a target gene family. Most of these were proprietary compounds, such as favored leads from discovery projects, representatives from in-house combinatorial libraries, novel hits from HTS screens, and analogs of known substrates or inhibitors. All of these compounds were selected manually with input from a variety of chemists and modelers and emphasis was placed on compounds with high solubility, good PK characteristics, available analogs, and activity against multiple members of the gene family.

Follow-up libraries consisted of compounds purchased or synthesized around hits from a primary SHAPES screen. The most desirable hits have known chemistry, a clear relationship between structure and binding affinity, and no competing patents. It has been shown that the best results are obtained when the search for analogs is as broad as possible. This is partly because of the inherent limitations of similarity searches using bit strings, which tend to exaggerate the diversity of small molecule libraries and produce counter-intuitive similarities (for example, failing to recognize topologically similar molecules with different heteroatom substitution patterns) [86, 87]. For this reason, it is always wise to manually edit the results of similarity searches. In addition, broad searches are more likely to avoid the previously described local minimum problem. Also, the number of compounds must be large enough to give a reasonable likelihood of finding hits above the detection limit of the follow-up assay. Given an HTS hit rate around 0.3%, a 500-compound general screening library has a 76% chance of yielding one or two hits. As per their experience, a 500-compound follow-up library generally produced HTS-detectable hits, since HTS screens of SHAPES follow-up libraries usually produce hit rates around ten fold higher than random compound libraries. Large improvements in potency demanded large follow-up libraries, since linear improvements in potency require geometrically more compounds [88].

If a model is available for the active site of the protein, then virtual screening methods may be used to prioritize the choice of follow-up compounds. For example, hits from the SHAPES Linking Library may be followed up by conceptually breaking the compounds apart at the linker and docking commercially available reagents to replace each portion of the molecule [76]. Docking programs are currently limited in their ability to predict binding modes and relative affinities *a priori* because of difficulty accounting for flexibility and water bound in the active site. The addition of knowledge about which scaffolds bind and proximity restraints from chemical shift perturbation or NOE experiments promises to improve the predictive ability of these docking models.

SHAPES screening has been implemented in several different ways in order to contribute to projects at various stages of development. It was found that NMR screening has little impact if run concurrently with HTS, since the resources available for follow-up are invariably directed toward the more potent HTS hits. Because SHAPES screening needs no assay development and requires only a few milligrams of protein, new targets can be screened long before HTS. It was found that even with follow-up screens of only a few hundred compounds it is common place to find inhibitors with affinities ranging from micromolar to nanomolar. When a high-throughput assay is unavailable, such inhibitors can be also used as potential probe molecules for use in assay development.

SHAPES screening can also be run in "target rescue" mode to find new leads after HTS screens have failed to generate sufficient viable leads or to identify scaffolds that are clear of intellectual property conflicts. It was found that SHAPES screening often identifies classes of compounds that were tested in the original HTS screen, but were not identified as hits at 30 μM. Why are these classes missed by HTS? One possible explanation is that compounds in the HTS general screening library are typically more complex, which reduces the likelihood of binding. The simpler SHAPES screening compounds are more likely to bind and this binding can be observed by NMR even when too weak to be detected under standard HTS screening conditions.

Thus Vertex has recently implemented SHAPES screening as part of a "chemo-genomics"- based approach [89] by screening proprietary scaffolds against multiple members of a gene family. This permitted them to find new indications for existing scaffolds, determine intrinsic selectivity patterns, and assess the "drugability" of new targets.

NMR Screening Libraries

Objective and purpose is important in the selection of compounds for screening. There are four basic types of libraries: general screening, targeted, focused, and follow-up. General screening libraries are the most common out of four libraries used for primary screens to identify novel ligands. In many cases, the hits from such a library can be readily pursued by purchasing commercially available analogs. Analogs are usually purchased rather than synthesized because the hits are relatively weak and unless a very large library is screened it is unlikely that there will be enough hits within a given compound class to generate meaningful SAR information.

Molecular descriptors are used to define molecular diversity for a given set of compounds. Descriptors used should be based on correlation of the molecule with its biological activity. This is to select the compounds based on important factors based on relevant descriptors. Thus unimportant descriptor based section of compounds should be carefully eliminated. Structural keys and 2D fingerprints help much for the selection of compounds rather than traditional QSAR descriptors. QSAR descriptors would not be helpful in finding much diverse search of compounds among several compounds with weak affinities. In addition, pharmacophores and molecular steric information can be used to design analog libraries for ill defined series of compounds.

A number of different approaches have been used for selecting diverse set of compounds. One of the methods is selection of compounds with maximum dissimilarity, in which each new compound is chosen to be as dissimilar as possible to those already selected. This reduces library size drastically without reducing the likelihood of discovering classes of active compounds. Possible drawback of this approach is that selection is biased towards unique compounds, so close analogues are unavailable for follow-up. Commercially available clustering software can be useful to identify groups of similar compounds for the follow-up screen in turn it helps choosing additional compounds from the active clusters. There are different clustering softwares (Nonhierarchical Jarvis-Patrick clustering, Hierarchical clustering by Ward's method) can be used for the selection of neighborhood compounds for finding hits. Hierarchical clustering by Ward's method is reportedly best for separating active and inactive compounds [90]. Different keys as descriptors (MACCS, BCUT) can be used to select the compounds from clusters.

Different types of libraries are designed as per applicability and follow-ups discussed here under.

Compounds categorized under *Targeted* libraries consist are subjected for binding with family of closely related proteins. Similarities between the active target sites of homologous proteins lead to identify the compounds with higher hit rates. The proteins selected for binding studies may be of substrates, cofactors, or known inhibitors. The targeted library must represent the principal classes of known binders while conforming to experimental requirements, and the compounds satisfying these criteria are limited in number and novelty. Since analogs of hits are not available commercially, synthesis of expected compounds is required for follow-up.

Combinatorial chemistry offers a route to large targeted libraries at a reduced cost per compound. When screening combinatorial libraries by NMR, it is important to ensure that they can be made in sufficient quantities as discrete, purified compounds and that they are sufficiently soluble. Whenever any mixture of compounds is to be screened using ligand detection then it must be possible to resolve among individual compounds through NMR spectra. Usually Combinatorial libraries have been targeted against enzyme classes by incorporating recognition elements (such as transition state mimics) with known affinity for the target class into the molecules.

Focused libraries are directed against a specific protein and the target is preparing antagonists. These known inhibitors against a specific protein are used to build library. First step in developing these compounds is identifying the possible active target through Insilico screening. Insilico screening typically refers to computational docking of molecules in the active site of the protein in order to identify the most promising candidates [79, 81, 91-96]. In one published report, comparison of docking and NMR screening results for 3300 commercially available compounds against FKBP-12 [97] showed that virtual screening could reduce the number of compounds to be screened by two- to three-fold without eliminating any that actually bound. Based on the results obtained through virtual screening enzymatic assays are performed by using combinatorial reagents [98-100]. Virtual screening must significantly enhance hit rates in order to justify the expense of constructing new focused libraries for every primary screening target. In practice, they are rarely used for primary NMR screens and are more commonly created to follow up primary screening hits.

The construction of *follow-up* libraries is a critical element of elaboration and variation strategies. General libraries give primary screening information with few hits obtained within a given scaffold class. The limitation with general screening library is that no detailed structural information is possible to obtain. Thus, the first step in following up primary screening hits is typically to search databases of readily available compounds for analogs to screen, with the intent of discerning a relationship between chemical structure and activity. Functional groups are added or modified to increase the number of interactions between the molecule and the target, thereby increasing the potency, or replaced with isosteres [101-103] to broaden the lead class. The molecular core is permuted in order to discover equipotent alternative scaffolds with significantly different structures [104].

The compounds with similar structures are predicted to be with similar SAR and are the basis for same biological property. This assumption is invariably taken for granted since in its absence there is essentially no other rationale by which to follow up hits! In practice testing compounds similar to weak hits usually results in more weak hits. So, to find ligands with more affinity, efficacy and potency and to move out of the local minimum of confined search requires testing more disparate compounds, which dramatically increases the size of the follow-up library. For this reason, it is important to use an intelligent search strategy when seeking follow-up compounds to elaborate upon simple scaffolds. Analogs of primary screening hits are easily found by searching databases of compounds using commercially available similarity matching [105] or nearest-neighbor searching [106] tools. However, a conventional global similarity search, in which the entire molecule is matched to the target structure, may fall prey to the local minimum problem because the compounds matched cannot be significantly larger than the original hit. A substructure similarity search, [107] which matches only the target substructure is preferable, because it finds compounds that contain scaffolds resembling the primary screening hit while allowing free variation elsewhere in the molecule.

The correlation of molecular descriptors with biological activity helps us understanding the term "neighborhood behavior" (relative similarity of scaffolds) of the series of compounds. The compounds with close or similar biological activity come under same umbrella. Thus compounds with same neighborhood have similar biological activities. This similarity in biological activity can be measured using well known Tanimoto index and MACCS keys as descriptors. A uniformly distributed set of compounds avoids overlap between neighborhoods and so is more efficient for discovering active compounds than a random set. When hits are found, the likelihood of finding the hits is highest for compounds within the neighborhood of the original hit. In addition to the commonly used MACCS structural fragment keys and Daylight or UNITY fingerprints, similarity searching can be carried out around a well-defined compound class using local descriptors such as atom pairs [108, 109] or topomeric shape [100, 111]. Also, ligand-based pharmacophore searches are able to identify follow-up compounds that are less obvious and more diverse than similarity searches [95, 112-116].

OPTIMIZING MOLECULAR COMPLEXITY

Molecular complexity arises with increase in the bulk of the molecule result in more interactions. Simple molecules are with less in the complexity and are more likely to bind to the target. But simple molecules usually have lower affinities with less selectivity. To improve the affinities of the ligands simple molecules are to be constructed in to larger molecules by combining them. As bulk of the molecule is increased apart from improved binding with target several unwanted interactions of ligands with proteins are more likely to occur. At the same time, models suggest that the overall probability of finding a complementary match between molecule and receptor declines exponentially with increasing molecular complexity [117]. Consequently, there is an optimum window in which molecules are complex enough to give detectable inhibition but not so complex as to make hits too rare in the first place. Hence the designed molecules should fall in this optimum window without exceeding the limits of complexity.

NMR screening is sometimes much superior over HTS screening in eliminating false positives probably because the latter are much more susceptible to compounds interfering with the assay system. NMR can detect compounds with multiple binding modes and even very weak binders.

So, how large should screening compounds be? They should be as large as necessary for detectable binding but no larger. Since NMR can detect millimolar binders, this size can be very small indeed. For molecules with up to 10 heavy atoms, the binding free energy contribution per heavy atom can be as high as-1.5 kcal/mol, so even very small compounds can, in principle, bind very tightly [118]. In practice, this is extremely rare. Typical results are exemplified by a library of combinatorial monomers used at GlaxoSmithKline, in which compounds of 100-250 Da yield activities of 0.2-10 mM [117]. By comparison, most drugs have molecular weights between 150 and 500. [119-122] In order to balance the need to find a reasonable number of hits against the need for specificity, it is advisable to screen compounds across the intermediate range of ca. 100-350 Da.

In addition to molecular weight, percentage saturation of the standard Daylight fingerprint has been used to describe the degree of molecular functionality [123]. Compounds with saturation outside a certain range (e.g., <10% and >60% saturation) can be excluded in order to remove overly simple or complicated molecules that would otherwise pass a simple molecular weight cutoff.

SELECTING FOR DRUG-LIKE CHARACTER

In recent years, it has become common practice to bias screening libraries in favor of molecules possessing characteristics similar to those of known drugs. Drug-like compounds are presumed to possess more attractive toxicological and ADME properties, and this advantage is thought to outweigh the reduced likelihood of finding novel, as yet undiscovered classes of drugs. Three approaches have been used to predict drug-likeness: selection based on physicochemical properties, functional group filtering, and topology based classification. Physicochemical profiling of drug databases produced the popular "rules of five" [119], which state that most orally available compounds possess a molecular weight 500, 5 hydrogen bond donors (OH and NH), 10 hydrogen bond acceptors (O and N), and log P 5 or Astex rule of 3 (three or fewer hydrogen bond acceptors, three or fewer hydrogen bond donors and ClogP $\leq$ 3). Other studies have further extended and refined the descriptor set. These descriptors are readily implemented as filters that remove molecules with undesirable properties from lists of candidates. Alternatively, physicochemical descriptors may be used as the basis for machine-learning programs that separate drugs from non-drugs [120, 123, 124] and predict ADME and toxicity properties [125-129]. Chemists and biologists have long known that certain chemical moieties are likely to produce false positives in biochemical assays because of their chemical reactivity [130]. Software filters are now routinely used at Vertex [79, 80] and other companies to flag compounds containing functional groups known empirically to contribute to reactivity, insolubility, toxicity, or poor ADME.

Topologically based methods compare the molecular structures (atom type and connectivity) of drugs and nondrugs to identify distinguishing features. The molecular features used for classification range from small to large, including atom types [82, 131, 132], molecular fragments and functional groups [77, 133-135], scaffolds and building blocks [78, 136] and whole molecules [137]. The classification methods include scoring functions [82, 136], decision trees [133] and neural networks [124, 132, 137-139]. These methods correctly classify 80-90% of known drugs, while classifying 10-30% of non-drugs as drugs [123, 131, 132, 134]. Although this performance is impressive (claimed to be comparable to the intuition of a medicinal chemist), non-specialists often prefer easy-to-use filtering methods.

It has been pointed out recently that compounds selected on the basis of drug-like properties may actually be sub-optimal for use as drug discovery leads [140]. When drug-like compounds are screened, the resulting hits typically have molecular weights (350- 450) and log P values (3-4.5) at the high end of the preferred "rules of five" ranges [140]. In the process of optimizing the potency of these drug-like leads, the molecules become more complex and lipophilic (median molecular weight increase of 69 Da, log P increase of 0.43, 1 ring and 2 rotatable bonds typically added), adversely affecting their pharmacokinetic properties. Compounds with molecular weight (100-350) and log P (1-3) at the low end of the preferred range offer better starting points, since addition of lipophilic groups promises to improve both potency and oral bioavailability [140]. Libraries consisting of small, polar, lead-like molecules (particularly those with a single charge at physiological pH) are thus better than drug-like libraries for producing useful leads [140]. Such compounds also tend to be highly soluble, which is beneficial for NMR experiments.

SOLUBILITY REQUIREMENTS

Aqueous solubility of compounds is a critical issue, since NMR screens must be run at relatively high compound concentrations, anywhere from millimolar for NOE, chemical shift perturbation or affinity. NMR methods [63, 64, 72, 141-145] to ca. 50 M for saturation- transfer methods [72] or screens run using a cryoprobe [146]. The ability to identify potentially insoluble compounds would obviously be very useful when designing screening libraries.

It is difficult to accurately predict aqueous solubility from chemical structure, because it involves disruption of the crystal lattice as well as solvation of the compound. Simple methods based on log P and melting temperature have been widely used [147]. Recently, various prediction methods have been reported [148-158] that are able to predict aqueous solubility to within ca. 0.5 log units (roughly a factor of 3 in concentrations). Although these predictors may not be precise or robust enough to select final compounds, they can be used as rough filters for narrowing the list of candidates (Table **2**).

Table 2. Solubility Influencing Parameters

Molecular weight:	150 to 450
Number of atoms:	10 to 70
H-bond donors:	0 to 3
Rotatable bonds Polar bonds:	1 to 8
Number of rings:	1 to 4
Ring size (number of atoms):	3 to 6
Unsubstituted cyclic atoms:	1 to 8
Log P:	0.5 to 5
Molar refractivity:	40 to 130
2K_ shape index:	2 to 12

There is currently no substitute for experimentally testing the solubility of all compounds before screening. Even this simple measurement is not trivial. The common practice of diluting DMSO stock solutions into buffer and measuring solubility based on apparent absorption (A) or, more precisely, extinction at 650 nm is very unreliable. In general, $A650$ measurements grossly overestimate solubility, since solutions that appear optically clear often contain high molecular weight aggregates (detectable by more sensitive light-scattering instruments and by nephelometry). Furthermore, the precipitation of compounds is often kinetically controlled [119], so that a compound diluted from DMSO into buffer will precipitate slowly over many hours. The result can be disastrous for the NMR screen, since aggregates can produce both false negatives and false positives regardless of screening method. Hence preference is to test all compounds for aggregation prior to screening using the sign of cross peaks in NOESY spectra to assess the rotational correlation time.

DESIGNING MIXTURES

Screening compounds in mixtures rather than singly has obvious advantages of higher throughput and reduced protein consumption. There are some drawbacks, however, which can be partly alleviated by careful mixture design:

Deconvolution of mixtures to identify individual hits is necessary for large, spectrally unresolved mixtures and for any screens using protein detection. The total number of data points required to screen and deconvolute mixtures in one step is minimized when the number of compounds per mixture equals 1/(hit rate)1/2 [159]. When the hit rate is relatively low, so that mixtures are unlikely to contain more than one hit, it is possible to avoid deconvolution by assaying each compound in duplicate, with each replicate in an orthogonal mixture [160]. When the hit rate is relatively high, the optimal mixture size becomes so small (e.g., mixtures of 3 for a 10% hit rate) that it may be more convenient to screen individual compounds [161]. Otherwise, for mixtures of more than four compounds, it is more efficient to carry out a two-step deconvolution, in which active mixtures of N compounds are divided into smaller mixtures of n compounds and re-tested; active n-compound mixtures are then tested singly [159]. The efficiency is highest when n is the square root of N [159]. When a ligand-detected screening method is used, deconvolution can be avoided by designing the mixtures so that there are resolved NMR resonances for every compound in the mixture [63].

Competition between components of a mixture becomes more likely as the hit rate and pool size increase. For a 10% hit rate, competing hits will be found in 26% of mixtures of ten compounds, but only 2.7% in mixtures of three [76]. Unless mixtures are deconvoluted competition can lead to false negatives when the competing compounds have different affinities. For screens using protein detection, competition can also produce false positives due to the additive effect of multiple weak binders. Competition can be controlled by limiting the size of mixtures and pooling dissimilar compounds to reduce the likelihood of two compounds binding at the same site.

Reactions between compounds in mixtures are more common than is generally recognized. Mixtures of 10 compounds stored in DMSO show evidence of chemical reactions between components in 25% of random mixtures, but this is reduced to 9% in mixtures strategically designed to separate acids from bases and electrophiles from nucleophiles [122]. Care must be taken in designing strategic mixtures, first to designate the compound types correctly (especially challenging for compounds of unknown pKa) [122], and then to adjust for the additive effect of combining multiple acids or bases into one mixture (requiring high buffer concentrations to control pH) [162]. Decreasing mixture size can dramatically reduce the likelihood of inter-component reactions. For example, decreasing mixture size from 10 to 5 reduces the number of potential pair wise interactions by 4-5 fold [76]. Although the screening of random 100-compound mixtures has been reported [147], the high likelihood of interactions between components (>90% of mixtures) makes the practicality of such large mixtures rather questionable. In practice, most laboratories screen pools of 4-10 compounds.

APPLICATIONS OF MULTIDIMENSIONAL NMR IN DRUG DESIGN

As already mentioned, there has been good utility of multidimensional NMR in designing novel compounds for various targets. A few examples representing various NMR techniques have been listed below.

Compounds with nanomolar affinities for the FK506 binding protein (FKBP) were developed by tethering two ligands with micromolar affinities. In this experiment ^{15}N-HSQC spectrum of uniformly ^{15}N labeled FKBP was used. To determine the binding interactions of the developed compounds with FKBP, intermolecular nuclear overhauser effects were measured. This technique has advantage over fluorimetric or calorimetric assay as ^{15}N-HSQC spectra do not show any signals of ligands which enables to detect the binding even in higher concentrations where as former methods gives rise to large background signals. In addition it has the ability to rapidly determine the different binding site locations of the fragments, which is critical for interpreting structure-activity relationships and guides the synthesis of linked compounds [163].

Although the transferred NOE experimental information has been often used to determine the structure of the ligand when bound to the target, for screening purposes it provides a very sensitive and highly reliable diagnostic for binding as well. Two mixtures of oligosaccharides were screened against *Aleuria aurantia* agglutinin. From the 2D spectra, a single binding ligand from each mixture was readily identified and characterized. This provides an excellent example for utilization of transferred NOE information to screen libraries of small molecules against a target protein [164].

Likewise two potent inhibitors (< 25 nM) for stromelysin, a matrix metalloproteinase were designed using the above technique. It provides a stepwise description of the design strategy of the inhibitors and also it is an excellent source of information for scientists working in the field [165]. Similarly fragments were discovered that bound region specifically at two different sites of Human Papiloma Virus E$_2$ protein. Ligands were then optimized for the site deemed most relevant, the DNA-binding site. Ultimately inhibitors of DNA binding in the low micromolar range were synthesized based on the screening hits. This work is of special interest as lead compounds were obtained for a target, HPV E$_2$-protein for which little chemistry was known [166].

The 2D transferred NOE or 2D ^{1}H-^{15}N heteronuclear correlation experiments consumes lot much of time. This drawback was overcome to certain extent by NMR line broadening and pulsed field gradient diffusion techniques. Like the NOE, other NMR observables that exhibit a marked dependence on molecular mass include the transverse relaxation time (T$_2$) and the translational diffusion coefficient (D). In relaxation editing, binding of a small molecule to a large protein will cause the ligand T$_2$ values to decrease and the signals to broaden. The degree of broadening increases with the increase in affinity of the ligand to the macromolecule. This line broadening is measured by comparing 1D NMR spectra of the ligand with and without the protein. Further it is a known fact that the diffusion of small, free molecule in solution is several orders faster than a large protein bound molecule which obviously reduces the diffusion coefficient. The extent of diffusion coefficient is the function of its affinity to the protein which can be measured using standard 1D pulsed field gradient NMR. Literature reveals that these methods were successfully implemented. This has several advantages along with less time consumption like it won't require ^{15}N labeled protein and allows one to examine targets with no molecular weight limitation. Both the NOE and relaxation data has been extended to know the TGFα residues involved in binding to the epidermal growth factor (EGF) receptors [167, 168].

Though the NOE is a successful method for defining the molecular requirements of complex formation but it suffers from severe limitations. First there are many factors like internuclear distance, mixing time and differential cross-relaxation and exchange rates in the free and bound ligand states that influences the intensity of NOE cross peaks. Second TGFα is dynamic in nature as the structural flexibility present in some regions is concurrent with rigidity in the β-sheets and hairpin regions of the ligand. The consequence of the variation in structural plasticity is the large difference in the number of NOEs / residue in different parts of the molecule. Thus it does not yield an accurate representation of the extent of contribution of individual amino acids to binding interaction. Hence to overcome this problem the interaction of TGFα with EGF receptors has been further envisaged by using heteronuclear single quantum correlation spectroscopy (HSQC) experiment [169]. As it overcomes the limitations found in NOE technique it gives more conclusive and informative data on the residues comprising the binding determinants. This in turn helps in designing the various agonists and antagonists which goes a long way in curing the diseases. This approach has also been utilized for studying the interaction between EncZ and OmpR proteins [170].

Recently the interligand nuclear overhauser effect (Interligand NOE or ILOE's) [171, 172] in conjunction with two dimensional [1]H - [15]N TROSY has been utilized to design small molecules that interact with Bid, a protein involved in neurodegenerative diseases, cerebral ischemia or brain trauma [173]. This technique is increasingly tackling very challenging drug targets which were otherwise not possible so easily by other techniques.

Interactions of lipopolysaccharides with polymixins were studied effectively by isotope filtered nuclear overhauser effect spectroscopy (NOESY) experiments along with restrained molecular dynamic simulation and chemical shift mapping. This is a recently developed technique which has several advantages over the other techniques described above [174].

Competition Saturation Transfer Difference, Water-LOGSY, Heteronuclear 2D ^{15}N HSQC and surface plasmon resonance experiments were used to validate four point pharmacophore generated by computational methods for heparinase enzyme. The experimental studies confirmed the reliability of pharmacophore model, its applicability to in silico databases in order to reduce the number of compounds to be experimentally screened and the possibility of generating fragment libraries enriched in heparinase inhibitors [175].

An excellent review of fifty potent inhibitors designed by using fragment based drug design, some of them in clinical trials, developed by thirteen different organization has been published. [176] Further, twenty three published fragment based design studies have been analyzed recently [177].

LIMITATIONS OF MULTIDIMENSIONAL NMR AND FRAGMENT BASED DRUG DESIGN

Impurities present with the ligands or targeted proteins for binding studies may interfere with the results for interpretation. Utmost purification of compounds and isolation of target proteins before subjecting for binding studies is important to get accurate results. Stability of the compounds is important at varied temperatures.

Size of the molecule is another important criteria, where the size of the compound and protein should fall in the optimum window. Since more the complex of the molecule structures more the interactions with different sites of target protein that may give false positives some times.

Solubility is another prerequisite for NMR screening of the compounds.

The time consumed to realize success by fragment based screens will be relatively longer than other lead generation approaches as fragment hits need to be given sufficient consideration and may require more cycles for optimization [178]. Further the compound collections used for fragment based drug discovery is a hybrid of a true fragment library and a typical high throughput screen collection. As the size of the fragment increases, the size of the library needed increases and as complexity grows the chances of an H-bonding mismatch or steric clash between the fragment and the target protein rapidly increases reducing the chance of finding hits [179].

FUTURE OF NMR BASED DRUG DESIGN

Biological mechanisms are almost always based - at the molecular level - on more or less specific intermolecular interactions involving biomacromolecules (e.g. receptor-hormone; transporter-substrate; enzyme-inhibitor; antibody-antigen; nucleic acids-regula-tory proteins). Understanding these interactions is a prerequisite for a modern approach to drug design, since the majority of newly developed drugs specifically target one or more of these systems in order to suppress, enhance or modify the intermolecular interactions involved.

NMR is uniquely able to provide information about the underlying interactions on various levels, from a merely qualitative indication of complex formation up to high-resolution structures and kinetic data. Therefore, NMR can make important contributions to drug design directed at molecular targets and to helping to understand the affinities and specificity of intermolecular interactions on a structural basis.

Based on long-established techniques as well as on very recent developments, NMR has a lot to offer in the field of drug discovery and development. Its applicability to ever larger biomacromolecules has made it one of the premier methods for solving three-dimensional structures of biologically active molecules and complexes in general. Structures of proteins and protein-ligand complexes (and increasingly also information about their dynamics) are needed for modeling the interactions between drugs and their macromolecular targets, and NMR spectroscopy can deliver valuable information in these areas. In addition, specific NMR techniques have been developed to directly yield useful information for drug development purposes, such as NMR-based screening and lead compound improvement. In the fascinating field of intermolecular interactions, NMR will continue to make important contributions to both our theoretical understanding of biomolecular mechanisms and its application to a modern, structure-based drug design.

When the crystal structure of the protein is not available or the size of the protein is beyond 30 KDa or protein is insoluble the identification of the hotspots in the protein has been assessed by computational methods like HOOK [180], CAVEAT [181], Re-core [182], Confirm [183] MED-SuMo [184, 185] etc. But the computational methods are not as reliable as experimental methods unless it is validated. Hence these techniques are used as augments for further lead generation.

FUTURE DIRECTIONS

Multitarget Drugs

Due to the redundancy in biological networks, modulating the function of multiple protein targets simultaneously can be beneficial for treating complex diseases [186, 187]. Many currently marketed drugs act via multiple targets, but the discovery of their multiple mechanism of action was usually serendipitous and retrospective. Compounds that are designed to modulate functions of several targets are in general larger and more lipophilic than compounds in the clinic or in the market [188, 189]. The reason for this is the current 'framework combination strategy', where two selective ligands are combined into one dual ligand, which most likely will contain features that are important for one of the target proteins only. The result of this strategy is contradictory to the notion that larger compounds are more selective than smaller compounds, [188, 190] i.e. multitarget compounds should typically be smaller than target-selective compounds. Fragment-based screening is the ideal approach to find a core scaffold capable of binding to two or several target proteins. The core scaffold could then be optimized to a compound with appropriately balanced affinities between the target proteins. Naturally, the greatest likelihood of success would be when the target proteins share a conserved binding site, e.g. kinases. The RAMPED-UP NMR method [191], where several differently labeled target proteins can be screened for binders simultaneously, will probably prove to be useful in finding multitarget core scaffolds.

Membrane Protein Targets

Currently published examples of fragment-based screening by NMR are only applied to soluble target proteins. There are, however, NMR techniques that have been used to detect the binding of small molecules to integral membrane proteins reconstituted in liposomes. STD has been used to detect the binding of peptides to integrin embedded in liposomes [39] and the bound conformation of a GPCR-bound peptide was determined by transferred NOE experiments [192]. The problem is rather the difficulty in producing and successfully reconstituting milligram amounts of therapeutically interesting integral membrane proteins, e.g. GPCRs, membrane-bound enzymes (e.g. tyrosine kinases) and ion channels. In recent years, the number of crystal structures of integral membrane proteins has increased dramatically; especially notable is a second high-resolution GPCR structure [193]. This is mainly a consequence of the impressive progress in the protein production process [194] and leaves good hope that integral membrane protein targets will be subjected to fragment-based screening by NMR and structure-based drug design in the very near future. The integral membrane proteins that will prove most suitable for fragment screening by NMR can be predicted (i) to be possible to produce in milligram quantities and reconstitute in liposomes and (ii) not to contain large amounts of detergents in the binding pocket of interest, i.e. the binding pocket should be hydrophilic. If the binding pocket is lipophilic, the fragments will probably not be able to displace the detergent molecules that could bind with relatively high affinity. A method that appears to be very promising for target proteins that are difficult to produce or that are insoluble, such as membrane proteins, is called TINS (target-immobilized NMR screening) [195, 196]. Binding is detected by comparing ^{1}H 1D spectra in the presence and absence of the target protein that is immobilized on a solid support. The compounds are pumped through a dual flow cell and binding is detected as a simple reduction in ligand peak height in the cell with target protein present. The binders are then washed off and the experiment is repeated with the next fragment cocktail using the same immobilized protein. Hence only a single protein sample is required to screen the fragment library. The use of paramagnetic spin labels attached either to the target protein in the vicinity of the binding pocket of interest or to a known ligand, can also be predicted to play a role in fragment screening of membrane proteins.

CONCLUSION

As instrumentation and experimental methods improve, the role of NMR screening in drug discovery continues to evolve. Of the various criteria for selecting compounds, it is fair to say that solubility remains the most stringent, but the use of probes with cryogenically cooled preamplifiers [197] may mitigate this problem by allowing screens to be run at lower concentrations. Given that advances continue to be made in competing methods for detecting ligand binding (such as mass spectroscopy, surface plasmon resonance, capillary electrophoresis, calorimetry, etc.), such improvements may be essential if NMR is to remain an attractive method for screening.

Like other screening methods, NMR screening is subject to the "garbage in, garbage out" rule; the results are only as good as the compounds that are tested. It has been observed that hit rates by this method are 10 - 1000 times higher than HTS method. It continues to be challenging to design screening libraries that will provide meaningful information, and in particular to design follow-up libraries that bridge the gap between weakly binding NMR screening hits and leads with affinities high enough to be detected by conventional enzymatic assays. The increased use of combinatorial libraries should increase the speed with which follow-up libraries can be synthesized. At the same time, the increased use of genomic information will make gene-family-targeted libraries more useful. Regardless of future advances, thoughtful selection of a screening strategy and intelligent design of primary screening and follow-up libraries will continue to be essential.

As a whole we would like to conclude that one single technology isolated will be never enough for drug discovery and development, as this process is horribly complex and can fail for highly diverse reasons. Hence fragment based drug design approach emerges as one of the most accomplished methodologies in which multidimensional NMR can augment.

ACKNOWLEDGEMENTS

We profusely thank Sri. K. V. Vishnu Raju, Chairman of Sri Vishnu Educational Society, Hyderabad and Sri. Srinivas Reddy, Dean of Sri Vishnu Educational Society, Hyderabad for their financial and moral support respectively. The infrastructural facilities created by Dr. D. Basava Raju, Principal, Shri Vishnu College of Pharmacy to carry out our work is highly appreciated. The Grant in Aid in the form of Career Award for Young Teachers, Fast Track Fellowship for Young Scientists and Minor Research Grant by All India Council for technical Education, Department of Science & Technology and Council for Scientific and Industrial Research respectively are duly acknowledged. We are thankful to Mr. Anand Kumar Rotte,

University of Tubingen, Germany and Mr. Venkat Ratnam Devadasu, University of Strathclyde, UK for providing us the literature required to accomplish this article successfully.

REFERENCES

[1]   Phiel CJ, Zhang F, Huang EY, Guenther MG, Lazar MA, Klein PS. Histone deacetylase is a direct target for valproic acid, a potent anticonvulsant, mood stabilizer and teratogen. J Biol Chem 2001; 276: 36734-41.
[2]   Gottlicher M. Valproic acid: An old drug newly discovered as inhibitor of histone deacetylases. Ann Hematol 2004; 83(Suppl. 1): S91-2.
[3]   Hopkins AL, Groom CR. The druggable genome. Nat Rev Drug Discov 2002; 1: 727-30.
[4]   Braun P, LaBaer J. High through protein production for functional proteomics. Trends Biotechnol 2003; 21: 383-8.
[5]   Webb TR. Current directions in the evolution of compound libraries. Curr Opin Drug Discov Devel 2005; 8: 303-8.
[6]   Posner BA. High-throughput screening-driven lead discovery: meeting the challenges of finding new therapeutics. Curr Opin Drug Dicov Devel 2005; 8: 487-94.
[7]   Betz UA. How many genomics targets can a portfolio afford? Drug Discov Today 2005; 10: 1057-63.
[8]   Betz UA, Farquhar R, Ziegelbauer K. Genomics: Success or failure to deliver drug targets? Curr Opin Chem Biol 2005; 9: 387-91.
[9]   Teague SJ, Davis AM, Leeson PD, Oprea T. The design of lead like combinatorial libraries. Angew Chem Int Ed Engl 1999; 38: 3743-8.
[10]   Hann MM, Leach AR, Harper G. Molecular complexity and its impact on the probability of finding leads from drug discovery. J Chem Inf Comput Sci 2001; 41: 856-64.
[11]   Lipinski CA. Drug-like properties and the causes of poor solubility and poor permeability. J Pharmacol Toxicol Methods 2000; 44: 235-49.
[12]   Oprea T, Davis AM, Teague SJ, Leeson PD. Is there a difference between leads and drugs? A historical perspective. J Chem Inf Comput Sci 2001; 41: 1308-15.
[13]   Kassel DB. Applications of high-thoughput ADME in drug discovery. Curr Opin Chem Biol 2004; 8: 339-45.
[14]   Macarron R. Critical review of the role of HTS in drug discovery. Drug Discov Today 2006; 11: 277-9.
[15]   Jacoby E, Schuffenhuer A, Popov M, *et al*. Key aspects of the Novartis compound collection enhancement project for the compilation of a comprehensive chemogenomics drug discovery screening collection. Curr Top Med Chem 2005; 5: 397-411.
[16]   Leach AR, Hann MM. The in silico world of virtual libraries. Drug Discov Today 2000; 5: 326-36.
[17]   Erlanson DA, McDowell RS, O'Brien T. Fragment-based drug discovery. J Med Chem 2004; 47: 3463-82.
[18]   Carr RA, Congreve M, Murray CW, Rees DC. Fragment-based lead discovery: leads by design. Drug Discov Today 2005; 10: 987-92.
[19]   Graffinity. Graffinity's fragment based drug discovery process - RAISE. http://www. graffinity. com/t_elements.php
[20]   Huber W. A new strategy for improved secondary screening and lead optimization using high-resolution SPR characterization of compound-target interactions. J Mol Recognit 2005; 18: 273-81.
[21]   Blundell TL, Jhoti H, Abell C. High-throughput crystallography for lead discovery in drug design. Nat Rev Drug Discov 2002; 1: 45-54.
[22]   Silverstein RM, Webster FX. Spectrometric identification of Organic Compounds. 6th ed. New York, John Wiley & Sons Inc., 1998: pp. 189-191, 217, 234, 237.
[23]   Pavia DL, Lampmann GM, Kriz GS. Introduction to Spectroscopy. 3rd ed. Thomson Brooks/Cole 2007; pp. 174-6.
[24]   Ni F. Recent developments in transferred NOE methods. Prog Nucl Magn Res 1994; 26: 517-606.
[25]   Li D, DeRose E, London RE. The inter-ligand Overhauser effect: a powerful new NMR approach for mapping structural relationships of macromolecular ligands. J Biomol NMR 1999; 15: 71-6.
[26]   Becattini B, Pellecchia M. SAR by ILOEs: an NMR-based approach to reverse chemical genetics. Chemistry 2006; 12: 2658-62.
[27]   Clore GM, Gronenborn AM. Theory and application of the transferred nuclear Overhauser effect to the study of the conformations of small ligands bound to protein. J Magn Reson 1982; 48: 402-17.
[28]   Mayer M, Meyer B. Characterization of ligand binding by saturation transfer difference NMR spectroscopy. Angew Chem Int Ed 1999; 38: 1784-8.
[29]   Hoffman RA, Forsén S. High resolution nuclear magnetic double and multiple resonances. Prog Nucl Magn Reson Spectrosc 1966; 1: 15-204.
[30]   Peter Schuck. Protein interactions: biophysical approaches for the study of complex reversible systems. NY10013, USA. Springer science + business media, LLC, 233, Springer st 2007; 5: 201-2.
[31]   Cutting B, Shelke SV, Dragic Z, Wagner B, Gathje H, Kelm S, Ernst B. Sensitivity enhancement in saturation transfer difference (STD) experiments through optimized excitation schemes. Magn Reson Chem 2007; 45: 720-4.
[32]   Mayer M, James TL. Detecting ligand binding to a small RNA target via saturation transfer difference NMR experiments in D2O and H2O. J Am Chem Soc 2002; 124: 13376-7.
[33]   Jayalakshmi V, Krishna NR. Complete relaxation and conformational exchange matrix (CORCEMA) analysis of intermolecular saturation transfer effects in reversibly forming ligand-receptor complexes. J Magn Reson 2002; 155: 106-18.
[34]   Yan J, Kline AD, Mo H, Shapiro MJ, Zartler ER. The effect of relaxation on the epitope mapping by saturation transfer difference NMR. J Magn Reson 2003; 163: 270-6.
[35]   Mayer M, Meyer B. Group epitope mapping by saturation transfer difference NMR to identify segments of a ligand in direct contact with a protein receptor. J Am Chem Soc 2001; 123: 6108-17.
[36]   Sandstrom C, Berteau O, Gemma E, Oscarson S, Kenne L, Gronenborn AM. Atomic mapping of the interactions between the antiviral agent cyanovirin-N and oligomannosides by saturation-transfer difference NMR. Biochemistry 2004; 43: 13926-31.
[37]   Hajduk PJ, Mack JC, Olejniczak ET, Park C, Dandliker PJ, Beutel BA. SOS-NMR: a saturation transfer NMR-based method for determining the structures of protein-ligand complexes. J Am Chem Soc 2004; 126: 2390-8.
[38]   Benie AJ, Moser R, Bauml E, Blaas D, Peters T. Virus-ligand interactions: identification and characterization of ligand binding by NMR spectroscopy. J Am Chem Soc 2003; 125: 14-15.

[39]   Meinecke R, Meyer B. Determination of the binding specificity of an integral membrane protein by saturation transfer difference NMR: RGD peptide ligands binding to integrin alphaIIbbeta3. J Med Chem 2001; 44: 3059-65.

[40]   Claasen B, Axmann M, Meinecke R, Meyer B. Direct observation of ligand binding to membrane proteins in living cells by a saturation transfer double difference (STDD) NMRspectroscopy method shows a significantly higher affinity of integrin alpha(IIb)beta3 in native platelets than in liposomes. J Am Chem Soc 2005; 127: 916-9.

[41]   Dalvit C, Pevarello P, Tato M, Veronesi M, Vulpetti A. Identification of compounds with binding affinity to proteins via magnetization transfer from bulk water. J Biomol NMR 2000; 18: 65-8.

[42]   Dalvit C, Fogliatto G, Stewart A, Veronesi M, Stockman B. WaterLOGSY as a method for primary NMR screening: practical aspects and range of applicability. J Biomol NMR 2001; 21: 349-59.

[43]   Wolfgang J, Daniel AE. Fragment based approaches in drug discovery. Germany: Wiley VCH; 2006; PP. 209.

[44]   Dalvit C. Homonuclear 1D and 2D NMR experiments for the observation of solvent-solute interactions. J Magn Reson B 1996; 112: 282-8.

[45]   Dalvit C, Fogliatto G, Stewart A, Veronesi M, Stockman B. Water-LOGSY as a method for primary screening: practical aspects and range of applicability. J Biomol NMR 2001; 21: 349-59.

[46]   Lepre CA, Moore JM, Peng JW. Theory and applications of NMR-based screening in pharmaceutical research. Chem Rev 2004; 104: 3641-76.

[47]   Johnson EC, Feher VA, Peng JW, Moore JM, Williamson JR. Application of NMR SHAPES screening to an RNA target. J Am Chem Soc 2003; 125: 15724-5.

[48]   Di Micco S, Bassarello C, Bifulco G, Riccio R, Gomez-Paloma L. Differential-frequency saturation transfer difference NMR spectroscopy allows the detection of different ligand DNA binding modes. Angew Chem Int Ed 2006; 45: 224-8.

[49]   Feeney J, Batchelor JG, Albrand JP, Roberts GCK. The effects of intermediate exchange processes on the estimation of equilibrium constants by NMR. J Magn Reson 1979; 33: 519-29.

[50]   Meiboom S, Gill D. Modified spin-echo method for measuring nuclear relaxation times. Rev Sci Instrum 1958; 29: 688-91.

[51]   Hajduk PJ, Olejniczak ET, Fesik SW. One-dimensional relaxation- and diffusion-edited NMR methods for screening compounds that bind to macromolecules. J Am Chem Soc 1997; 119: 12257-61.

[52]   van Dongen MJP, Uppenberg J, Svensson S, Lundback T, Kerud T, Wikstrom M , Schultz J. Structure-based screening as applied to human FABP4, a highly efficient alternative to HTS for hit generation. J Am Chem Soc 2002; 124: 11874-80.

[53]   van Dongen M, Weigelt J, Uppenberg J, Schultz J, Wikstrom M. Structure-based screening and design in drug discovery. Drug Discov Today 2002; 7: 471-8.

[54]   Dalvit C, Flocco M, Knapp S, Mostardini M, Perego R, Stockman BJ, Veronesi M, Varasi M. High-throughput NMR-based screening with competition binding experiments. J Am Chem Soc 2002; 124: 7702-9.

[55]   Jahnke W, Perez LB, Paris CG, Strauss A, Fendrich G, Nalin CM. Second-site NMR screening with a spin-labeled first ligand. J Am Chem Soc 2000; 122: 7394-5.

[56]   Jahnke W, Rüdisser S, Zurini M. Spin label enhancedNMRscreening. J Am Chem Soc 2001; 123: 3149-50.

[57]   Jahnke W, Florsheimer A, Blommers MJ, Paris CJ, Heim J, Nalin CM, Perez LB. Second-site NMR screening and linker design. Curr Top Med Chem 2003; 3: 69-80.

[58]   Peng JW. Cross-correlated 19F relaxation measurements for the study of fluorinated ligand- receptor interactions. J Magn Reson 2001; 153: 32-47.

[59]   Dalvit C, Flocco M, Veronesi M, Stockman BJ. Fluorine-NMR competition binding experiments for high-throughput screening of large compound mixtures. Comb Chem High Throughput Screen 2002; 5: 605-11.

[60]   Dalvit C, Dalvit C, Fagerness PE, Hadden DTA, Sarver RW, Stockman BJ. Fluorine-NMR experiments for high-throughput screening: theoretical aspects, practical considerations and range of applicability. J Am Chem Soc 2003; 125: 7696-03.

[61]   www.cryst.bbk.ac.uk/PPPS2/projects/schirra/html/1Dnmr/2Dnmr/3Dnmr.htm

[62]   Protein Data Bank, Brookhaven National (http://www.pdb.bnl.gov); currently being transferred to the research collaborator for structural bioinformatics, RCSB (http://www.rcsb.org).

[63]   Fejzo J, Lepre CA, Peng JW, Bemis GW, Ajay, Murcko MA, Moore JM. The SHAPES strategy: an NMR-based approach for lead generation in drug discovery. Chem Biol 1999; 6: 755-69.

[64]   Shuker SB, Hajduk PJ, Meadows RP, Fesik SW. Discovering high-affinity ligans for proteins: SAR by NMR. Science 1996; 274: 1531-4.

[65]   Hajduk PJ, Sheppard G, Nettesheim DG, Olejniczak ET, Shuker SB, Meadows RP, Steinman DH, Carrera GM JUN, Marcotte PA, Severin J, Walter K, Smith H, Gubbins E, Simmer R, Holzman TF, Morgan DW, Davidsen SK, Summers JB, Fesik SW. Discovery of Potent Nonpeptide Inhibitors of Stromelysin Using SAR by NMR. J Am Chem Soc 1997; 119: 5818-27.

[66]   Medek A, Hajduk PJ, Mack J, Fesik SW. The Use of Differential Chemical Shifts for Determining the Binding Site Location and Orientation of Protein-Bound Ligands. J Am Chem Soc 2000; 122: 1241-2.

[67]   Ross A. Automation of NMR measurements and data evaluation for systematically screening interactions of small molecules with target proteins in NMR: Drug discovery and design Post-Genomic Analysis. McLean, Virginia: Cambridge Health Tech. Institute 2000.

[68]   Boehm HJ, Boehringer M, Bur D, et al. Novel Inhibitors of DNA Gyrase: 3D Structure Based Biased Needle Screening, Hit Validation by Biophysical Methods, and 3D Guided Optimization. A Promising Alternative to Random Screening. J Med Chem 2000; 43: 2664-74.

[69]   Hajduk PJ, Augeri DJ, Mack J, Mendoza R, Yang J, Betz SF, Fesik SW. NMR-Based Screening of Proteins Containing ^{13}C-Labeled Methyl Groups. J Am Chem Soc 2000; 122: 7898-904.

[70]   Pellecchia M, Meininger D, Shen AL, et al. SEA-TROSY (Solvent Exposed Amides with TROSY): A Method to Resolve the Problem of Spectral Overlap in Very Large Proteins. J Am Chem Soc 2001; 123: 4633-34.

[71]   Weigelt J, Van Dongen M, Uppenberg J, Schultz J, Wikström M. Site-Selective Screening by NMR Spectroscopy with Labeled Amino Acid Pairs. J Am Chem Soc 2002; 124: 2446-7.

[72]   Meyer B, Weimar T, Peters T. Screening mixtures for biological activity by NMR. Eur J Bio Chem 1997; 246: 705-9.

[73]   Mayer M, Meyer B. Characterization of Ligand Binding by Saturation Transfer Difference NMR Spectroscopy. Angew Chem Int Ed Engl 1999; 38: 1784-8.

[74]   Meyer B. STD NMR to screen libraries and characterize binding. In NMR: drug discovery and design - post genomic analysis. McLean, Virginia: Cambridge Health Institute 2000.

[75]   Meinecke R, Meyer B. Determination of the Binding Specificity of an Integral Membrane Protein by Saturation Transfer Difference NMR: RGD Peptide Ligands Binding to Integrin $\alpha_{IIb}\beta_3$. J Med Chem 2001; 44: 3059-65.
[76]   Lepre CA. Library design for NMR-based screening. Drug Discov Today 2001; 6: 133-40.
[77]   Bemis GW, Murcko MA. Properties of Known Drugs. 2. Side Chains. J Med Chem. 1999; 42: 5095-9.
[78]   Bemis GW, Murcko MA. The Properties of Known Drugs. 1. Molecular Frameworks. J Med Chem 1996; 39: 2887-93.
[79]   Walters WP, Stahl MT, Murko MA. Virtual screening - an overview. Drug Discov Today 1998; 3: 160-78.
[80]   Walters WP, Murcko MA. Library filtering systems and prediction of drug like properties, in virtual screening for bioactive molecules. New York. Wiley-VCH: 2000; 15-32.
[81]   Leach AR, Hann MM. The *in silico* world of virtual libraries. Drug Discov Today 2000; 5: 326-6.
[82]   Wang J, Ramnarayan K. Toward Designing Drug-Like Libraries: A Novel Computational Approach for Prediction of Drug Feasibility of Compounds. J Comb Chem 1999; 1: 524-33.
[83]   Jarvis RA, Patric EA. Clustering using a similarity measure based on shared near neighbors. IEEE Trans Comput. 1973; C-22: 1025-34.
[84]   Bayada DM, Hamersma H, Van Geerestein VJ. Molecular Diversity and Representativity in Chemical Databases. J Chem Inf Comput Sci 1999; 39: 1-10.
[85]   Menard PR, Lewis RA, Mason JS. Rational Screening Set Design and Compound Selection: Cascaded Clustering. J Chem Inf Comput Sci 1998; 38: 497-505.
[86]   Dixon SL, Koehler RT. The Hidden Component of Size in Two-Dimensional Fragment Descriptors: Side Effects on Sampling in Bioactive Libraries. J Med Chem 1999; 42: 2887-900.
[87]   Flower DR. On the Properties of Bit String-Based Measures of Chemical Similarity. J Chem Inf Comput Sci. 1998; 38: 379-86.
[88]   Young SS, Sheffield CF, Farmen M. Optimum Utilization of a Compound Collection or Chemical Library for Drug Discovery. J Chem Inf Comp Sci 1997; 37: 892-9.
[89]   Caron PR, Mullican MD, Mashal RD, Wilson KP, Su MS, Murcko MA. Chemogenomic approaches to drug discovery. Curr Opin Chem Biol 2001; 5: 464-70.
[90]   Brown RD, Martin YC. Use of Structure−Activity Data to Compare Structure-Based Clustering Methods and Descriptors for Use in Compound Selection. J Chem Inf Comput Sci 1996; 36: 572-84.
[91]   Bohm H-J, Stahl M. Structure-based library design: molecular modelling merges with combinatorial chemistry. Curr Opin Chem Biol 2000; 4: 283-6.
[92]   Drewry DH, Young SS. Approaches to the design of combinatorial libraries. Chemometer Intell Lab 1999; 48: 1-20.
[93]   Vanm Drie JH, Lajiness MS. Approaches to virtual library design. Drug Discov Today 1998; 3: 274-83.
[94]   Higgs RE, Bemis KG, Watson IA, Wikel JH. Experimental Designs for Selecting Molecules from Large Chemical Databases. J Chem Inf Comput Sci 1997; 37: 861-70.
[95]   Good AC, Lewis RA. New Methodology for Profiling Combinatorial Libraries and Screening Sets: Cleaning Up the Design Process with HARPick. J Med Chem 1997; 40: 3926-36.
[96]   Gramer RD, Pattersn DE, Clarck RD, Soltanshahi F, Lawless MS. Virtual Compound Libraries: A New Approach to Decision Making in Molecular Discovery Research. J Chem Inf Comput Sci 1998; 38: 1010-23.
[97]   Muegge I, Martin YC, Hajduk PJ, Fesik SW. Evaluation of PMF Scoring in Docking Weak Ligands to the FK506 Binding Protein. J Med Chem 1999; 42: 2498-503.
[98]   Haque TS, Skillman G, Lee CE, *et al*. Potent, Low-Molecular-Weight Non-Peptide Inhibitors of Malarial Aspartyl Protease Plasmepsin II. J Med Chem 1999; 42: 1428-40.
[99]   Reich SH, Johnson T, Wallace MB, *et al*. Substituted Benzamide Inhibitors of Human Rhinovirus 3C Protease: Structure-Based Design, Synthesis, and Biological Evaluation. J Med Chem 2000; 43: 1670-83.
[100] Brown RD, Martin YC. Designing Combinatorial Library Mixtures Using a Genetic Algorithm. J Med Chem 1997; 40: 2304-13.
[101] Kubinyi H. Similarity and Dissimilarity: A Medicinal Chemist's View. Perspect Drug Discov Des 1998; 9: 225-2.
[102] Lipinski CA, Ed. Bioisosterism in drug design. Ann Rep Med Chem ed. R. C. Allen. 21, Academic press, Inc. 1986.
[103] Patni GA, Lavoie EJ. Bioisosterism: A Rational Approach in Drug Design. Chem Rev 1996; 96: 3147-76.
[104] Schneider G, Neidhart W, Giller T, Schmidt G. Scaffold-Hopping by Topological Pharmacophore Search: A Contribution to Virtual Screening. Angew Chem Int Ed Engl 1999; 38: 2894-6.
[105] Willet P, Barnard JM, Downs GM. Chemical Similarity Searching. J Chem Inf Comput Sci 1998; 38: 983-96.
[106] Stanton DT, Morris TW, Roychoudhury S, Parker CN. Application of Nearest-Neighbor and Cluster Analyses in Pharmaceutical Lead Discovery. J Chem Inf Comput Sci 1999; 39: 21-7.
[107] Hagadone TR. Molecular substructure similarity searching: efficient retrieval in two-dimensional structure databases. J Chem Inf Comput Sci 1992; 32: 515-21.
[108] Carhart RE, Smith DH, Venkataraghavan R. Atom pairs as molecular features in structure-activity studies: definition and applications. J Chem Inf Comput Sci 1985; 25: 64-73.
[109] Sheridan RP, Miller MD, Underwood DJ, Kearsley SK. Chemical Similarity Using Geometric Atom Pair Descriptors. J Chem Inf Comput Sci 1996; 36: 128-36.
[110] Cramer RD, Poss MA, Hermsmeier MA, Caulfield TJ, Kowala MC, Valentine MT. Prospective Identification of Biologically Active Structures by Topomer Shape Similarity Searching. J Med Chem 1999; 42: 3919-33.
[111] Andrews KM, Cramer RD. Toward General Methods of Targeted Library Design: Topomer Shape Similarity Searching with Diverse Structures as Queries. J Med Chem 2000; 43: 1723-40.
[112] Pickett SD, Mason JS, Mclay IM. Diversity Profiling and Design Using 3D Pharmacophores: Pharmacophore-Derived Queries (PDQ). J Chem Inf Comput Sci 1996; 36: 1214-23.
[113] McGregor MJ, Muskal SM. Pharmacophore Fingerprinting. 1. Application to QSAR and Focused Library Design. J Chem Inf Comput Sci 1999; 39: 569-74.
[114] Marriott DP, Dougall IG, Meghani P, Liu Y-J, Flower DR. Lead Generation Using Pharmacophore Mapping and Three-Dimensional Database Searching: Application to Muscarinic M_3 Receptor Antagonists. J Med Chem 1999; 42: 3210-6.
[115] Pickett SD, McLay IM, Clark DE. Enhancing the Hit-to-Lead Properties of Lead Optimization Libraries. J Chem Inf Comput Sci 2000; 40: 263-72.
[116] Mason JS, Hermsmeier MA. Diversity assessment. Curr Opin Chem Biol 1999; 3: 342-9.

[117] Hann MM, Leach AR, Harper G. Molecular Complexity and Its Impact on the Probability of Finding Leads for Drug Discovery. J Chem Inf Comput Sci 2001; 41: 856-64.

[118] Kuntz ID, Chen K, Sharp KA, Kollman PA. The maximal affinity of ligands. Proc Natl Acad Sci USA 1999; 96: 9997-10002.

[119] Lipinski CA, Lombardo F, Dominy BW, Feeney PJ. Experimental and computational approaches to estimate solubility and permeability in drug discovery and development settings. Adv Drug Deliv Rev 1997; 23: 3-25.

[120] Ghose AKm, Viswanadhan VN, Wendoloski JJ. Knowledge-Based Approach in Designing Combinatorial or Medicinal Chemistry Libraries for Drug Discovery. 1. A Qualitative and Quantitative Characterization of Known Drug Databases. J Comb Chem 1999; 1: 55-68.

[121] Gillet VJ, Willet P, Bradshaw J. Identification of Biological Activity Profiles Using Substructural Analysis and Genetic Algorithms. J Chem Inf Comput Sci 1998; 38: 165-79.

[122] Oprea TI. Property distribution of drug-related chemical databases. J Comput -Aid Mol Des. 2000; 14: 251-64.

[123] Hann M, Hudson B, Lewell X, Rob Lifely, Miller L, Ramsden N. Strategic Pooling of Compounds for High-Throughput Screening. J Chem Inf Comput Sci 1999; 39: 897-902.

[124] Ajay A, Walters WP, Murcko MA. Can we learn to distinguish between "Drug-like" and "Nondrug-like" Molecules? J Med Chem 1998; 41: 3314-24.

[125] Ajay A, Bemis GW, Murcko MA. Designing Libraries with CNS Activity. J Med Chem. 1999; 42: 4942-51.

[126] Clark DE, Picket SD. Computational methods for the prediction of 'drug-likeness'. Drug Discov Today 2000; 5: 49-58.

[127] Mitchell T, Showell GA. Design strategies for building drug-like chemical libraries. Curr Opin Drug Discov Develop 2001; 4: 314-8.

[128] Blake JF. Chemoinformatics - predicting the physicochemical properties of 'drug-like' molecules. Curr Opin Bitechnol 2000; 11: 104-7.

[129] Cronin MTD. Computational methods for the prediction of drug toxicity. Curr Opin Drug Discov Dev 2000; 3: 292-297.

[130] Stenberg P, Luthman K, Artursson P. Virtual screening of intestinal drug permeability J Contr Rel 2000; 2: 231-43.

[131] Rishton GM. Reactive compounds and in vitro false positives in HTS. Drug Discov Today 1997; 2: 382-84.

[132] Sadowski J, Kubinyi H. A Scoring Scheme for Discriminating between Drugs and Nondrugs. J Med Chem 1998; 41: 3325-29.

[133] Frimurer TM, Bywater R, Naerum L, Lauritsen LN, Brunak S. Improving the Odds in Discriminating "Drug-like" from "Non Drug-like" Compounds. J Chem Inf Comput Sci 2000; 40: 1315-24.

[134] Lewell XQ, Judd DB, Watson SP, Hann MM. RECAP Retrosynthetic Combinatorial Analysis Procedure: A Powerful New Technique for Identifying Privileged Molecular Fragments with Useful Applications in Combinatorial Chemistry. J Chem Inf Comput Sci 1998; 38: 511-22.

[135] Wagener M, van Geerestein VJ. Potential Drugs and Nondrugs: Prediction and Identification of Important Structural Features. J Chem Inf Comput Sci 2000; 40: 280-92.

[136] Muegge I, Head SL, Brittelli D. Simple Selection Criteria for Drug-like Chemical Matter. J Med Chem 2001; 44: 1841-6.

[137] Xu J, Stevenson J. Drug-like Index: A new approach to measure drug-like compounds and their diversity. J Chem Inf Comput Sci 2000; 40: 1177-87.

[138] Burden FR, Winkler DA. New QSAR Methods Applied to Structure–Activity Mapping and Combinatorial Chemistry. J Chem Inf Comput Sci 1999; 39: 236-42.

[139] Sadowski J. Optimization of chemical libraries by neural networks. Curr Opin Chem Biol 2000; 4: 280-2.

[140] Sadowski J. Database profiling by neural networks, in virtual screening for bioactive molecules, H.-J. Bohm, Schneider, Eds. Wiley- VCH: New York 2000; p. 117-29.

[141] Teague SJ, Davis AM, Leeson PD, *et al.* The Design of Leadlike Combinatorial Libraries. Angew. Chem Int Ed Engl 1999; 38: 3743-7.

[142] Chen A, Shapiro MJ. NOE Pumping: A Novel NMR Technique for Identification of Compounds with Binding Affinity to Macromolecules. J Am Chem Soc 1998; 120: 10258-9.

[143] Lin M, Shapiro MJ. Mixture Analysis in Combinatorial Chemistry. Application of Diffusion-Resolved NMR Spectroscopy. J Org Chem 1996; 61: 7617-9.

[144] Lin ML, Shapiro MJ, Wareing JR. Diffusion-Edited NMR–Affinity NMR for Direct Observation of Molecular Interactions. J Am Chem Soc 1997; 119: 5249-50.

[145] Lin M, Shapiro MJ, Wareing JR. Screening Mixtures by Affinity NMR. J Org Chem 1997; 62: 8930-1.

[146] Dalvit C, Pevarello P, Tato M, Veronesi M, Vulpetti A, Sundström M. Identification of compounds with binding affinity to proteins via magnetization transfer from bulk water. J Biomol NMR 2000; 18: 65-8.

[147] Meylan WM, Howard PH. Atom/fragment contribution method for estimating octanol-water partition coefficients. J Pharm Sci 1995; 84: 83-92.

[148] Meylan WM, Howard PH, Boethling RS. Improved method for estimating water solubility from octanol/water partition coefficient. Environ Toxicol Chem 1996; 15: 100-6.

[149] McElroy NR, Jurs PC. Prediction of Aqueous Solubility of Heteroatom-Containing Organic Compounds from Molecular Structure. J Chem Inf Comput Sci 2001; 41: 1237-47.

[150] Jorgensen WL, Duffy EM. Structure-activity relationships of trans-3,5-disubstituted pyrrolidinylthio-1β-methylcarbapenems. Part 2: J-111, 225, J-114, 870, J-114, 871 and related compounds. Bioorg Med Chem Lett 2000; 10: 1155-8.

[151] Abraham MH, Le J. The correlation and prediction of the solubility of compounds in water using an amended solvation energy relationship. J Pharm Sci 1999; 88: 868-80.

[152] Huuskonen J, Salo M, Taskinen J. Neural network modeling for estimation of the aqueous solubility of structurally related drugs. J Pharm Sci 1997; 86: 450-4.

[153] Huuskonen J. Aqueous Solubility Prediction of Drugs Based on Molecular Topology and Neural Network Modeling. J Chem Inf Comput Sci 1998; 38: 450-6.

[154] Huuskonen J. Estimation of Aqueous Solubility for a Diverse Set of Organic Compounds Based on Molecular Topology. J Chem Inf Comput Sci 2000; 40: 773-7.

[155] Huuskonen J. Comb Chem High Throughput Screen 2001; 4: 311-6.

[156] Klopman G, Zhu H. Estimation of the Aqueous Solubility of Organic Molecules by the Group Contribution Approach. J Chem Inf Comput Sci 2001; 41: 439-45.

[157]   Mitchell BE, Jurs PC. Prediction of Aqueous Solubility of Organic Compounds from Molecular Structure. J Chem Inf Comput Sci 1998; 38: 489-96.
[158]   Jain N, Yalkowsky SH. Estimation of the aqueous solubility I: Application to organic nonelectrolytes. J Pharm Sci 2001; 90: 234-52.
[159]   Ran Y, Jain N, Yalkowsky SH. Prediction Of Aqueous Solubility Of Organic Compounds By The General Solubility Equation (Gse). J Chem Inf Comput Sci 2001; 41: 1208-17.
[160]   Teixido J, Michelotti EL, Tice CM. Ruminations Regarding the Design of Small Mixtures for Biological Testing. J Comb Chem 2000; 2: 658-74.
[161]   Snider M. Screening of Compound Libraries. Consommé or Gumbo? J Biomol Screen 1998; 3: 169-70.
[162]   Ross A, Schlotterbeck G, Klaus W, Senn H. Automation of NMR measurements and data evaluation for systematically screening interactions of small molecules with target proteins. J Biomol NMR 2000; 16: 139-46.
[163]   Shuker SB, Hajduk PJ, Meadows RP, Fesik SW. Discovering high affinity ligands for proteins: SAR by NMR. Science 1996; 274: 1531-4.
[164]   Meyer B, Weimar T, Peters T. Screening mixtures for biological activity by NMR. European J Biochem 1997; 246: 705-9.
[165]   Hajduk PJ, Sheppard G, Nettesheim DG, et al. Discovery of potent non-peptide inhibitors of stromelysin using SAR by NMR. J Am Chem Soc 1997; 119: 5818-27.
[166]   Hajduk PJ, Dinges J, Miknis GF, et al. NMR-based discovery of lead inhibitors that block DNA binding of the human papilloma virus E2 protein. J Med Chem 1997; 40: 3144-50.
[167]   Hoyt DW, Harkins RN, Debanne MT, O'Connor-McCourt M, Sykes BD. Interaction of transforming growth factor alpha with the epidermal growth factor receptor: Binding kinetics and differential mobility within the bound TGFα. Biochemistry 1994; 33: 15283-92.
[168]   McInnes C, Hoyt DW, Harkins RN, et al. NMR study of transforming growth factor alpha (TGFα) epidermal growth factor receptor complex: Visualization of human TGFα binding determinants through nuclear overhauser enhancement analysis. J Biol Chem 1996; 271: 32204-11.
[169]   McInnes C, Grothe S, O'Connor-McCourt M, Sykes BD. NMR study of the differential contributions of residues of transforming growth factor alpha to association with receptor. Prot Eng 2000; 13: 143-7.
[170]   Rajagopal P, Waygood EB, Reizer J, Saier MH, Jr, Klevit RE. Demonstration of protein-protein interaction specificity by NMR chemical shift mapping. Protein Sci 1997; 6: 2624-7.
[171]   Becattini B, Pellecchia M. SAR by ILOE's: an NMR-based approach to reverse chemical genetics. Chemistry 2006; 12: 2658-62.
[172]   Becattini B, Sareth S, Zhai D, Crowell K, Leone M, Reed J, Pellecchia M. Targeting apoptosis via chemical design: Inhibition of Bid-Induced Cell Death by Small Organic Molecules. Chem Biol 2004; 11: 1107-17.
[173]   Becattini B, Culmsee C, Leone M, et al. Structure activity relationships by interligand NOE-based design and synthesis of antiapoptotic compounds targeting. Bid Proc Natl Acad Sci 2006; 103: 12602-6.
[174]   Mares J, Kumaran S, Gobbo M, Zerbe O. Interactions of lipopolysaccharide and polymixin studied by NMR spectroscopy. J Biol Chem 2009; 284: 11498-506.
[175]   Gozalbes R, Mosulen S, Carbajo RJ, Pineda-Lucena A. Development and NMR validation of minimal pharmacophore hypothesis for the generation of fragment libraries enriched in heparinase inhibitors. J Comput Aided Mol Des 2009; 23: 555-69.
[176]   Hajduk PJ, Greer JA. Decade of fragment based drug design: strategies, advances and lessons learned. Nat Rev Drug Discov 2007; 6: 211-9.
[177]   De Kloe GE, Bailey D, Leurs R, de Esch IJP. Transforming fragments into candidates: small becomes big in medicinal chemistry. Drug Discov Today 2009; 14: 630-46.
[178]   McCarthy DJ. Challenges of fragment screening. J Comput Aided Mol Des 2009; 23: 449-51.
[179]   Wendy AW. Fragment based drug discovery. J Comput Aided Mol Des 2009; 23: 453-8.
[180]   Eisen MB. HOOK: A program for finding novel molecular architectures that satisfy the chemical and steric requirements of a macromolecule binding sites. Proteins 1994; 19: 199- 221.
[181]   Lauri G, Bartlett PA. CAVEAT - A program to facilitate the design of organic molecules. J Comput Aided Mol Des 1994; 8: 51-66.
[182]   Maass P. Re-Core: A fast and versatile method for scaffold hopping based on small molecule crystal structure conformations. J Chem Inf Model 2007; 47: 390-9.
[183]   Thompson DC. Confirm: Connecting fragments in receptor molecules. J Comput Aided Mol Des 2008; 22: 761-72.
[184]   Moriaud F. Computational fragment based approach at PDB scale by protein local similarity. J Chem Inf Model 2009; in press.
[185]   Ross A, Senn H. Automation of measurements and data evaluation in biomolecular NMR screening. Drug Discov Today 2001; 6: 583-93.
[186]   Csermely P, Agoston V, Pongor S. The efficiency of multi-target drugs: the network approach might help drug design. Trends Pharmacol Sci 2005; 26: 178-82.
[187]   Morphy R, Rankovic Z. Fragments, network biology and designing multiple ligands. Drug Discov Today 2007; 12: 156-160.
[188]   Morphy R, Rankovic Z. The physicochemical challenges of designing multiple ligands. J Med Chem 2006; 49: 4961-70.
[189]   Morphy R. The influence of target family and functional activity on the physicochemical properties of pre-clinical compounds. J Med Chem 2006; 49: 2969-78.
[190]   Hopkins AL, Mason JS, Overington JP. Can we rationally design promiscuous drugs?, Curr Opin Struct Biol 2006; 16: 127-36.
[191]   Zartler ER, Yan J, Mo H, Kline AD, Shapiro MJ. RAMPED-UP NMR: multiplexed NMR-based screening for drug discovery. J Am Chem Soc 2003; 125: 10941-6.
[192]   Inooka H, Ohtaki T, Kitahara O, et al. Conformation of a peptide ligand bound to its G-protein coupled receptor. Nat Struct Biol 2001; 8: 161-5.
[193]   Cherezov V, Rosenbaum DM, Hanson MA, et al. High-resolution crystal structure of an engineered human beta2-adrenergic G protein-coupled receptor. Science 2007; 318: 1258-65.
[194]   Lundstrom K. Structural genomics and drug discovery. J Cell Mol Med 2007; 11: 224-38.

[195]　Vanwetswinkel S, Heetebrij RJ, Van Duynhoven J, Hollander D, Filippov P, Hajduk G, Siegal. TINS, target immobilized NMR screening: an efficient and sensitive method for ligand discovery. Chem Biol 2005; 12: 207-16.

[196]　Marquardsen T, Hofmann M, Hollander JG, *et al.* Development of a dual cell, flow-injection sample holder and NMR probe for comparative ligand-binding studies. J Magn Reson 2006; 182: 55-65.

[197]　Fragments 2009 conference organized by the Royal Society of Chemistry Biological and Medicinal Chemistry Group at Astra Zeneca, Cheshire, England, 2009; Mar 4-5.

[198]　Fejzo J, Fejzo J, Lepre CA, *et al.* The SHAPES strategy: an NMR-based approach for lead generation in drug discovery. Chem Biol 1999; 6: 755-69.

[199]　Lin M, Shapiro MJ, Wareing JR. Diffusion-edited NMR - affinity NMR for direct observation of molecular interactions. J Am Chem Soc 1997; 119: 5249-50.

Investigation of Drug Delivery Behaviors by NMR Spectroscopy

Soo-Jin Park[*] and Ki-Seok Kim

*Department of Chemistry, Inha University, 253 Yonghyun-dong,
Nam-gu Incheon, 402-751, Korea*

Abstract: Nuclear magnetic resonance (NMR) spectroscopy is the most widely applicable method for drug discovery and analysis. This technique provides a highly specific tool for identifying a drug substance containing impurities and residual solvents and their metabolites in biological media. It also provides a suitable analytical technique for their absolute quantification. In recent years, NMR spectroscopy has been increasingly used to monitor the cumulative drug release, drug dissolution, and diffusion coefficient of drugs from drug delivery systems *in vitro* and *in vivo*. Furthermore, this technique provides a better understanding of the release behaviors of drugs from drug delivery systems based on diffusion, dissolution, and osmosis mechanisms. Although early studies have been mainly qualitative in nature, these techniques can offer considerable information on release processes at the molecular level. Moreover, NMR spectroscopy has been used to detect structural changes that occur in drug delivery systems during the dissolution process. This review focuses on an overview of drug delivery systems and NMR spectroscopy and the application of NMR spectroscopy to drug release behaviors in drug delivery systems.

Keywords: NMR spectroscopy, drug delivery system, drug release, drug dissolution, diffusion coefficient.

1. INTRODUCTION

Conventional dosage forms, such as oral delivery and injection, are the predominant routes for drug administration. However, such dosage forms cannot easily control the rate of drug delivery or the target area of a given drug and are often associated with immediate or rapid drug release. Controlled drug delivery systems (CDDS), i.e., those that release the drug with a controlled profile to maintain an appropriate concentration for desired periods of time, are of great interest in the pharmaceutical industry. CDDS offer various advantages and more effective therapies through elimination of both under- and overdosing, decreased indosing frequency and improved patient compliance, minimized *in vivo* fluctuation of drug concentrations, localized drug delivery, drug protection against degradation in the body and reduced side effects. In addition, site-specific controlled release systems offer many distinct advantages compared with classical drug delivery methods [1-4].

For a long time, a number of CDDS, such as oral, transdermal, and injectable drug delivery systems, have been investigated. For CDDS, various delivery systems or devices, such as micro- or nanoparticles, hydrogels, and osmotic pumps, etc., have been studied and different types of drug delivery systems such as continuous or pulsatile release and pH, temperature, light, ionic strength, and electro-sensitive systems have also been studied during the last decade. An appropriately designed CDDS can serve as a major advance toward solving the problems associated with existing drug delivery systems. Thus, a number of research studies are being developed [5-7].

Generally, most CDDS involve one or more release mechanisms for rate control and the drug release rates from these delivery systems can vary at different release stages. The mechanisms driving drug delivery systems can be classified into three types: diffusion, chemical reaction, and erosion/or degradation. Prevalent drug delivery mechanisms for drug delivery devices are resorption of the drug carrier material onto specific sites and then diffusion [8-10]. The resorption of these devices may, however, cause an inflammatory tissue response that interferes with the treatment sought with the molecules. Therefore, ideal materials for drug delivery systems are biodegradable with high biocompatibility. And much research has also been carried out to obtain a desired release profile through the control of various factors, such as drug loading, morphology, and structure for controlled release systems [11-13].

*Corresponding author: Fax: +82-32-860-8438; E-mail: sjpark@inha.ac.kr

Atta-ur-Rahman / M. Iqbal Choudhary (Eds.)

In recent years, nuclear magnetic resonance (NMR) spectroscopy methods have been increasingly used to monitor the drug release from drug delivery systems *in vitro* and *in vivo* through surface characteristics such as porosity and its evolution, and also a transport of the release medium within the void space. Generally, three distinct sub-disciplines of NMR are solid-state NMR, NMR imaging (MRI), and solution-state NMR, which is the most widely used in drug research. NMR allows measurement of the diffusivity of various molecules such as liquids and drugs within a void space over length scales of a few microns. It can also be used to measure pore size at the nanometer scale and pore distribution through observation of the freezing and melting point depression of fluids imbibed within pores [14-16]. In addition, MRI allows observation of larger length-scale transport processes in drug delivery systems. The advantages of *in vitro* applications of MRI can determine the hydration, swelling, and erosion of polymer matrixes and this technique also has unique preclinical applications in monitoring drug efficiency. In addition, MRI is sensitive to mobile protons and can selectively observe liquids diffused within a porous polymer matrix. Thus it is a useful technique for monitoring liquid penetration into or out of, and liquid distribution within, porous polymers. The NMR technique including MRI also benefits from its non-destructive and non-invasive feature under certain circumstances, allowing experiments to be conducted in situ. Thus it is suitable for studying the time course of the distribution of liquid within a drug delivery system [17-19].

This article focuses on monitoring the release behaviors of drugs from various drug delivery systems such as polymer particles and hydrogels using an analytical technique that incorporates both NMR and MRI. The first part surveys the advantages of NMR and MRI methods in drug research. The second part deals with various drug delivery devices. The third part reports on some examples of studying various factors on how various factors affecting drug release behaviors have been studied using NMR. Finally, the fourth part reports on hydration and diffusion mechanisms for drug release from polymer matrixes in a release medium using MRI.

2. BASIC PRINCIPLE OF NMR

NMR spectroscopy technique was first developed in 1946 by research groups of Stanford and M.I.T. It has been important analysis tool in various fields of the industry, since the development of the high-resolution NMR spectrometer in the 1950s. Many researchers began to apply the NMR technique to their research fields including chemistry and physics with advanced hardware. Generally, last 50 years, NMR technique was developed into the organic chemistry to determine the detailed chemical structure of the natural or synthesized materials. Also, as well-known, another application of NMR technique was the magnetic resonance image (MRI), which is used extensively in the medical field to obtain the information on the tissue of the human body [20, 21].

The principle of NMR technique is based on the magnetic properties of atomic nuclei and these characteristics can be used to obtain chemical and structural information of the organic compounds. Basically, organic compounds are composed of the major elements, such as hydrogen (H), carbon (C), phosphorus (P), nitrogen (N), and oxygen (O). Also, there are the fluorine (F), chlorine (Cl), bromine (Br), and iodine (I) and some metal atoms. Atomic nuclei are composed of nucleons including protons and neutrons and these subatomic particles have 'spin (I)' property. It behaves like an angular momentum and therefore each of these elements can be detected by the NMR. In many atoms, such as ^1H, ^{13}C, ^{31}P, ^{15}N, and ^{19}F, the nucleus does possess an overall spin, whereas in some atoms, such as ^{12}C, ^{16}O, and ^{32}S, the spins are paired and there are no overall spin [22, 23]. This spin interacts with an external magnetic field and the spin of nucleus one can determine as following below rules:

1. If the number of neutrons and protons are both even, the nucleus has no spin ($I=0$).

2. If the number of neutrons plus the number of protons is odd, then the nucleus has a half-integer spin ($I=1/2, 3/2, 5/2$).

3. If the number of neutrons and protons are both odd, then the nucleus has an integer spin ($I=1, 2, 3$).

And, the following features lead to the NMR phenomenon. A spinning charge generates a magnetic field and the resulting spin-magnet has a magnetic moment (μ) proportional to the spin. In the presence of an external magnetic field (B_0), two spin states exist, i.e., +1/2 (α) and -1/2 (β). The magnetic moment of the lower energy state (+1/2) is aligned with the external field, but that of the higher energy spin state (-1/2) is oppose to the external field.

The difference in energy between two spin states is dependent on the external magnetic field strength, and is always very small. The diagram represents that the two spin states have the same energy when the external field is zero and then changed as the increase of field strength. At a field equal to B_x a formula for the energy difference is given.

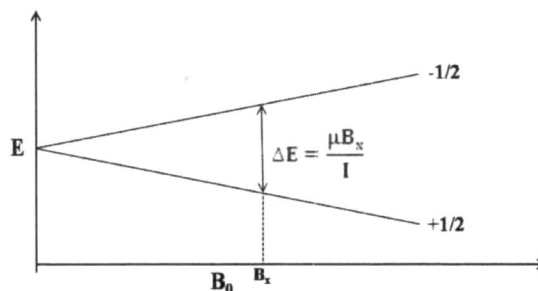

$$\Delta E = \frac{\mu B_x}{I}$$

For NMR spectroscopy, strong magnetic fields are essential. For example, the magnetic field of earth is not constant and shows approximately 10^{-4} T at ground level. However, NMR spectrometers are commonly used with strong magnets of about 1 to 20 T. Even though high magnetic fields, the energy difference between the two spin states is less than 0.1 cal/mole. For NMR, this small energy difference (ΔE) is usually given as a frequency of the range from 20 to 900 Mz and depended on the given magnetic field strength. A sample irradiated by radio frequency (RF) energy, which is corresponded to the spin state separation of nuclei will cause excitation of those nuclei in the +1/2 state to the higher -1/2 spin state. Therefore, NMR is the energetically mildest method used to examine the structure of molecules. For spin 1/2 nuclei, the energy difference between the two spin states at a given magnetic field strength will be proportional to their magnetic moments. For the four common nuclei noted above, the magnetic moments are: ^{1}H=2.7927, ^{19}F=2.6273, ^{31}P=1.1305, and ^{13}C=0.7022. These moments are in nuclear magnetons, which are 5.05078×10^{-27} JT^{-1}.

3. CLASSIFICATION OF NMR INSTRUMENTS

Generally, two types, i.e., continuous wave and Fourier transform, of NMR instrument are used. The continuous wave type is used in an early stage for the experiment and Fourier transform type is mainly used, since 1970s.

In a few decades, NMR spectrometer is known as continuous wave spectroscopy. The principle of continuous wave type is similar to optical-scan spectrometers. The sample is held in a strong magnetic field, and the frequency of the source is slowly scanned. However, the efficiency of continuous wave spectroscopy is lower than that of Fourier transform spectroscopy because this technique probes the NMR response at individual frequencies in succession. Therefore, this system is hardly used except for a few wideline experiments through specialty solid-state NMR.

Commonly, most applications of NMR involve full NMR spectra, which is the intensity of the NMR signal as a function of frequency. The magnitude of the energy changes in NMR spectroscopy is very small, indicating that sensitivity can be a limitation at very low concentrations. To increase sensitivity, one of the methods is the recording many spectra and adding them together. However, to collect the spectra, a lot of time is needed, when continuous wave NMR used. Compared to continuous wave NMR, in Fourier transform NMR, all frequencies in a spectral width are collected simultaneously with a radio frequency pulse and the collection time of the spectra is very short. Fourier transform NMR spectrometers use a pulse of radiofrequency (RF) radiation and the frequency width of the RF pulse is wide enough to simultaneously excite nuclei in all environments. The nuclei reemit RF radiation at their respective resonance frequencies and then a time domain emission signal, known as free induction decay (FID) is recorded by the instrument. The frequencies are obtained by Fourier transformation of the FID of the time-based data and fast decay give broad line and slow decay give sharp line.

4. APPLICATIONS OF NMR TECHNIQUES

NMR spectroscopy is useful for the study of molecules by recording the interaction of radiofrequency electromagnetic radiation with the nuclei of molecules placed in a strong magnetic field. NMR can provide a powerful toolkit to probe the molecular structure of organic- and bio-chemical compounds and uniquely structural content, such as chemical shifts, multiplicity, integrals, intramolecular relationships, and a wide range of dynamic processes, including molecular motions in solution, chemical exchange, and ligand binding [24, 25]. NMR has found various fields of chemistry, physics, biology, and materials science. Together with X-ray crystallography, NMR spectroscopy is one of the two leading technologies for the structure determination of molecules at atomic resolution [26]. Some of the applications of NMR spectroscopy are listed below:

1. *Molecular Structure and Dynamics*

   To determine atomic-resolution structure of molecules and quantify motional properties of molecules.

2. *Ionization State*

   To determine the chemical properties of functional groups in molecules, such as the ionization states of ionizable groups.

3. *Weak Intermolecular Interactions*

   To study weak functional interactions between molecules in the micromolar to millimolar range.

4. *Hydrogen Bonding*

   To detect hydrogen bonding interactions.

5. *Drug Screening, Design, and Metabolite Analysis*

   A powerful technology for identifying drug leads and determining the conformations of the compounds bound to enzymes, receptors, and other proteins and for metabolite analysis.

6. *Chemical Analysis*

   A powerful tool for chemical identification and conformational analysis of synthetic or natural materials.

7. *Material Science*

   A useful technique for the research of polymer chemistry and physics.

5. APPLICATION OF NMR TECHNIQUE FOR DRUG DELIVERY RESEARCH

NMR techniques have long been an effective analytical method in the chemical industry to verify synthesis and compound characterization. Also, NMR has gained widespread acceptance in recent years as a most powerful tool for drug delivery research that has many advantages, as mentioned above [27-29].

Indeed, the last 30 years have seen considerable development of biological NMR since it is the only physical method used routinely at the molecular level for direct study of biological samples, from biofluids, cell or tissue extracts, excised tissues, packed intact cells to isolated living cells or isolated perfused organs, and finally animal models and human subjects.

Recently, NMR techniques have become a major tool for many applications not only in chemistry but also in other fields such as materials science, biology, and medicine. Indeed in recent years, NMR techniques have been increasingly used in drug discovery and to monitor drug delivery systems *in vitro* and *in vivo*. It can provide detailed information about the form of the drug within the matrix, and can identify and quantify the changes in form that may have occurred during processing. NMR spectroscopy can also study drug–excipient interactions by detecting differences in the chemical shifts of the API or the excipient upon formulation. Also, NMR is unique in its ability to permit analysis of the metabolism of both endogenous and xenobiotic compounds such as drugs [30-33].

MRI is also a useful tool to obtain a powerful image that provides internal images of materials or living organisms at a macroscopic scale. It is one of the few techniques for observing internal phases inside materials in situ and can produce 2D and 3D images. However, it has seen only limited application in pharmaceutical research studies, even though it has widespread use and many advantages [34]. In recent pharmaceutical research, MRI studies have focused on drug release mechanisms in drug delivery systems and have been used for a wide range of potential research on topics such as bioadhesion and tablet properties. Although MRI study is currently focused on the diffusion of liquid or drugs, polymer swelling, and the erosion/dissolution of conventional oral dosage forms, only a few papers have examined the various dosage forms for drug delivery to the eyes, skin, and vagina. Also, it is a potential analysis method for monitoring the *in vivo* fate of micro- or nano-sized drug carriers for gene delivery and drug targeting to specific sites [35, 36].

Recently, there have been relatively few reports of the use of NMR technique to characterize drugs in polymeric matrices. Table **1** gives some examples of NMR spectroscopy applied for drug delivery researches. And, in this section we present pulsed-gradient spin-echo NMR and MRI to characterize the drug delivery system.

5.1. Pulsed-Gradient Spin-Echo NMR

For drug delivery systems, several forms are available including liposomes, micelles, hydrogels, and micro- or nanoparticle. NMR technique can be used to quantify the diffusion of the small molecules in numerous drug delivery systems.

The pulsed-gradient spin-echo NMR (PGSE-NMR) offers many advantages like other NMR techniques. The most merit of PGSE-NMR is the chemical specificity, indicating that PGSE-NMR can identify the characteristics of each component within the mixture in a single experiment [37]. However, there is limitation for NMR experiment. The analysis samples have to contain a spin-active nucleus, whereas ^{1}H and ^{19}F are particularly viable for biological samples. The PGSE-NMR has been recently used to study the polymer solutions, gels, supramolecular, and diffusion of small molecular [38, 39]. The main feature of the PGSE-NMR is the application of magnetic field-gradients G(r) that encodes into the NMR signal through the frequency.

Table 1.   Some Examples of NMR and MRI for the Determination of the Drug Delivery Behaviors

No.	Substrate	Active Materials	Objective	Techniques	Refs.
1	PLGA	5-fluoro-5'deoxyuridine	Pore geometry	PFG-NMR, Cryoporometry	[14]
2	Cellulose beads		Pore geometry	Spin echo NMR	[16]
3	GMS:Paraffin	Theophylline	Relaxation time	PGSE-NMR	[17]
4	Mucus		SDC	PGSE-NMR	[44]
5	Biscuit dough		Water mobility	^{1}H NMR	[94]
6	Microemulsions	Vinpocetine	Permeation rate	PFG-NMR	[112]
7	PVA		Molecule diffusion	CONVEX NMR	[113]
8	Phospholipid	Bisphenol A	Drug delivery site	^{13}C NMR	[114]
9	PVA/PVP	Asparagines	Interaction of drug/matrix	^{13}C NMR	[115]
10	Fish oil/EPA	Ketoprofen	Skin permeation	^{1}H NMR	[116]
11	HPMC, HEC, HPC	Sodium salicylate	SDC	PFGSE-NMR	[117]
12	Liposome	PEG-lipid	Structural information	^{31}P NMR	[118]
13	Microemulsions	Prilocaine HCl	SDC	PGSE-NMR	[120]
14	HPMC	5-fluorouracil	SDC	^{19}F NMR	[128]
15	Casein		Water mobility	BT-NMR	[152]
16	PEO	Ketoprofen	Miscibility of polymer/drug	Solid state NMR	[153]
17	HPMC	Antipyrine	Swelling kinetics	MRI	[129]
18	PEO		Swelling kinetics	MRI	[130]
19	Pulsatile capsule	Propranolol HCl	Capsule hydration	MRI	[132]
20	HPMC		Swelling kinetics	MRI	[133]
21	HPMC	Naproxen, naproxen sodium	Water mobility	MRI	[134]
22	Starch		Swelling kinetics	MRI	[136]

(Table 1) Contd.....

No.	Substrate	Active Materials	Objective	Techniques	Refs.
23	Starch/Lactose/PVP	Paracetamol	Disintegration	MRI	[139]
24	Osmotic systems	Isradipine	Swelling kinetics	Benchtop-MRI	[140]
25	PGLA	Goserelin	Swelling kinetics	MRI	[141]
26	HPC, HPMC		SDC	MRI	[142]
27	PolyHEMA		SDC	MRI	[143]
28	PVA		SDC	MRI	[144]
29		Gd-DTPA	Drug permeability	DCE-MRI	[149]
30	Episclera implant	Gd-DTPA	Pharmacokinetics	MRI	[151]
31	Polysaccharides	Acetaminophen, ciprofloxacin	SDC	MRI	[154]
32	Phospholipid		Drug delivery site	MRI	[155]
33	Mycobacteria	Ethionamide	Metabolism	HRMAS NMR	[156]
34		Cyclophosphamide	Metabolism	^{31}P NMR	[157]
35		5-fluorouracil	Metabolism	^{19}F NMR	[158]

-SDC: Self-diffusion coefficients
-HPMC: hydroxypropyl methyl cellulose
-HPC: hydroxypropylcellulose
-HEC: hydroxyethylcellulose
-PVP: poly(vinyl pyrrolidone)
-PEO: poly(ethylene oxide)
-PVA: poly(vinyl alcohol)
-PGLA: poly(d,l-lactide-co-glycolide)
-PolyHEMA: Poly(hydroxyethyl methacrylate)
-GMS: glycomonosaccharide

Generally, diffusion experiments determine a displacement of the molecules during evolution time and are presented by pseudo-2D dataset. The field-gradients are pulsed at the start and end of the evolution time to record the NMR signal with the spatial information. Many pulse sequences have been used to quantify the diffusion for a range of different sample circumstances [40, 41] and the Stejskal-Tanner [42] is commonly used, as well known the spin echo sequence obtained with pulsed gradients, resulting in improved sensitivity to diffusion compared to the steady state gradients [43].

The diffusion of the molecular is usually determined by the calculation of diffusion coefficient. To quantify the diffusion coefficient, there are various analysis methods including the integral of a peak arised from the specific molecule and fitting of the exponential attenuation of this signal. For example, diffusion behaviors of monodisperse molecule show a uniform and exponential decay (Fig. **1a**), whereas polydispersity or slow exchange of the molecule lead to non-exponential decay (Fig. **1b** and **c**). In addition, non-exponential behaviors may arise from the attenuation of a signal arising from similar functional groups in the components, resulting in the spectral overlap.

Recently, there is advanced methods to determine diffusion coefficients of an unknown species in a mixture as well as the known species. One of the most commonly used methods is the diffusion ordered spectroscopy (DOSY) NMR technique, which provide a 2D representation of the diffusion coefficient versus spectral characteristic. Thus, the diffusion coefficients offer dynamical and structural information by the size and shape of the diffusing species in the medium. It is representing that the diffusion coefficients are highly affected to morphologies and surface features of the diffusing species. Also, the physical or chemical interactions between diffusing species are main factor on the diffusion behaviors, resulting in the decrease of diffusion rate by aggregation of diffusing species. If many aggregates of diffusion species are formed, there are two diffusion behaviors, such as the faster rate associated with the non-aggregated species and slower rate associated with aggregates. Therefore, the diffusion of small molecule is dependant to the degree of aggregation and NMR technique is more useful to study diffusion behaviors of small molecules in condition of the non-aggregate and weak interaction between diffusion species.

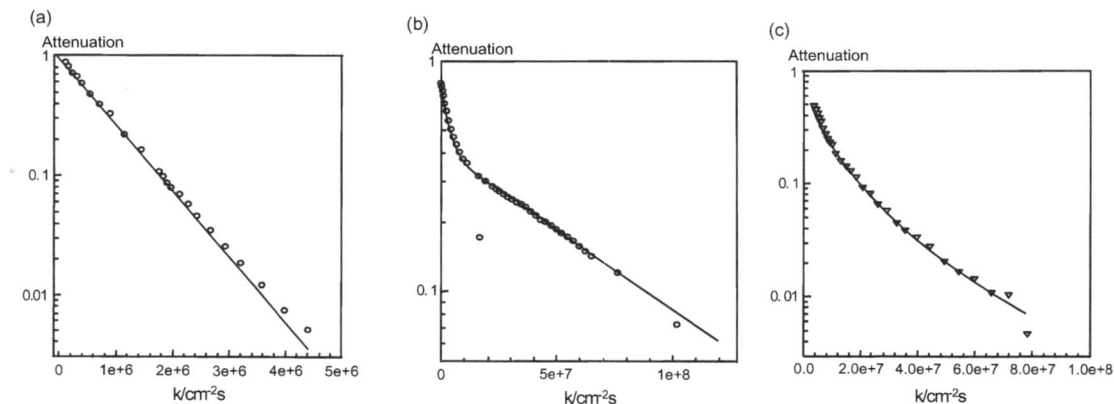

Fig. (1). Attenuation functions arising from the integral of a simple polymer in solution and in a polymer blend showing (**a**) monodisperse, (**b**) bi-exponential and (**c**) polydisperse behaviors [44].

5.2. Magnetic Resonance Images

As one of the NMR techniques, MRI is a non-invasive imaging technique and widely used in medicine and drug delivery system [45, 46]. MRI are formed from the NMR signal by certain nuclei, such as ^1H, ^{19}F, ^{31}P, and ^{13}C in strong magnetic field and irradiated with radio waves. These nuclei have a magnetic moment, leading to the alignment with an applied magnetic field and the NMR signal is proportional to the applied magnetic field. Therefore, MRI is generally applied to samples containing certain nuclei in high concentrations.

In the field of drug delivery systems, MRI can be used to measure the distribution of water in the magnetic field. MRI is generally based on using magnetic field gradients to record the NMR signal to obtain spatial information. The frequency of the NMR signal is depended on water position within the sample. The water mobility and concentration at a specific position within the sample can be determined by the difference of amplitude of the NMR signal.

After excitation by a radio-frequency pulse, the nuclear magnetisation becomes equilibrium state by relaxation. This behavior can be verified by two relaxation times, such as the spin lattice relaxation time (T_1) and the spin-spin relaxation time (T_2). The signal intensity of an MRI is proportional to the concentration of water within the sample. In addition, MRI can be used to observe the change of relaxation times with the application of appropriate sequences of radio-frequency pulse, because the relaxation times strongly depend upon the local molecular environment.

MRI is also used to study the molecular self-diffusion coefficient in the drug delivery systems. The investigation of diffusion behaviors is valuable in the pharmaceutical dosage forms with magnetic field gradients. In this case, a pulsed field gradient is used to tag the position of molecules due to the position dependence on the change of the NMR signal. This offers the prevention of the phase shift of molecules between gradient pulses, leading to restoring the NMR signal. However, the rephasing of molecules between pulses is incomplete, resulting in the signal attenuation. The signal attenuation can be related to the molecular self-diffusion coefficient.

6. CONTROLLED DRUG DELIVERY SYSTEMS

The past two decades have seen rapid growth in controlled drug delivery systems in the field of modern medication and the pharmaceutical industry since controlled release can affect the drug pharmacokinetic, drug bioavailability, safety, and efficacy of the drugs [47, 48]. This increasing interest in drug delivery is due to the increasing need for safe drugs capable of reaching the target with minimal side effects [49, 50]. Over the years, through controlled release research, different systems, ranging from coated tablets and gels to biodegradable micro- and nanoparticles and osmotic systems, have been explored experimentally and computationally to obtain predesigned release profiles. The delivery systems currently available enlist carriers that are either simple, soluble macromolecules such as monoclonal antibodies, natural and synthetic biodegradable polymers, polysaccharides, and more complex particulate multicomponent structures such as microcapsules, nanoparticles, cells, lipoproteins, liposomes, etc. Among a wide number of materials, micelles, liposomes, and polymeric and co-polymeric nanoparticles are widely employed as drug carriers for sustained/controlled delivery [51-53].

6.1. Micro- and Nanoparticles for Controlled Drug Delivery Systems

6.1.1. Microparticles

Microcapsule systems including microcapsules, microparticles, and microspheres for drug delivery have found wide application in pharmacy. Microcapsules are small containers with diameters in the 0.1–100 μm range, containing active compounds in their structures, as shown in Fig. (2). They have been used for a wide range of applications, such as drug delivery systems, foods, cosmetics, and agricultural purposes due to their advantages of greater convenience, physicochemical stability, and the controlled release of active compounds [54-57].

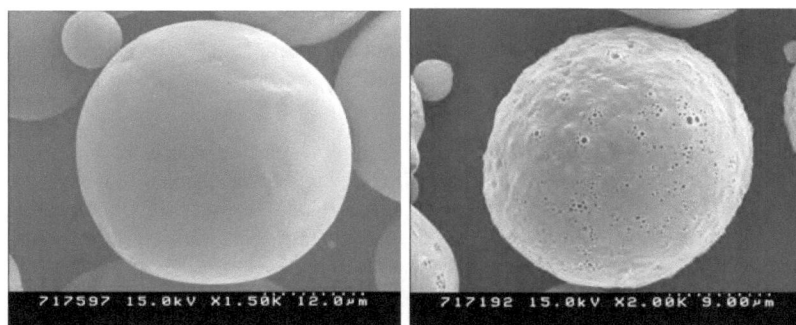

Fig. 2. SEM images of different microcapsules [Ref 57].

For drug delivery systems, biodegradable (or biocompatible) polymeric microcapsules have recently attracted some attention because of their potential application in controlled drug delivery. It has been shown that polymeric microcapsules can be used, intravenously, to administer peptides and other drugs. Using polymeric microcapsules could increase availability, decrease possible associated adverse effects and avoid surgical implantation in some cases. As biodegradable polymers, synthetic aliphatic polyester, such as poly(lactic acid) (PLA), poly(glycolic acid) (PGA), and poly(ε-caprolactone) (PCL), are often used in biomedical applications because they are biocompatible and non-toxic [58-60]. Over the past decade, various technologies have proposed biodegradable microcapsules which include solvent evaporation [61, 62], phase separation [63], and spray drying [64, 65]. The morphologies of different types of microcapsules are shown in Fig. (3).

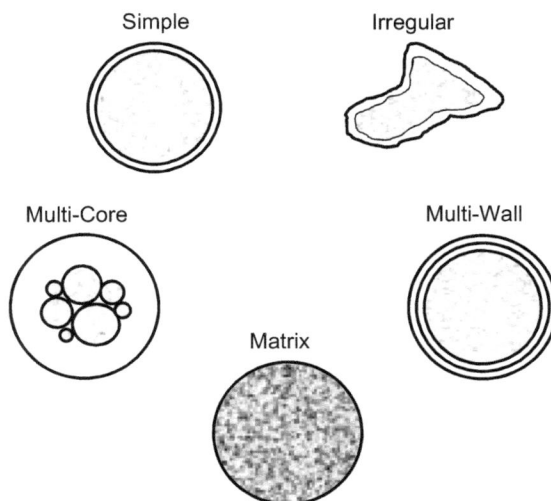

Fig. (3). Morphology of different types of microcapsules [Ref 66].

6.1.2. Nanoparticles

Over the years, nanoparticles have shown potential application in the biotechnology, medical, and pharmaceutical industries. In recent years, several studies have focused on the use of nanotechnology in drug delivery systems because it offers suitable advantages such as site-specific targeting and controlled delivery of drugs and other active materials [67-69]. Nanoparticle drug delivery systems are nanometeric carriers used to deliver drugs or biomolecules. Generally, nanometeric carriers, which are classified as nanoparticles (or nanospheres) and nanocapsules (Fig. **4**), also vary in size from 10 nm and 1 μm with various morphologies [70, 71].

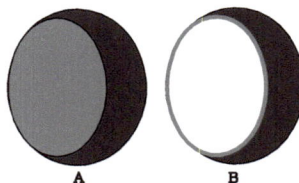

Fig. (4). Schematic representation of a nanosphere (**A**) and nanocapsules (**B**). In nanospheres, the whole particle consists of a continuous polymer network. Nanocapsules present a core-shell structure with a liquid core surrounded by a polymer shell [Ref. 72].

These systems offer many advantages in drug delivery, particularly with regard to improved drug safety, efficacy, and utility. They do so by providing targeted and controlled drug delivery due to the biodegradability, pH, ion and/or temperature sensitivity of the materials; improving bioavailability; extending drug or gene effects in the target tissues; improving the stability of therapeutic agents against chemical/enzymatic degradation; and reducing toxic side effects. In addition, they can pass through micro- or nanoscale capillaries because of their high tiny volume. They can also avoid rapid clearance by phagocytes so that their duration in the bloodstream is greatly prolonged and they can penetrate cells and tissue gaps to arrive at target organs such as the liver, spleen, lung, spinal cord and lymph nodes. The nanoscale size of these delivery systems provides the basis for all these advantages [73-75].

Several types of nanoparticulate systems have been attempted as potential drug delivery systems, including biodegradable polymeric nanoparticles, polymeric micelles, solid nano-particles, lipid-based nanoparticles, nanoliposomes, inorganic nanoparticles, dendrimers, magnetic nanoparticles, and quantum dots, etc., as shown in Fig. (**5**). As drug delivery systems, nanoparticles can entrap drugs or biomolecules into their interior structures and/or absorb drugs or biomolecules onto their exterior surfaces. The drug is dissolved and then entrapped, encapsulated or attached to a nanoparticle matrix and depending upon the method of preparation, nanoparticle carriers can be obtained. Presently, nanoparticles have been widely used to deliver drugs, polypeptides, proteins, vaccines, nucleic acids, genes, and so on [76-78].

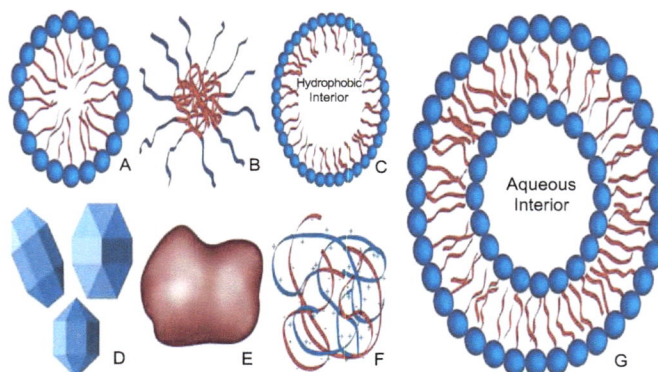

Fig. 5. Schematic illustrations of various nanoparticles; **A, B:** Micelle, **C:** Nanoemulsion, **D:** Crystalline nanoparticles. **E:** Amorphous polymeric nanoparticle. **F:** Condensed ionic oligomers, **G:** Single-walled liposome consisting of an amphiphilic bilayer surrounding an aqueous core [Ref. 78].

For drug delivery systems, several drug release mechanisms of the micro- and nanoparticles may be responsible for the release of the therapeutic agent: diffusion of the drug through the polymer matrix; erosion/degradation of the polymer matrix; and dissolution of the drug in the surrounding medium. The relative importance of each mechanism varies from system to system [79].

In diffusion-controlled systems, the drug carrier retains its structural features even after the drug is exhausted in the release medium. Also, degradation of the polymer matrix occurred throughout the drug release process depending on the drug release time. The rate of diffusion in these systems was controlled by the following factors: 1) solubility and hydrophilicity/hydrophobicity of the drug and polymers in the release medium; 2) structural characteristics such as porosity, tortuosity, surface area, and shape of the system; 3) loaded drug concentration; 4) chemical interaction between the drug and the polymer; and 5) polymer properties such as glass transition temperature and molecular weight [80, 81].

Erosion, which refers to the dissolution or degradation of polymers, means the progressive weight loss of the polymer matrix and subsequent loss of the carrier structure during the release process. Understanding the drug release mechanism of polymer carriers by erosion is very important to determine their release characteristics. A number of parameters affecting drug release may be altered in controlling the drug's erosional release. However, the degradation of the polymer matrix means the bond cleavage and shortening of the polymer chain length, resulting from hydrolysis, an enzymatic reaction or both. Therefore, in degradable polymer systems, polymer degradation is faster than the erosion of the carrier system. In general, polymer degradation is influenced by the following factors: 1) chemical structure and molecular weight of the polymers; 2) polymer concentration; 3) polymer morphologies such as shape, size, and porosity; and 4) environmental conditions such as pH and temperature; and 5) sterilization [82, 83].

6.2. Hydrogels for Controlled Drug Delivery Systems

Since the introduction of hydrophobic gels in the early 1960s, hydrogels have attracted interest as important materials for pharmaceutical application, particularly in recent years. Conventionally, hydrogels are classified as hydrophobic/hydrophilic polymers and non-crosslinked/crosslinked polymers. Recently, many studies of hydrophilic hydrogels have been undertaken with a view to application in drug delivery systems, building upon earlier hydrophobic hydrogel research. They have been used in various applications such as diagnostic, therapeutic, and implantable devices, biosensors, artificial skin, and controlled drug delivery systems because of their biocompatibility with the human body [84-86].

Hydrogels, which are polymeric networks with a three-dimensional configuration, can absorb high amounts of biological fluid, a capability attributed to the presence of hydrophilic groups such as –OH, –COOH, –CONH–, –CONH$_2$–, and –SO$_3$H on polymers. These functional groups lead to different degrees of hydration of the polymers (more than 90%wt) [87, 88]. In addition, functional groups play an important role in drug diffusion from these polymers by controlling ionization and swelling ratio through reactions with the external environment including temperature, ionic strength of the swelling agent, pH or a combination of two or more factors [89-91].

Generally, hydrogels can be classified based on various characteristics such as the side groups of polymer, chemical, physical and structural features, as well as responsiveness to environmental conditions. For pharmaceutical and biological applications, the polymers for hydrogels are obtained from natural sources, synthesis or a combination of natural and synthesized materials [92-94]. Typical examples of hydrophilic polymers used for hydrogels are summarized in Table **2**.

Table 2. Hydrophilic Polymers Used in Preparation of Hydrogels [Ref. 96]

Natural polymers and their derivatives
Anionic polymers: HA, alginic acid, pectin, carrageenan, chondroitin sulfate, dextran sulfate
Cationic polymers: chitosan, polylysine
Amphipathic polymers: collagen (and gelatin), carboxymethyl chitin, fibrin
Neutral polymers: dextran, agarose, pullulan
Synthetic polymers
Polyesters: PEG–PLA–PEG, PEG–PLGA–PEG, PEG–PCL–PEG, PLA–PEG–PLA, PHB, P(PF-co-EG)6acrylate end groups, P(PEG/PBO terephthalate)
Other polymers: PEG-bis-(PLA-acrylate), PEG6CDs, PEG-g-P(AAm-co-Vamine), PAAm, P(NIPAAm-co-AAc), P(NIPAAm-co-EMA), PVAc/PVA, PNVP, P(MMA-co-HEMA), P(AN-coallyl sulfonate), P(biscarboxy-phenoxy-phosphazene), P(GEMA-sulfate)
Combinations of natural and synthetic polymers
P(PEG-co-peptides), alginate-g-(PEO–PPO–PEO), P(PLGA-co-serine), collagen-acrylate, alginate-acrylate, P(HPMA-g-peptide), P(HEMA/Matrigel®), HA-g-NIPAAm

For application in drug delivery systems, hydrogels prepared using natural polymers can offer various advantages as they are usually non-toxic, biocompatible, and possess remarkable physicochemical properties. In addition, synthetic polymers can form well-tailored structures, leading to enhance control of the water absorption rate, degradation kinetic, and physicochemical properties and degradation of the kinetic and physicochemical properties [95]. Accordingly, hydrophilic hydrogels capable of absorbing high amounts of water in their structures show distinctive properties compared to hydrophobic polymeric hydrogels.

The drug release mechanism from hydrophilic hydrogels is generally defined by diffusion of the drug from hydrogel networks in the release medium and hydrophilic hydrogels show highly different drug release behaviors compared with hydrophobic hydrogels due to their hydrophilic nature. In addition, from various studies on the drug release from hydrogels, drug release mechanisms from hydrogels can be categorized as: 1) diffusion controlled, 2) swelling controlled, and 3) chemically controlled. Fick's law indicates that diffusion is the main mechanism for the drug release from hydrogels and is affected by various parameters such as solubility, the degree of crosslinking, chemical structure and composition as well as the release environment [97, 98]. In the swelling mechanism, drug diffusion is significantly increased with hydrogel swelling by absorption of the release medium. Therefore, swelling is considered an important factor in controlling release behaviors [99, 100]. Finally, chemically controlled release is determined by chemical reactions occurring within the hydrogel, including polymeric chain cleavage *via* hydrolytic or enzymatic degradation, or reversible/irreversible reactions occurring between the drug and the hydrogel [101].

7. NUCLEAR MAGNETIC RESONANCE FOR DRUG RELEASE STUDIES

Recently, NMR has become a useful tool for determining the molecular structure and has been used in various applications such as materials science, biology, and medicine since proton nuclear magnetic resonance (^{1}H NMR) in the liquid and the paraffin was studied by Bloch and Purcell. Over the last 30 years, NMR in biological research has been developed in order to analyze the release behaviors of drug delivery systems from *in vitro* and *in vivo* studies at the molecular level [102]. Currently, the use of NMR to investigate drug release behaviors from various dosage forms has drawn increasing interest.

Griffiths *et al.* [44] reviewed the use of pulsed-gradient spin-echo NMR (PGSE-NMR) to quantify the diffusion of drugs (polyplexes, lipoplexes, particles) in mucosal systems. Mucus covers many epithelial surfaces in mammalian organs and prevents foreign particles that enter the body from accessing cells. However, the mucus layer also represents a potential barrier to the efficient delivery of nano-sized drug delivery systems to the underlying mucosal epithelium. Many studies have considered the ability of nano-sized particles and polymers to diffuse within the mucosal network using a range of different techniques, including multiple-particle tracking (MPT), diffusion chamber studies, and fluorescence recovery after photobleaching (FRAP). This article reviewed the current understanding of the interaction of the diffusion of nano-sized structures within mucosal networks and presents PGSE-NMR as a new tool to investigate the mobility of molecular species through mucosal networks and related biological gels.

PGSE-NMR is a non-invasive technique and can be used to quantify the diffusion of micro- and nano-sized drug delivery constructs in colloidal systems and biogels such as mucin glycoprotein solutions, cartilage, whey, and casein protein gels [103-109]. Lafitte *et al.* [110] used PFG-NMR diffusometry to study the effect of porcine gastric mucin on the diffusion of PEGs spanning a molecular weight range 1,020 gmol^{-1}<Mw<716,500 gmol^{-1} in 5wt.% mucin gels as functions of pH, ionic strength, and temperature. They concluded that the structure of the mucin molecules revealed a stronger dependence on PEG diffusion as a result of changes in the pH compared with changes in ionic strength and temperature. It was found that intermediate molecular weight PEG displayed a slower diffusion at pH 4 compared to pH 1 and 7, as shown in Fig. (**6**). The increase of diffusion coefficients between pH 1 and 4 was due to the mucin network being less homogeneous at pH 1. In addition, the diffusion of the PEGs was faster at pH 7 than at pH 4, indicating a decrease in flexibility of the PEG molecules within the network by stronger hydrophobic interactions as the pH decreased.

Assifaoui *et al.* [111] used a low-field ^{1}H NMR technique to characterize water proton mobility in biscuit dough containing starch, gluten, lipids, sugars, fats and water. From transverse relaxation times (T_2) and Carr-Purcell-Meiboom-Gill (CPMG) experiments, they indicated that this system can be distinguished as four populations. The first population measured by the FID sequence was observed at T_{2s}^*~11 μs due to

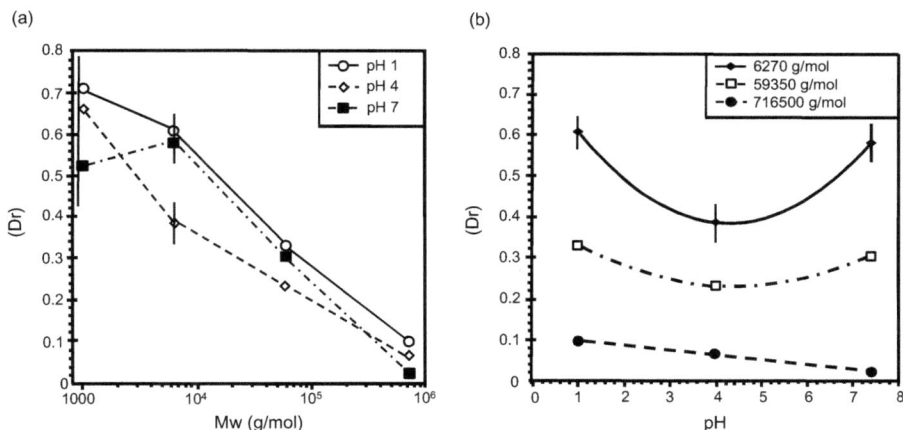

Fig. (6). Relative diffusion coefficients of PEG probe molecules in PGM solutions at 5 wt.% reported as function of the Mw of the polymers (**a**) and as function of the pH (**b**).

the presence of flexible protons associated with the crystalline phase of palm oil, starch, and gluten. The other three populations measured by the CPMG sequence were observed at $T_2(1)\sim2$ ms, $T_2(2)\sim12$ ms, and $T_2(3)\sim105$ ms, respectively (Fig. **7**). Population (1) corresponded to intra-granular protons, while

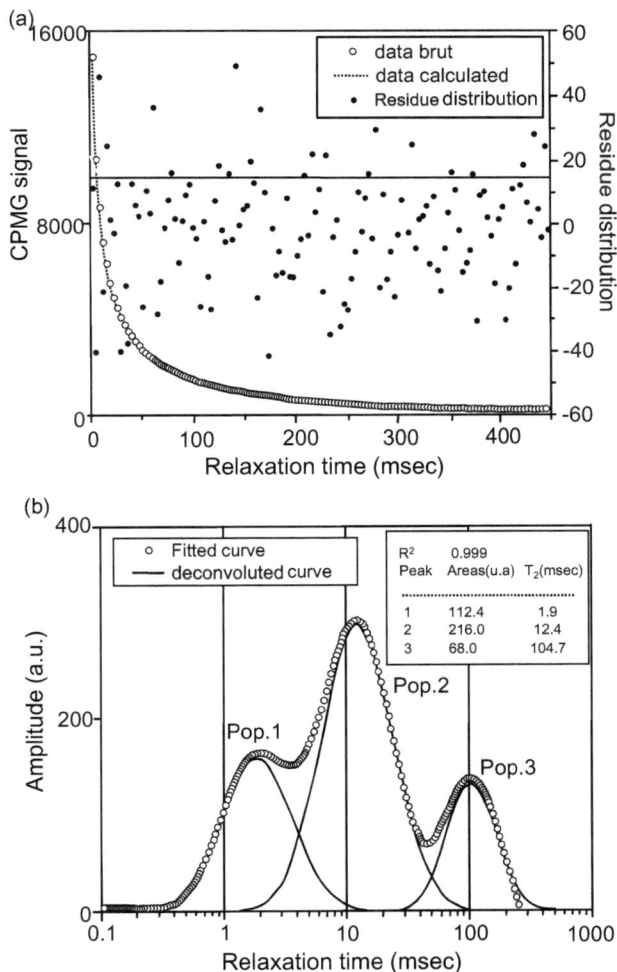

Fig. (7). (**a**) NMR CPMG signal for biscuit dough at 19.4% of moisture and residue distribution obtained from WinDxp software, (**b**) T_2 distribution from the fit; lines represent the deconvolution by three-Gaussian functions. Values on the table present the deconvolution results.

population (2) was very sensitive to water and sucrose content and to temperature. Finally, population (3) was associated with apolar protons due to the fat fraction present in the biscuit dough formula. These results can provide the homogeneity and physical state of biscuit dough systems necessary to understand the mechanisms of dough formation.

Hua *et al.* [112] investigated the correlation between the transdermal permeation rate and structural characteristics of vinpocetine microemulsion using pulsed-field gradient-nuclear magnetic resonance (PFG-NMR). They prepared a novel microemulsion to increase the solubility and the *in vitro* transdermal delivery of poorly water-soluble vinpocetine. The solubility and self-diffusion coefficients (SDC) of vinpocetine were investigated as a function of microemulsion composition with equal drug concentrations in their transdermal delivery. They observed that the SDC of Labrasol, oleic acid, and Transcutol P in all microemulsion systems were very slow (10^{-11} m^2s^{-1} range), while the water diffusion in all microemulsions was faster than that of oil and S_{mix}, as shown in Fig. (**8**). This suggests that water constitutes a continuous free phase in these systems. The diffusion of vinpocetine was very slow in all systems. This indicates that the vinpocetine was incorporated into the inner lipophilic of transdermal systems, which was attributed to its high solubility in the oil-S_{mix} mixture and low solubility in water.

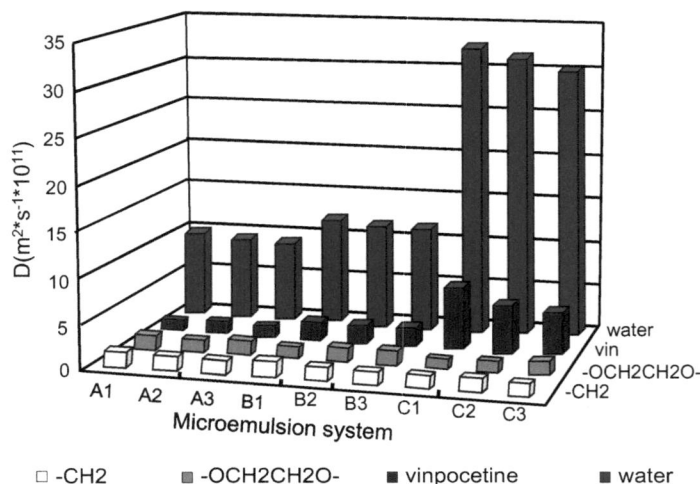

Fig. (8). The self-diffusion coefficients of different components ($-CH_2$, $-OCH_2CH_2O-$, vinpocetine, water) in microemulsions (A1, A2, A3, B1, B2, B3, C1, C2, C3) containing 1% vinpocetine at 25 ℃. The signals of the methylene groups ($-CH_2$) corresponded to multiple species, including Labrasol, oleic acid, and Transcutol P. The self-diffusion coefficients for the unique signal of Labrasol were determined from the resonance of the ethylenedioxy group (OCH_2CH_2O) of the polyethylene glycol (PEG) moiety.

Regan *et al.* [113] used NMR measurements to investigate the diffusion coefficient of the aromatic amino acid phenylalanine (Phe) from PVA hydrogels through a comparison of CONVEX and standard pulsed-gradient spin-echo (PGSE) methods. Generally, PFG-NMR measurements have the advantage to determine the molecular SDC directly, but this technique is difficult to perform in tissues and hydrated polymers. However, CONVEX is an advanced technique to solve this problem by means of NMR excitation-sculpting water suppression. They observed that CONVEX provides superior results, such as a flatter baseline, less phase distortion, and more linear Stejskal-Tanner plots compared with the PGSE method, and enabled a reliable comparison of relative accurate diffusion coefficients of Phe in hydrogels. They concluded that the CONVEX method is a particularly useful tool to determine aromatic low molecular weight compounds and measure the diffusion of aromatic-rich molecules.

Okamura *et al.* [114] monitored the delivery of Bisphenol A (Bis A) from water to phospholipid vesicles using the non-invasive ^1H and ^{13}C NMR technique. From chemical shift differences of the ring proton signals, they indicated that Bis A was trapped to the interface between the lipid head group and the hydrocarbon chain, and the delivery site was confirmed by the ring current effect of Bis A on phospholipid proton signals. In Fig. (**9**), the ^{13}C NMR spectra of the phospholipid vesicles provide complementary evidence for specifying the delivery site. The penetration of Bis A in the lipid interfacial part is typical of shallow penetration (type II). Also, to investigate endocrine disruptors (EDs), NMR is a useful tool and contributed to clarification of the molecular mechanism of the ED delivery into the lipid bilayer. Finally, it is also powerful to monitor the decomposition and the release of EDs accumulated in the membrane interior.

Lubach *et al.* [115] used solid-state NMR to study the state of both the drug and the matrix using two systems: 1) Bupivacaine-loaded microspheres composed of tristearin and encapsulated using a solid protein matrix; and 2) ^{13}C-labeled asparagines (Asn)-loaded poly(vinyl pyrrolidone) (PVP) and poly(vinyl alcohol) (PVA) matrices.

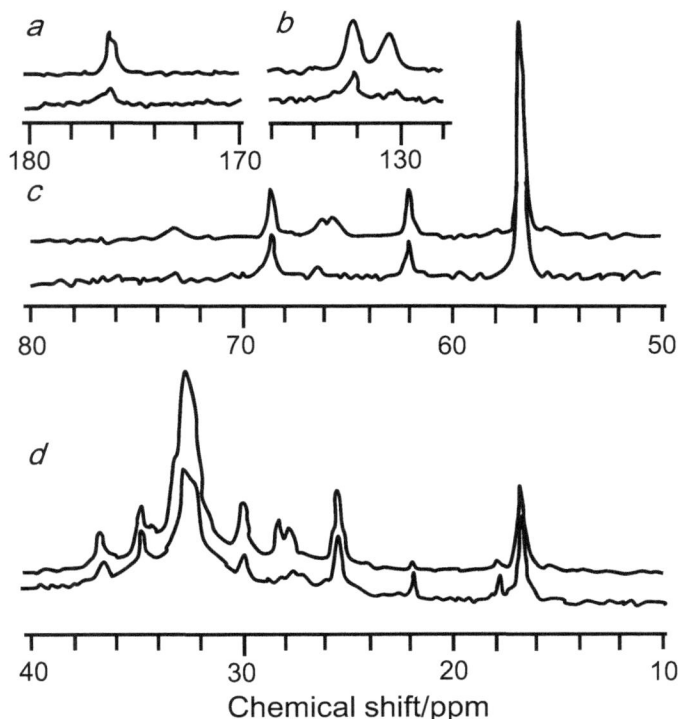

Fig. (9). ^{13}C NMR spectra of the EPC SUV in the presence and the absence of Bis A, at the EPC atom sites of (**a**) carbonyl, (**b**) olefinic, (**c**) choline and glycerol, and (**d**) methylene and methyl carbons. In each spectral region, the upper trace represents the EPC spectrum before the addition of Bis A, while the lower is that after the Bis A delivery. Asterisks denote the signals of the reference, DSS.

Although the solid-state NMR has only recently been used to study these systems, it is a potentially useful technique to obtain insight into the chemistry of solid polymeric pharmaceutical formulations. In their article, they determined the quantity of different forms in polymer matrices, the interaction between drugs and polymers, and the physical and chemical stability of the drugs and polymers using line width and relaxation times to reactivity obtained from solid-state NMR spectroscopy.

Thomasa *et al.* [116] determined the modulation of aromatic protons of ketoprofen using ^{1}H NMR spectra from different formulations containing varying concentrations of fish oil and a control saturated triglyceride. They observed that the changes in the chemical shift of aromatic protons in ketoprofen depend on the concentration of fatty acids. Molecular modeling results indicate that all complexes showed large binding energies from *ca.* 90 to 160 kJmol^{-1}, a finding attributed to strong hydrogen bonds in all cases, and the geometries of these complexes gave rise to regiospecifically solvated complexes. The complexes including ketoprofen with fish oils, EPA or DHA showed increased permeation. It is suggested that the more hydrophilic ketoprofen can aid the permeation of the triglyceride/ free fatty acid through the epidermis *via* the pull effect, indicating the synergistic permeation enhancement of these complexes.

Ferrero *et al.* [117] used pulsed-field-gradient spin-echo nuclear magnetic resonance (PFG-SE NMR) to investigate the release mechanism *via* self-diffusion of the model solute sodium salicylate in hydrogels prepared by various hydrophilic polymers, such as hydroxypropyl methylcellulose (HPMC), hydroxyethylcellulose (HEC), and hydroxypropyl-cellulose (HPC) of varying weight fractions and molecular weights in D$_2$O. Additionally, the extent of bound water and the presence of liquid crystals in the gels as drug diffusion were determined using differential scanning calorimetry (DSC) and polarized light microscopy, respectively. They observed that solute diffusivity is not significantly affected by the nature of cellulose derivative, whereas the SDC of the solute depended on polymer weight fraction, ascertaining the free-volume theory. They also indicated that the structural factor of the polymer matrix was a key point affecting the diffusion. However, the polymer molar mass of the cellulose derivative also did not affect solute self-diffusivity. It is expected that solute molecules can only diffuse in a void space by the solvent. Overall, the hydrated cellulose derivative showed similar retarding effects, which means that the type of swellable polymer did not affect the solute diffusivity in the hydrated gel layer.

Leal *et al.* [118] investigated a series of liposomal formulations as a function of size and PEGylation extent to control drug release using proton-detected NMR diffusion and ^{31}P NMR chemical shifts/bandwidth measurements. ^{1}H NMR was demonstrated to be a quick, non-invasive method to simultaneously determine the diffusional motion, the extent of PEG-lipid incorporation and liposome size. Advantageously, the measurements could be done at the formulation's original conditions of composition and concentration. Additional structural information was obtained by ^{31}P NMR, which is sensitive to aggregate size and shape. They concluded that as the PEGylation extent, self-diffusion ^{1}H NMR provides information about the size and diffusion coefficients of micelles and liposomes. The ^{31}P spectra showed information about the structural features of mixed micelles comprising both PEG-lipids and phospholipids as varying temperatures and PEG molecular weights. Consequently, the most efficient PEG-lipid incorporation in liposomes was achieved with lower molecular weight PEG (2000 Da *vs.* 5000 Da) and when the PEG-lipid acyl chain length matched the acyl chain length of the liposomal core phospholipid.

Perkins *et al.* [14] used NMR cryoporometry to investigate the nature and mechanisms of the structural evolution of the PLGA polymer microspheres that ultimately control drug release kinetics. In order to design polymer carriers for controlled drug delivery it is important to completely understand the nature of a polymer matrix and the mechanisms of drug release from polymer carriers. NMR cryoporometry is a useful tool to study drug release from polymer carriers. However, the study of cryodiffusometry including integrated cryoporometry and the PFG NMR method, as well as specialized cryoporometry techniques such as scanning loops, has limited application. In this work, cryoporometry scanning loops were used to determine the network geometry. Also, cryoporometry freezing curves and PFG NMR were used to investigate the evolution in the pore-scale connectivity and the larger-scale interconnectedness of the nanoporous void space following immersion of PLGA microspheres in an aqueous phase. The molecular weight did significantly affect the trajectory of the structural evolution of the polymer. They indicated that the scanning loops provided the delivery information as pore geometry and the pores of the microsphere are an important factor to control release of larger drug molecules.

Ek *et al.* [119] estimated the tortuosity of porous structures in cellulose beads from water self-diffusion studies by a spin echo NMR technique and direct release measurement. They observed that the tortuosity of the cellulose beads obtained by NMR was slightly higher (τ=2.7) compared with the value obtained from direct release measurements (τ=1.6-2.4). From the change of tortuosity in the release medium, it has been shown to be possible to predict drug release from the structural data of porous cellulose beads. Therefore, without any direct drug release experiments, one can conclude whether the drug was sufficiently sustained by the cellulose matrix itself or if film coating might be necessary. They concluded that the cellulose beads are capable for use in well-defined drug delivery systems and soluble drugs are released from the cellulose matrix, film coating might be necessary for sustained release. It is possible to regulate the drug release rate by varying surface characteristics such as porosity, particle size and drug loading in the cellulose beads.

Kreilgaard *et al.* [120] investigated the influence of structure and the composition of microemulsions on their transdermal delivery potential of a lipophilic lidocaine and a hydrophilic prilocaine hydrochloride as model drugs and microemulsion was compared with conventional vehicles to estimate the potential of microemulsions as drug delivery systems. SDC determined by pulsed-gradient spin-echo NMR spectroscopy and T_1 relaxation times were used to characterize the microemulsions using rat skin and Franz-type diffusion cells. The microemulsions showed increased transdermal delivery of a lipophilic model drug up to four times compared to a conventional o/w-emulsion vehicle, and delivery of a hydrophilic model drug almost 10 times compared to a hydrogel, depending on the microemulsion structure and drug load. The increased transdermal drug delivery from microemulsion was mainly due to the enhanced solubility of drugs, resulting in larger concentration gradients towards the skin and which were dependent on drug mobility in the microemulsion vehicle. Also, measuring SDC was valuable to optimize the fractional composition of a given microemulsion vehicle, in order to maximize drug delivery. The microemulsions did not perturb the skin barrier due to their low skin irritancy. From these results, the microemulsions were promising vehicles for future topical applications with low skin irritancy and PGSE

NMR combined with T relaxation time determinations provide valuable information about microemulsion structures and drug incorporation.

Collins *et al.* [121] used PFG-NMR techniques to characterize the surface-to-volume ratio and tortuosity of the pore structure of partially soluble pharmaceutical pellets. A one-shot Carr-Purcell-Meiboom-Gill sequence was used to determine the spin-spin (T_2) relaxation time of water trapped within the pellets and the pore size distribution of the pellets was analyzed by the Brownstein-Tarr model. They indicated that the pore structure was changed significantly by the diffusivity of water in the pellet matrix and that the surface-to-volume ratio and tortuosity of the pellets decreased with increasing immersion time. Also, in the drug-loaded pellets both the mean and modal pore size increased significantly with immersion time compared with placebo pellets, their values being 6.6 and 2.1 μm after 10 min immersion time (Fig. **10**). This indicated that the effects of immersion time and composition on the pellet pore structure could be used to predict drug release rates and enhance the design of controlled-drug delivery systems.

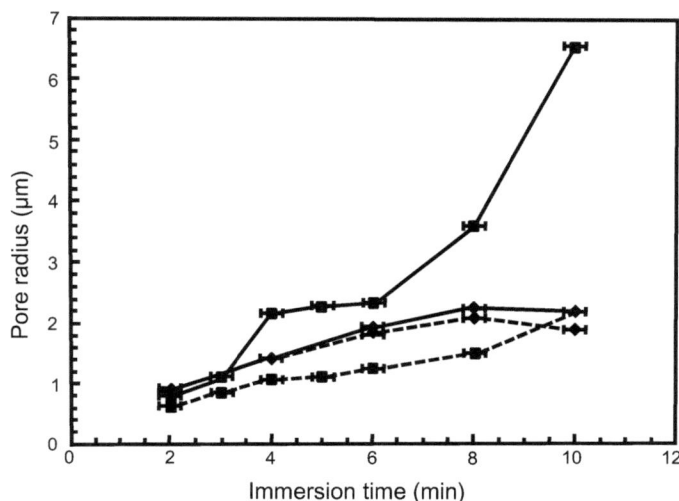

Fig. (10). Plot of the average mean (solid lines) and mode (dashed lines) of the PSD with immersion time for both placebo (◆) and durg-loaded pellets (■). The mean values for saturated pellets are 10.0 and 5.2 μm for the placebo and drug-loaded pellets, respectively. The modal pore size for saturated pellets 12.8 and 0.7 μm for the placebo and drug-loaded pellets, respectively.

8. MAGNETIC RESONANCE IMAGING FOR DRUG RELEASE STUDIES

MRI is a non-invasive technique and can provide cross-sectional images from solid pharmaceutical materials such as tablets or capsules. MRI has also been used to investigate the mechanisms of the drug release behaviors from dosage forms. Most studies have used MRI to observe the hydration of dosage forms as a consequence of biofluid penetration. In addition, MRI has been used to analyze the change of structure and behaviors of various dosage forms in biofluids as time course [122]. To investigate the drug release from dosage form, the use of MRI has increased, whereas there have been few reported studies. Therefore, in this section, we review the various MRI studies on drug release behavior from various dosage forms.

8.1. Oral Delivery Form

A number of studies have performed characterization of the formation of the gel layer and the influence of gel properties on drug release from dosage forms. Rajabi-Siahboomi *et al.* [123] used MRI imaging to investigate the swelling and gel layer formation of hydroxypropyl methylcellulose (HPMC) tablets, which have the grades of similar viscosity and various substituent levels. The kinetic development of the gel layer was clearly apparent in both the axial and radial planes, as shown in Fig. (**11**). They observed that the HPMC grade did affect the drug release, whereas the relationship between gel layer thickness and release kinetics was not identified. Also, the core of the HPMC matrix expanded remarkably in an axial direction with water diffusion as a function of immersion time, indicating the release of tablet compression forces.

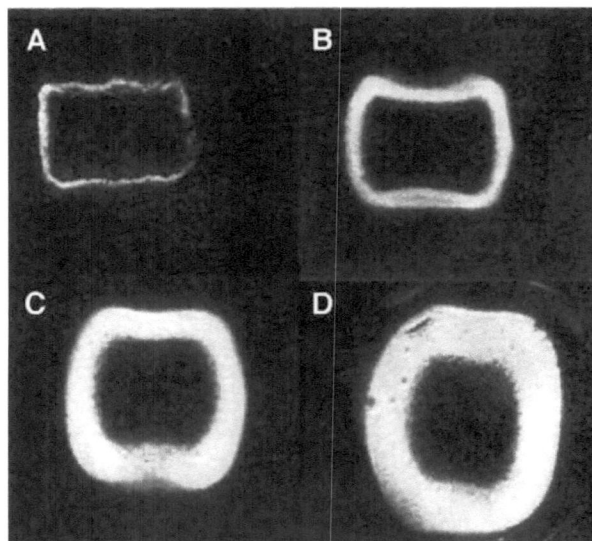

Fig. (11). Images of a K4M HPMC tablet undergoing hydration after (**A**) 10 min, (**B**) 30 min, (**C**) 60 min, and (**D**) 120 min.

The mobility of water within the gel layer using MRI imaging was investigated through the spatial distribution of SDC and the T_2 relaxation values for water, drug, and polymer. SDC is a measure of Brownian motion in the absence of a concentration gradient while T_2 is a measure of the translational and rotational freedom of the molecule in its local environment [124, 125].

Kojima and Nakagami [126] have studied the influence of cellulose ether types, such as hydrating low-substituted hydroxypropyl cellulose (LH41) tablets with or without a drug, hydroxypropylmethyl cellulose (HPMC), and hydroxypropyl cellulose (HPC) on the mobi-lity and diffusivity of water in the gel layer using MRI imaging. One-dimensional maps of the T_2 and water SDC were acquired and used to determine the position of the dry core, gel layer, and aqueous phase. Those maps showed that the extent of gel-layer growth in the tablets was in the order of HPC>HPMC>LH41, and a water mobility gradient existed across the gel layers of all three tablet formulations. The T_2 and SDC in the outer parts of the gel layers were close to those of free water. In contrast, these values in the inner parts of the gel layer decreased progressively, suggesting that water mobility and diffusivity around the core interface were highly restricted, as shown in Fig. (**12**). Furthermore, they observed a positive correlation between the T_2 of ^1H proton in the gel layer of the tablets and the drug release rate from the tablets.

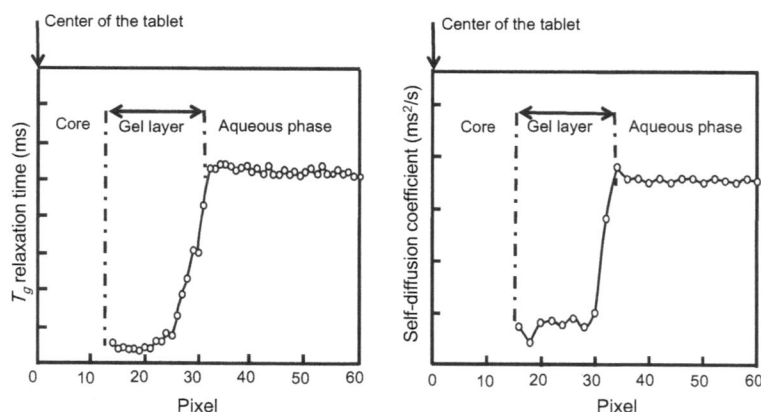

Fig. 12. One-Dimensional T_2 and ADC Maps of the LH41/PRAH tablet after 20 h hydration in water in the coronal direction.

Rajabi-Siahboomi *et al.* investigated an SDC of water in a USP2008 HPMC matrix using MRI imaging. They used a gradient to explain the retardation characteristics of HPMC matrices and indicated that their inner gel layer had a greater diffusional resistance to water. Fyfe CA and Blazek [127] used ^1H NMR spectroscopy to investigate the mobility of HPMC chains within the gel layer. From a relationship between the T_2 relaxation of bound water protons and polymer concentration, they indicated that HPMC chain mobility decreased substantially with increasing HPMC concentration, which resulted from the physical entanglement of polymer molecules with gel formation (Fig. **13**). They concluded that the gel layer consisted of three regions through T_2 weighted images of hydrated HPMC: 1) an outer layer composed of freely moving polymer chains by dissolution and diffusion in water; 2) the gel layer with polymer chain mobility reduced by inter-chain cohesive force; and 3) the inner layer of solid-like.

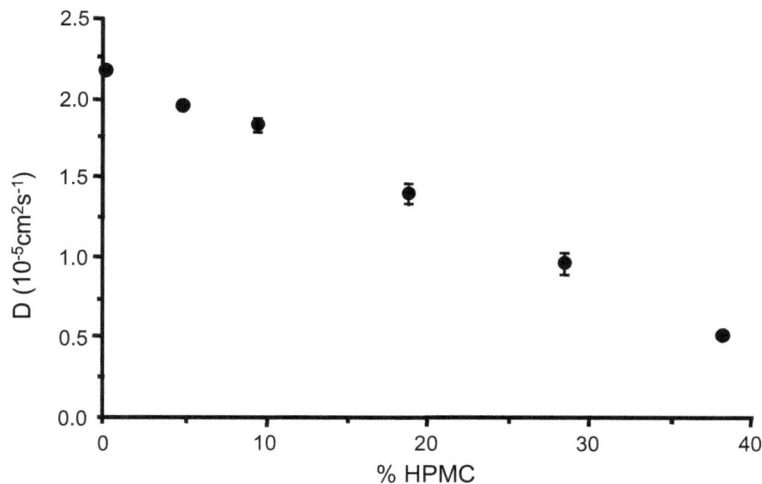

Fig. (13). Self-diffusion coefficients of the water component in the HPMC/water mixtures as determined with the PGSE experiment.

Fyfe and Blazek-Welsh [128] have studied the distribution and release of two model drugs, triflupromazine and 5-fluorouracil, from hydrating HPMC matrices using one-dimensional ^{19}F MRI imaging. From the SDC results they observed that the triflupromazine with low solubility showed a slow release rate from the matrix containing less than a 10%w/w HPMC concentration, indicating slow mobility within the gel layer of lowly hydrated regions. It is suggested that the release of the drug was associated with the erosion mechanism and the region of the matrix containing less than 10%w/w HPMC did not significantly affect gel strength, which is attributed to considerable chain disentanglement and dissolution into the bulk solution. However, the 5-fluorouracil showed fast release behaviors from the tablet containing up to 30%w/w HPMC, indicating free diffusion within the gel layer and. It is expected that a relatively soluble drug would depend on the release mechanism by diffusion. They concluded that understanding drug properties is useful to predict the drug release mechanism from various hydrophilic matrixes for drug delivery systems.

Dahlberg *et al.* [129] investigated the swelling characteristics of a hydroxypropyl methylcellulose (HPMC) matrix incorporating the hydrophilic drug antipyrine using MRI imaging. They observed the swelling behaviors and the drug release behaviors of HPMC tablets in the water. Using MRI imaging they confirmed the concentration released of the drug into the aqueous phase by swelling, which resulted from a diffusion mechanism. The distribution and hydration of the HPMC tablet during the swelling was evaluated by images shown in Fig. (**14**). At the initial stage, the tablet showed immobile images that corresponded with short T_2 values, leading to small image intensity. With further water penetration into the tablet, the tablets became more mobile and the T_2 value became longer, indicating an increase in their swelling. With increasing time the gel layer of the tablets was broadened and expanded and the HPMC distribution was fairly uniform after volume expansion of several times compared with that of the original tablet. This provides a significant understanding of the release mechanisms in HPMC tablets.

Hyde and Gladden [130] used one-dimensional NMR images to quantify the concentration profile of water and polymers in hydrating poly(ethylene oxide) discs. From spin-lattice relaxation time (T_1)-weighted imaging they measured the quantification of polymers, water concentration profiles, and penetrant and swelling kinetics. This indicates that differences exist in the T_1 relaxation time of water and polymers, quantification of the distribution of polymers and water across the hydrating gel layer. The T_1 of the water in the gel was proportional with water concentration, whereas the T_1 of the polymer in the gel was relatively constant. This suggested that the mobility of the polymer chain segments was relatively constant once the polymer was in contact with water. They concluded that both water penetration and polymer swelling were dependent on the diffusion mechanism.

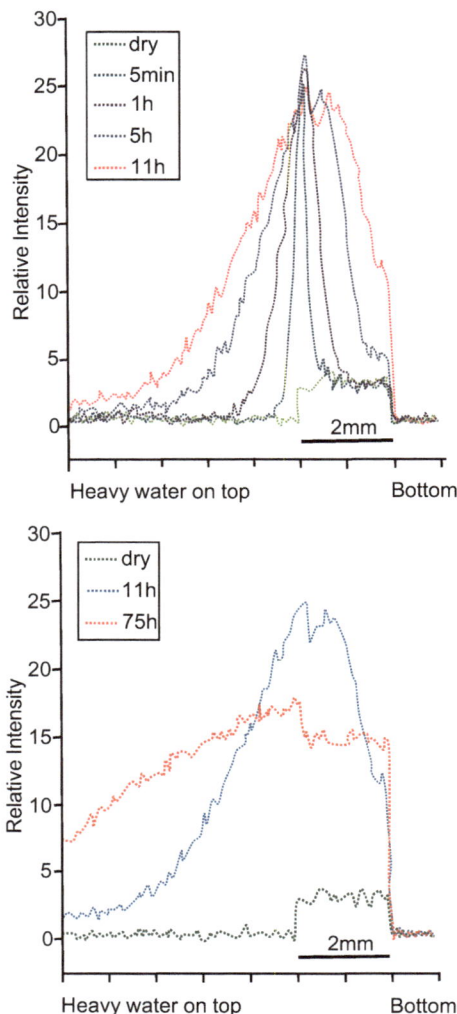

Fig. (14). 1D images illustrating the evolution of the distribution of the hydrated and therefore mobilized HPMC polymers during the swelling process for one of the tablets undergoing D2O penetration from the top. The signal amplitude is a combined measure of the polymer concentration and spin relaxation times as provided by Eq. (1). The polymer profiles are shown for the dry tablet and 5 min, 1 h, 5 h, and 11 h (**A**) and for the dry tablet and 11 h and 75 h (**B**) after the initiation of hydration. The reference tablet is outside the displayed range (to the right).

Messaritaki *et al.* [131] used PFG-NMR and confocal microscopy techniques to study the structural evolution and drug release profile of poly(d,l-lactide-co-glycolide) (PLGA) microspheres prepared by varying the drying process during immersion in an aqueous phase. They indicated that the cavities within the PLGA sphere grew over a period of 24-48 h and the variation of the drying process for the PLGA microspheres significantly affected the degree of permeability of the spheres to water by the changes in the pore structure of the microspheres. From confocal microscopy studies, the drug was distributed in all regions within a PLGA microsphere and the spatial distribution was heterogeneous, indicating a heterogeneous drug release process. Also, the release of a drug from the PLGA microspheres indicated that the rate of drug release was associated with the swelling rate of the PLGA matrix. This means that the mechanism of the drug was dependent on a percolation process, as shown in Fig. (**15**).

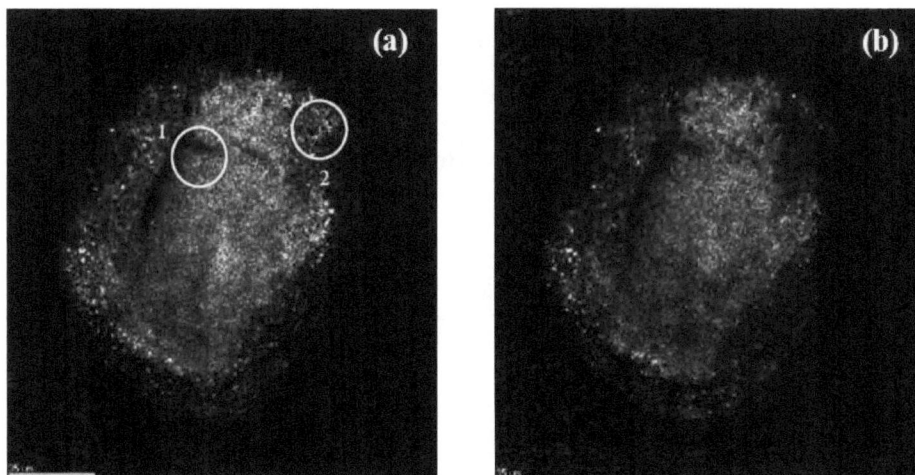

Fig. (15). Confocal micrographs of the dequatorialT plane of a microsphere containing model drug that has been immersed in flowing water for (**a**) 15 min and (**b**) 55 min. Whiter regions contain a higher concentration of drug. The circles in panel (a) denote the regions of interest (ROI) for which individual release profiles were obtained.

Sutcha *et al.* [132] used MRI imaging to investigate the effect of coating-dependent release mechanisms from chronopharmaceutical capsules. They observed that such capsules, which are coated with ethyl-cellulose to prevent water penetration, show clearly different release behaviors from those coated by organic or aqueous processes. Organic coating on the capsules provided a more delayed release, whereas aqueous-coated capsules showed less delayed and irregular release behaviors, as shown in Fig. (**16**). It is

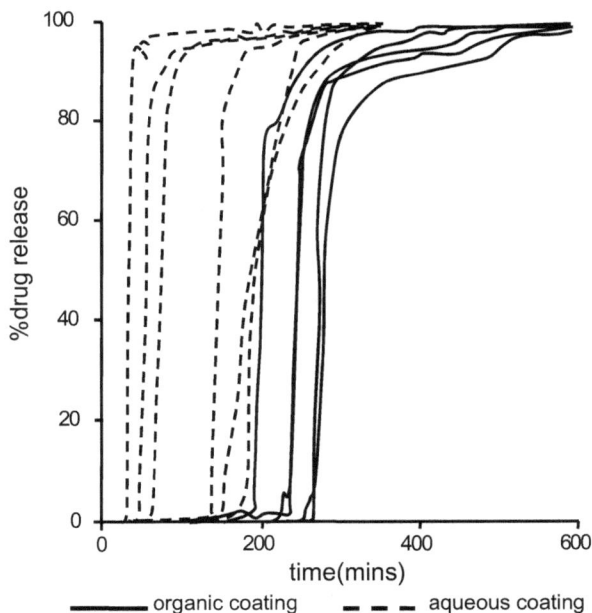

Fig. (16). Dissolution profiles for chronopharmaceutical capsules with lactose/dibasic calcium phosphate plugs coated with either an aqueous- or organic-based coating process.

clear that the rapid hydration of aqueous-coated capsules caused rapid influx of water into the capsule, leading to the swelling of the excipient containing the drug inside the capsule. In addition, MRI images showed that the seal between the capsule shell and plug was a key route for water penetration into the capsules (Fig. **17**). This indicates that the capsule plug was probably a key factor in controlling the drug release from the capsules. They concluded that MRI imaging is a powerful tool to determine the mechanisms of hydration and drug release in this novel dosage form. Tracking the mobility of water in the dosage form is important to understand the mechanism of failure in the aqueous-coated capsule and to clarify the action and kinetics of the organic-coated capsules.

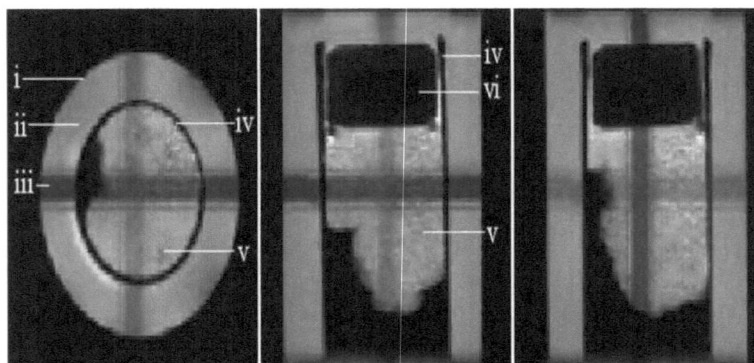

Fig. (17). Reference scans of chronopharmaceutical capsules (left to right), axial, coronal and sagittal views showing (**i**) tube wall, (**ii**) water around capsule, (**iii**) shadow artefact caused by interleaved acquisition of the three imaging planes, (**iv**) capsule edge, (**v**) hydrated material inside capsule and (**vi**) plug. The axial image is 6 mm from the top of the capsule.

Fyfe *et al.* [133] used MRI imaging to investigate the physical change and water penetration in solid dosage forms *via* a flow-through dissolution apparatus, USP Apparatus 4. Using NMR imaging, they illustrated the behaviors of diffusion, dissolution, and osmotic delivery systems, providing a better understanding of the processes in the release of drugs from drug delivery systems based on diffusion, dissolution and osmosis mechanisms. In addition, they indicated that MRI imaging is capable of complementing the drug release profiles obtained by the conventional USP method and is invaluable to the monitoring of drug delivery systems, providing direct information on the nature of molecular processes.

Harding *et al.* [17] used MRI imaging to study the drug release process from small (sub-mm) lipophilic matrix theophylline beads. They indicated that NMR imaging can be used to measure quantitatively the penetration of a liquid into drug-loaded spherical beads with sub-mm diameters. NMR imaging shows that the liquid ingress penetrated from front to center of the bead with increasing exposure time, as shown in Fig. (**18**). Also, both liquid penetra-tion and drug release rates increased as the glycomonosaccharide (GMS):paraffin ratio in the bead matrix formulation increased, a finding attributed to the increase in the concentration of liquid absorbed in the beads with increasing GMS content. It was also supported by the decrease in the liquid T_2 relaxation with increasing GMS content due to stronger interaction with the matrix material.

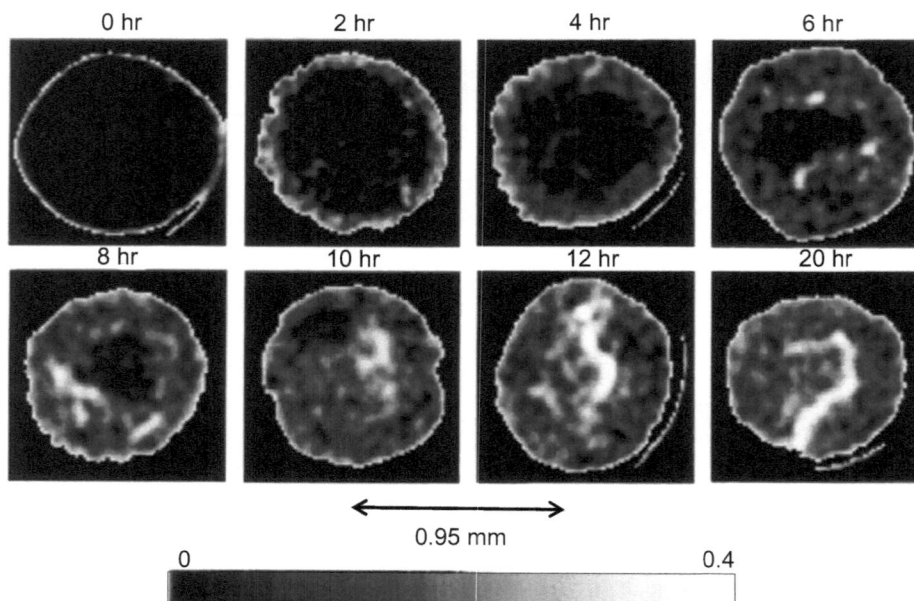

Fig. (18). Typical images acquired using the multi-slice imaging experiment showing the ingress of dissolution solution into beads of GMS:paraffin ratio (**a**) 1.5:1, (**b**) 1:3 and (**c**) 1:5, after varying exposure times in the dissolution bath as indicated in the figure. The intensity scale is relative to the intensity of free dissolution solution. The signal from the solution surrounding the beads has been removed to ease interpretation.

Many researchers have investigated the effect of the molecular weight of the polymer matrix on the drug release. Katzhendler *et al.* [134] studied the mechanism of drug release through microstructure, mobility, internal pH and the state of water within the gel layer of hydrated HPMC matrices containing naproxen and naproxen sodium, which have different molecular weights using nuclear magnetic resonance (NMR), electron paramagnetic resonance (EPS), and differential scanning calorimetry (DSC). They indicated that the HPMC of various viscosity grades showed similar microviscosity values in spite of the difference in their molecular weights and that the higher molecular weights of HPMC showed higher water absorption capacity and higher swelling. However, the drug release behaviors did affect as the difference in the solubility of the model drugs, naproxen sodium and naproxen, incorporated into the HPMC matrix. The less soluble naproxen-loaded HPMC shows how slow the release rate was compared with the HPMC matrices containing naproxen sodium of high solubility. This means that the low drug solubility led to the erosion-controlled mechanism, whereas the high drug solubility led to the diffusion mechanism.

Masaro *et al.* [135] studied the self-diffusion of water and poly(ethylene glycol) of a molecular weight of 600 (PEG-600) in aqueous systems of selected hydrophilic polymer matrixes, such as poly(vinyl alcohol) (PVA), hydroxypropyl methyl cellulose (HPMC), poly(*N,N*-diethylacrylamide) (PNNDEA), and poly(*N*-isopropylacrylamide) (PNIPA) using the pulsed-gradient spin-echo NMR technique. From the results, the diffusion of both water and PEG-600 in the same class of polymers was rather similar and did not vary significantly with molecular weight or small variations in the degree of hydrolysis of the polymer matrix used. It was expected that the major effect on diffusion in hydrophilic polymers would be the formation of hydrogen bonds as exemplified by the differences observed for PNNDEA and PNIPA. In PNIPA and PNNDEA, the formation of hydrogen bonds was not as easy as in PVAs and HPMCs. Therefore, the variations in the diffusion data should be attributed to the quality of the solvent, as shown in Fig. (**19**).

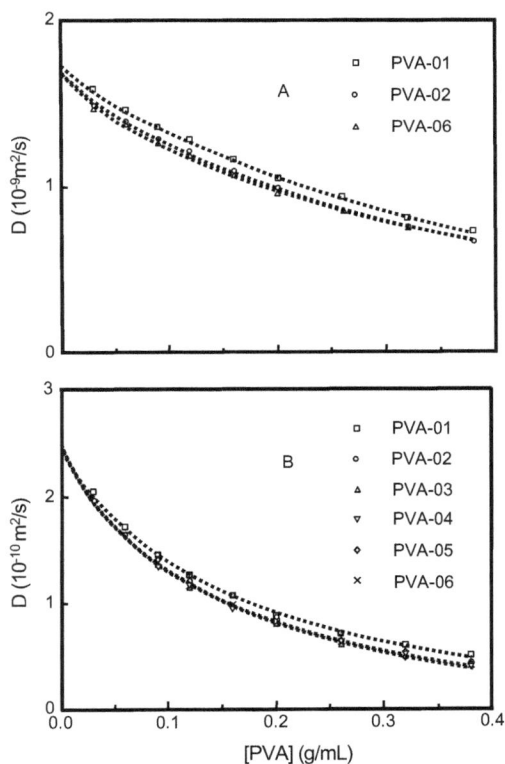

Fig. (19). Plot of the self-diffusion coefficient of water (**A**) and PEG-600 (**B**) as a function of the polymer (PNNDEA and PNIPA) concentration for at 25 °C. Dashed lines are fits to eq 3.

Baille *et al.* [136] used MRI imaging to investigate the effect of temperature on the swelling and water uptake of modified high-amylose starch matrices. Pharmaceutical tablets made of modified high-amylose starch have a hydrophilic polymer matrix into which water can penetrate with time to form a hydrogel. The tablets immersed in water were imaged at different time intervals on a 300 MHz NMR spectrometer. Radial images clearly showed the swelling of the tablets and the water concentration profile. The rate constants for water diffusion and the tablet swelling were extracted from the experimental data. At 25°C, the tablet showed proportionately increased water penetration with hydration time, indicating Case II diffusion kinetics, whereas at 37°C water penetration was not proportional (Fig. **20**). This change corresponded with the gelatinization temperature of the starch at approximately 37°C.

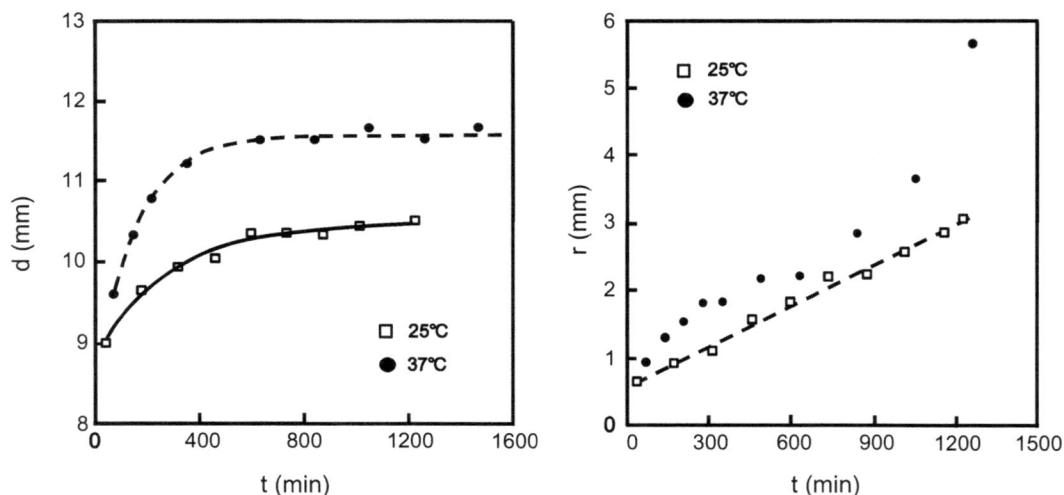

Fig. (20). Swelling of high-amylose starch tablet as a function of time at 25 and 37 °C (**a**) and Advance of solvent front expressed as r, the distance of the solvent front from the edge of the tablet toward the center of the tablet, at 25 and 37 °C as a function of immersion time (**b**).

Tritt-Goc and Piślewski *et al.* [137] determined the diffusion kinetics of water through the hydration of hydroxypropylmethylcellulose (HPMC) matrices in different pH environ-ments using 2D mapping of T_2 relaxation times and proton density *via* MRI imaging. The measurements were performed at two pH values of water, 2 and 6, and two temperatures, 25 and 37°C. They observed that water penetration into HPMC exhibit relaxation controlled (Case II kinetics) at pH=2 and Fickian behavior at pH=6. It was also observed that radial swelling was larger for the system composed of HPMC and water at pH=6 than at pH=2. Therefore, in the release medium environment, the change in pH value had an effect similar to the change in temperature. Consequently, the change of the diffusion mechanism with a pH value can give valuable information as to the application of the hydroxypropyl-methylcellulose matrix in the production of controlled release drugs.

Fahie *et al.* [138] investigated the drug release characteristics from porous and non-porous tablets using NMR imaging. The NMR images of tablets made of two different formulations were taken at various time intervals. They observed that a porous coated tablet shows a faster drug release, which was attributed to the higher exposed core surface area in the dissolution medium and this caused an increase in the rate of dissolution of the core. These results indicated that the porosity of the matrix can affect the drug release and control drug release behaviors, leading to the development of a successful controlled-dosage form.

Tritt-Goc and Kowalczuk *et al.* [139] studied the disintegration behavior of paracetamol tablets as a function of temperature by MRI imaging *via* the Snapshot FLASH method. The disintegration behaviors were carried out under a pH 2 gastric conditions and the total duration of the single experiment was 425 ms. They observed that the Snapshot FLASH MRI method can be used to successfully follow the *in vitro* behavior of the disintegration of paracetamol tablets varied at different temperatures. The distribution of protons of 4-(*N*-acetyl)aminophenol within the paracetamol tablet was shown to be homogeneous and the disintegration behavior was dependent on the temperature of the release medium, as shown in Fig. (**21**). These results indicated that MRI technique may serve to predict the disintegration behavior of paracetamol in the stomach after oral administration.

Malaterre *et al.* [140] studied the drug release mechanism from push–pull osmotic systems (PPOS) by MRI imaging using a new benchtop apparatus. The signal intensity profiles of both PPOS layers were monitored non-invasively over time to characterize the hydration and swelling kinetics. The drug release behaviors were well-correlated to the hydration kinetics. They indicated that (i) hydration and swelling critically depended on the tablet core composition, such as the presence of osmotic active agents, and

19°C

| 60s | 270s | 390s | 960s |

37°C

| 90s | 120s | 150s | 180s |

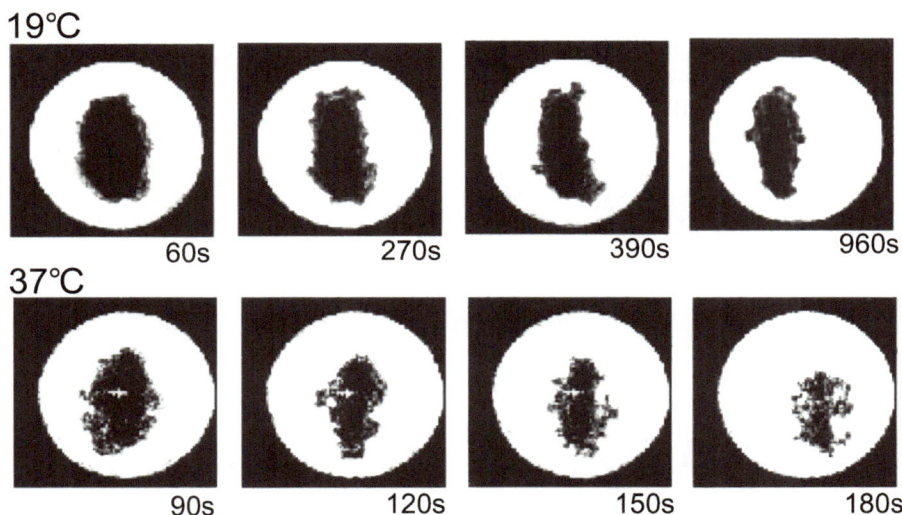

Fig. (21). 2D proton Snapshot FLASH representative images of the disintegration of the paracetamol tablet from Glaxo–Wellcome (Poznań, Poland) studied at two temperatures, 19 and 37 °C. The slice thickness was 2 mm and with an in-plane pixel resolution of 117×117 mm.

molecular weight of the matrix; (ii) high osmotic pressure developed by the push layer may lead to bypassing the drug layer and incomplete drug release; and (iii) the hydration of both the drug and the push layers needs to be properly balanced to efficiently deliver the drug. Therefore, MRI is a powerful tool for obtaining a better understanding of the drug release mechanism of push–pull osmotic systems, as shown in Fig. (**22**).

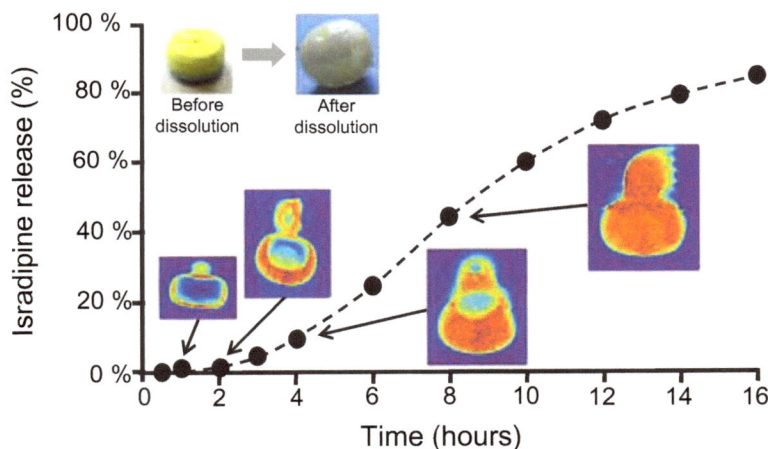

Fig. (22). Dissolution kinetics and corresponding T_1-weighted 1H-NMR images of DynaCircCR 5 mg tablets.

Biodegradable polymer-based drug delivery systems are becoming increasingly popular for the delivery of drugs with a high molecular weight. Hyde *et al.* [141] examined the liquid transport kinetics and morphology of the biodegradable polymer as a drug delivery system made of high molecular weight peptide in a 50:50 poly(glycolide-co-DL-lactide) using NMR imaging. They indicated that the NMR technique may be used to determine the effect of the presence of a drug on the liquid uptake kinetics, morphology, and the drug distribution within the polymer matrix. NMR images have shown that the incorporation of a high molecular peptide in poly(glycolide-co-DL-lactide) affected the physical structure of the polymer matrix and that the strong interaction between the drugs and the polymers led to slow diffusion compared with the drug-free polymer system. This indicates that strong interaction between the drug and the polymer affected the chain configuration and mobility of the polymer and the transport of buffer molecules in the polymer.

Kojima *et al.* [142] studied swelling and water mobility in direct compressed tablets of micronized low-substituted hydroxypropylcellulose (LH41), hydroxypropylcellulose (HPC), and hydroxypropylmethylcellulose (HPMC) tablets by MRI imaging. They observed that the contrast of the outer moiety of the LH41 tablet was slightly brighter in the coronal and transverse planes and that swelling, deformation, and cracking occurred on the edge of the tablet. According to the MRI images, the swelling of the tablets decreased as follows: HPMC>HPC>>LH41. The T_2 of water in the LH41 tablet was smaller than that of free water due to the strong interaction between the water and the polymer. The apparent SDC of water into the LH41 tablet was smaller than that of free water, resulting from the restriction of water diffusivity in the polymer. The HPC and HPMC tablets showed similar T_2 and SDC tendencies. In addition, from further T_2 analysis, two types of water with different mobilities existed in the HPC and HPMC tablets, whereas LH41 tablet showed one type, indicating the different gel layer properties of three polymers.

Ghi *et al.* [143] investigated the diffusion of water into hydroxyethyl methacrylate (HEMA) and its copolymers with tetrahydrofurfuryl methacrylate (THFMA) using NMR imaging. They observed that the diffusion of water in polyHEMA and its copolymers with THFMA showed a predominantly Fickian-type. From the Fickian model, the value of the mass diffusion coefficient for polyHEMA was 1.5×10^{-11} m^{2}s^{-1} which was less than the value of 2.0×10^{-11} m^{2}s^{-1} from mass uptake at 37°C. Also, the presence of an annulus of cracks in the glassy polyHEMA matrix was confirmed by T_2-weighted image profiles. These cracks were found in the copolymer, but the degree of crack formation was lower than that of polyHEMA, a finding attributed to the ductile nature of the THFMA.

Narasimhan *et al.* [144] used MRI imaging to study changing microstructure and molecular motion during dissolution of poly(vinyl alcohol) (PVA) in water. They measured one-dimensional water concentration profiles as a function of distance from the polymer-solvent interface and the self-diffusion coefficient of water by diffusion-weighted profiles. These results indicate that the diffusion coefficient was lower at the edge of the glassy core of the polymer. Also, it decreased toward the core of the polymer as the molecular weight of the polymer increased, as shown in Fig. (23). SDC in the polymer increased with increasing dissolution time. The water concentration profiles predicted by the mathematical model were qualitatively similar with the experimentally obtained profile. Furthermore, the scaling laws proposed in the model for the diffusion coefficients were verified. These results can provide supporting evidence that phenomena such as reptation are important near the glassy rubbery interface while Zimm-type diffusion occurs near the polymer-solvent interface, indicating the existence of a change in the mode of diffusion as solvent penetrated into the polymer.

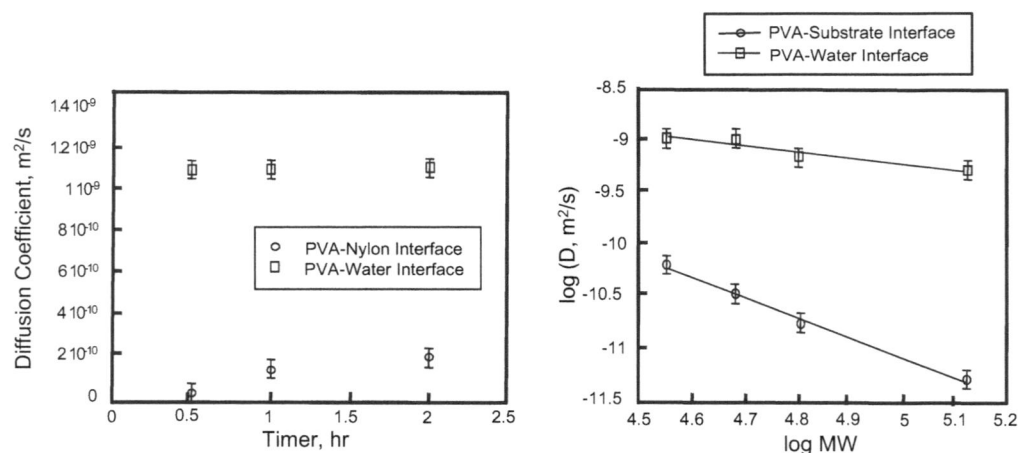

Fig. (23). Diffusion coefficient, *D*2, as a function of time at the PVA-substrate interface as well as at the PVA-water interface. The data are for a PVA sample of Mn=35,740 (a) and Molecular weight dependence of the polymer selfdiffusion coefficient measured near the PVA-substrate interface and the PVA-water interface. The measurements were made after 30 min of dissolution of the polymer. The slopes of the straight lines are -0.6 and -1.9, respectively.

8.2. Ocular Drug Delivery

Recently, MRI imaging has been used to investigate ocular drug delivery. Ocular drug delivery can be classified according to the anatomy of the eye: delivery to the anterior and posterior segments. Drug delivery to the anterior segment is the most common method, although it is ineffective and sometimes inconvenient. To improve topical drug delivery to the anterior segment of the eye, sustained release drug-delivery systems such as collagen shields, lenses, and emulsions have been investigated [145-147]. Another challenge is to deliver a drug to the posterior segment of the eye for the treatment of posterior eye disease. The advancement of non-invasive MRI has provided new opportunities for studying ocular drug-delivery systems compared with invasive conventional pharmacokinetic methods. In addition, MRI is useful to continuously monitor the distribution and elimination profile and the drug concentration during ocular drug delivery with a contrast agent or compound labeled with a contrast agent. Furthermore, MRI can provide insights into ocular pharmacokinetics without tissue perturbation and redistribution of the compound [148].

Berkowitz *et al.* [149] used GdDTPA and dynamic contrast-enhanced magnetic resonance imaging (DCE-MRI) to measure the permeability of the blood retinal barrier (BRB) in experimental diabetic retinopathy in rats. They indicated that DCE-MRI provides a sensitive, non-invasive, and linear assay without potential artifacts associated with death and enucleation, passive BRB permeability surface area (BRB PS) in experimental diabetes, indicating an accurate measurement method. The results showed that, with increasing time, sodium iodate-treated rats exhibited increased BRB PS compared to urethane-anesthetized control rats. This indicates that BRB became leaky, possibly causing an artificial increase in retinal permeability in methods. This suggested that the MRI method can provide useful evaluation of ocular drug delivery or drug treatment efficacy because of its noninvasiveness.

Alikacem *et al.* [150] studied the delivery of vascular endothelial growth factor (VEGF) from a poly(L-lactide-coglycolide)-based intravitreal implant through monitoring of blood ocular barriers such as the blood-retina barrier (BRB) using MRI imaging. They observed that sustained delivery of elevated amounts of VEGF in the vitreous cavity induced a BRB breakdown even earlier than 3 days after implantation, as shown in Fig. (**24**). This was achieved after a total sustained release of 7.42±0.54 mg/ml of VEGF. This retinal leakage regressed by more than half by the time the retinal neovascularization (NV) developed. Furthermore, a retinal detachment occurred after this retinal NV. These results are similar to proliferative diabetic retinopathy (PDR). The sustained elevation of VEGF in the vitreous cavity of rabbit eyes is potentially a good model to test the use of VEGF antagonists in the treatment or prevention of PDR in humans. The quantifiable change of a BRB breakdown by the contrast-enhanced MRI method is ideal to assess therapeutic intervention *in vivo* without killing the animal and may prove to be clinically useful in humans. In summary, MRI imaging can provide new possibilities in ocular drug-delivery research and will continue to assist the development and testing of ocular drug delivery.

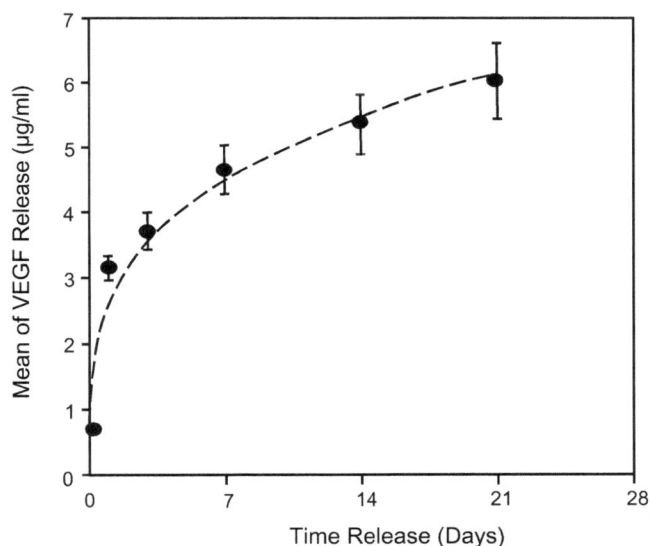

Fig. (24). Cumulative release of VEGF from the PLGA delivery device in 0.5 ml of 0.9% sodium chloride saline at 37°C. These data are fitted to a power function of time t, $f(t)=kt^n$, using a nonlinear square fit algorithm; k =2.57±0.031 and n= 0.29±0.05. This suggests that the VEGF release was governed by diffusion-like process.

Kim *et al.* [151] investigated drug delivery to the posterior segment of the eye from a sustained-release episclera implant containing gadolinium-DTPA (Gd-DTPA) and in the vitreous cavity using a contrast agent. Drug delivery was determined by the concentration profile, clearance, and pharmacokinetics of the contrast agent in the eye using MRI. Based on the pharmacokinetic data, they indicated that drug elimination from the subconjunctival space into the conjunctival lymphatics created a significant transscleral barrier for intraocular delivery of the contrast agent. Therefore, episcleral implants are not effective in delivering the contrast agent into the vitreous humor. It has been shown that dynamic three-dimensional MRI using Gd-DTPA or other contrast agents are useful tools to understand the spatial relationships of ocular drug distribution and clearance mechanisms in the eye.

9. CONCLUSIONS

Recently, NMR techniques have been increasingly used to monitor the cumulative drug release, drug dissolution, and diffusion coefficient of drugs from drug delivery systems *in vitro* and *in vivo*, although a few research papers have been published on such techniques in recent years, compared to other analysis methods. In addition, this technique is applicable as a highly specific tool, providing identification of a drug substance containing impurities and residual solvents and their metabolites in biological media as well as a suitable analytical technique for their absolute quantification.

In addition, the use of MRI imaging to study pharmaceutical systems *in vivo* is an emerging area of drug delivery research. The widespread distribution of clinical MRI imagers and the rapid technical progress in this field offer new avenues for many different kinds of *in vivo* experiments. Although current *in vitro* imaging of pharmaceutical dosage forms has focused on water penetration, diffusion, polymer swelling, and drug dissolution, future advances may allow the micro-environment of dosage forms to be characterized with respect to other physical parameters such as pH and viscosity. Also, the use of NMR spectroscopy to detect structural changes in drug delivery systems during dissolution has been demonstrated.

The few papers published so far show how major applications lie in the characterization of oral dosage form behaviors in the gastrointestinal tract and in drug delivery by routes, such as the eye, skin, and vagina. Monitoring the *in vivo* fate of colloidal drug carriers is also an area of great importance, and there are open questions in the field of gene delivery and drug targeting which might be answered by MRI.

Consequently, the use of both NMR and MRI to study the release behaviors of various drug delivery systems may provide a synergistic effect by the respective advantages and offer the possibility for fast, accurate understanding of the release mechanism from various drug delivery systems compared to using either alone.

REFERENCES

[1]   Sun Y, Peng Y, Chen Y, Shukla AJ. Application of artificial neural networks in the design of controlled release drug delivery systems. Adv Drug Deliv Rev 2003; 55: 1201-15.
[2]   Malaterre V, Metz H, Ogorka J, Gurny R, Loggia N, Mäder K. Benchtop-magnetic resonance imaging (BT-MRI) characterization of push–pull osmotic controlled release systems. J Control Release 2009; 133: 31-6.
[3]   Freiberg S, Zhu XX. Polymer microspheres for controlled drug release. Int J Pharm 2004; 282: 1-18.
[4]   Tao SL, Desai TA. Microfabricated drug delivery systems: from particles to pores. Adv Drug Deliv Rev 2003; 55: 315-28.
[5]   Fluri DA, Kemmer C, Baba MDE, Fussenegger M. A novel system for trigger-controlled drug release from polymer capsules. J Control Release 2008; 131: 211-9.
[6]   Jarvinen K, Akerman S, Svarfvar B, Tarvainen T, Viinikka P, Paronen P. Drug release from pH and ionic strength responsive poly (acrylic acid) grafted poly (vinylidenefluoride) membrane bags *in vitro*. Pharm Res 1998; 15: 802-5.
[7]   Andreani L, Cercená R, Ramos BGZ, Soldi V. Development and characterization of wheat gluten microspheres for use in a controlled release system. Mater Sci Eng C 2009; 29: 524-31.
[8]   Petitti M, Vanni M, Barresi AA. Controlled release of drug encapsulated as a solid core:Theoretical model and sensitivity analysis. Chem Eng Res Des 2008; 86: 1294-300.
[9]   Lu SM, Chen SR. Mathematical analysis of drug release from a coated particle. J Control Release 1993; 23: 105-21.
[10]   Liao YC, Lee DJ. Slow release from a coated sphere with slight deformations of coating film and drug matrix. J Pharm Sci 1997; 86: 92-100.
[11]   Birnbaum DT, Kosmala JD, Henthorn DB, Brannon-Peppas L. Controlled release of β-estradiol from PLAGA microparticles: the effect of organic phase solvent on encapsulation and release. J Control Release 2000; 65: 375-87.
[12]   Ko JA, Park HJ, Hwang SJ, Park JB, Lee JS. Preparation and characterization of chitosan microparticles intended for controlled drug delivery. Int J Pharm 2002; 249: 165-74.
[13]   Jeong JC, Lee J, Cho K. Effects of crystalline microstructure on drug release behaviour of poly(ε-caprolactone) microspheres. J Control Release 2003; 92: 249-58.
[14]   Perkins EL, Lowe JP, Edler KJ, Rigby SP. Studies of structure–transport relationships in biodegradable polymer microspheres for drug delivery using NMR cryodiffusometry. Chem Eng Sci 2010; 65: 611-25.
[15]   Dalvit C. Ligand- and substrate-based ^{19}F NMR screening: Principles and applications to drug discovery. Prog Nucl Magn Reson Spectrosc 2007; 51: 243-71.
[16]   Ek R, Gren T, Henriksson U, Nyqvist H, Nyström C, Ödberg L. Prediction of drug release by characterisation of the tortuosity in porous cellulose beads using a spin echo NMR technique. Int J Pharm 1995; 124: 9-18.
[17]   Hardinga S, Baumannb H, Grenc T, Seo A. NMR microscopy of the uptake, distribution and mobility of dissolution media in small, sub-millimetre drug delivery systems. J Control Release 2000; 66: 81-99.

[18] Metz H, Mäder K. Benchtop-NMR and MRI-A new analytical tool in drug delivery research. Int J Pharm 2008; 364: 170-5.
[19] Rajabi-Siahboomi AR, Bowtell RW, Mansfield P, Davies MC, Melia CD. Structure and behavior in hydrophilic matrix sustained release dosage forms: 4. Studies of water mobility and diffusion coefficients in the gel layer of HPMC tablets using NMR imaging. Pharm Res 1996; 13: 376-80.
[20] Levitt MH. Spin Dynamics: Basic Principles of NMR Spectroscopy. John Wiley & Sons, 2001.
[21] Holzgrabe U, Wawer I, Diehl B. NMR spectroscopy in drug development and analysis. Wiley-VCH, 1999.
[22] Holzgrabe U, Diehl BWK, Wawer I. NMR spectroscopy in pharmacy. J Pharm Biomed Anal 1998; 17: 557-616.
[23] Eisenreich W, Bacher A. Advances of high-resolution NMR techniques in the structural and metabolic analysis of plant biochemistry. Phytochemistry 2007; 68: 2799-815.
[24] Pellecchia M, Sem DS, Wüthrich K. NMR in drug discovery. Nature Rev 2002; 1: 11-219.
[25] Heller M, Kessler H. NMR spectroscopy in drug design. Pure Appl Chem 2001; 73: 1429-36.
[26] Powers R. Applications of NMR to structure-based drug design in structural genomics. J Struct Funct Genomics 2002; 2: 113-23.
[27] Martino R, Gilard V, Desmoulin F, Malet-Martino M. Fluorine-19 or phosphorus-31 NMR spectroscopy: A suitable analytical technique for quantitative *in vitro* metabolic studies of fluorinated or phosphorylated drugs. J Pharm Biomed Anal 2005; 38: 871-91.
[28] Jack RM, Smith JS, Villar HO, Sem DS, Coutts SM. Triad Therapeutics: integration of NMR structural determinations and smart chemistry to speed drug discovery. Drug Discov Today 2002; 7: 35-8.
[29] Holzgrabe U, Deubner R, Schollmayer C, Waibel B. Quantitative NMR spectroscopy-Applications in drug analysis. J Pharm Biomed Anal 2005; 38: 806-12.
[30] Corcoran O, Spraul M. LC-NMR-MS in drug discovery. Drug Discov Today 2003; 8: 624-31.
[31] Peng JW, Moore J, Abdul-Manan N. NMR experiments for lead generation in drug discovery. Prog Nucl Magn Reson Spectrosc 2004; 44: 225-56.
[32] Schachter DM, Xiong J, Tirol GC. Solid state NMR perspective of drug-polymer solid solutions:a model system based on poly(ethylene oxide). Int J Pharm 2004; 281: 89-101.
[33] Holzgrabe U, Wawer I, Diehl B. NMR spectroscopy in pharmaceutical analysis, Elsevier, 2008.
[34] Richardson JC, Bowtell RW, Mäder K, Melia CD. Pharmaceutical applications of magnetic resonance imaging (MRI). Adv Drug Delive Rev 2005; 57: 1191-209.
[35] Rajabi-Siahboomi AR, Melia CD, Davies MC, *et al.* Imaging the internal structure of the gel layer in hydrophilic matrix systems by NMR microscopy. J Pharm Pharmacol Suppl 1992; 44: 1062.
[36] Ashraf M, Iuorno VL, Coffin-Beach D, Evans CA, Augsburger LL. A novel nuclear magnetic resonance (NMR) imaging method for measuring the water front penetration rate in hydrophilic polymer matrix capsule plugs and its role in drug release. Pharm Res 1994; 11: 733-7.
[37] Griffiths P, Stilbs P. NMR self-diffusion studies of polymeric surfactants. Curr Opin Colloid Interface Sci 2002; 7: 249-52.
[38] Price WS. Recent advances in NMR diffusion techniques for studying drug binding. Aust J Chem 2003; 56: 855-60.
[39] Momot KI, Kuchel PW. PFG NMR diffusion experiments for complex systems. Concepts Magn Reson A 2006; 28A: 249-69.
[40] Tonning E, Polders D, Callaghan PT, Engelsen SB. A novel improved method for analysis of 2D diffusion-relaxation data-2D PARAFAC-Laplace decomposition. J Magn Reson 2007; 188: 10-23.
[41] Nilsson M, Morris GA. Improving pulse sequences for 3D DOSY: convection compensation. J Magn Reson 2005; 177: 203-11.
[42] Stejskal EO, Tanner JE. Spin diffusion measurements: spin echoes in the presence of a time-dependent field gradient. J Chem Phys 1965; 42: 288-92.
[43] Hrabe J, Kaur G, Guilfoyle DN. Principles and limitations of NMR diffusion measurements. J Med Phys 2007; 32: 34-42.
[44] Occhipinti P, Griffiths PC. Quantifying diffusion in mucosal systems by pulsed-gradient spin-echo NMR. Adv Drug Delive Rev 2008; 60: 1570-82.
[45] Bottomley PA. NMR in medicine. Comput Radiol 1984; 8: 57-77.
[46] Mäder K, Bacic G, Elmalak O, Langer R, Swartz HM. Non-invasive in vivo monitoring of drug release and polymer erosion from biodegradable polymers by EPR spectroscopy and NMR imaging. J Pharm Sci 1997; 86: 126-34.
[47] Pierigè F, Serafini S, Rossi L, Magnani M. Cell-based drug delivery. Adv Drug Delive Rev 2008; 60: 286-95.
[48] Griffith LG. Polymeric biomaterials. Acta Mater 2000; 48: 263-77.
[49] Huang X, Brazel CS. On the importance and mechanisms of burst release in matrix-controlled drug delivery systems. J Control Release 2001; 73: 121-36.
[50] Brazel CS, Peppas NA. Temperature- and pH-sensitive hydrogels for controlled release of antithrombotic agents. Mater Res Soc Symp Proc 1994; 331: 211-6.
[51] Yamada Y, Harashima H. Mitochondrial drug delivery systems for macromolecule and their therapeutic application to mitochondrial diseases. Adv Drug Delive Rev 2008; 60: 1439-62.
[52] Porter CJH, Pouton CW, Cuine JF, Charman WN. Enhancing intestinal drug solubilisation using lipid-based delivery systems. Adv Drug Delive Rev 2008; 60: 673-91.
[53] Gou ML, Li XY, Dai M, *et al.* A novel injectable local hydrophobic drug delivery system: Biodegradable nanoparticles in thermo-sensitive hydrogel. Int J Pharm 2008; 359: 228-33.
[54] Bussemer T, Dashevsky A, Bodmeier R. A pulsatile drug delivery system based on rupturable coated hard gelatin capsules. J Control Release 2003; 93: 331-9.
[55] Wyss A, Cordente N, von Stockar U, Marison IW. A novel approach for the extraction of herbicides and pesticides from water using liquid-core microcapsules. Biotechnol Bioeng 2004; 87: 734-42.
[56] Radt B, Smith TA, Caruso F. Optically addressable nanostructured capsules for controlled delivery. Adv Mater 2004; 16: 2184-9.
[57] Kim KS, Park SJ. Characterization and release behaviors of porous PCL/Eudragit RS microcapsules containing tulobuterol. Colloids Surf B: Biointerfaces, 2010; 76: 404-9.
[58] Jugminder SC, Mansoor MA. Biodegradable poly(o-caprolactone) nanoparticles for tumortargeted delivery of tamoxifen. Int J Pharm 2002; 249: 127-38.

[59]   Bouchemal K, Briançon S, Perrier E, Fessi H, Bonnet I, Zydowicz N. Synthesis and characterization of polyurethane and poly(ether urethane) nanocapsules using a new technique of interfacial polycondensation combined to spontaneous emulsification. Int J Pharm 2004; 269: 89-100.

[60]   Park SJ, Kim KS, Kim SH, Effect of poly(ethylene oxide) on the release behaviors of poly(ε-caprolactone) microcapsules containing erythromycin. Colloids Surf B: Biointerfaces, 2005; 43: 238-44.

[61]   Bodmeier R, McGinity JW. The preparation and evaluation of drug containing poly(DL-lactide) microspheres formed by the solvent evaporation. Pharm Res 1987; 4: 465-71.

[62]   Park SJ, Kim KS. Effect of oxygen plasma treatment on the release behaviors of poly(ε-caprolactone) microcapsules containing tocopherol. Colloids Surf B: Biointerfaces, 2005; 43: 138-42.

[63]   Ruiz JM, Tissier B, Benoit JP. Microencapsulation of peptide: a study of the phase separation of poly(D,L-lactic acid-co-glycolic acid)copolymers 50/50 by silicone oil. Int J Pharm 1989; 49: 69-77.

[64]   Bodmeier R, Chen H. Preparation of biodegradable polylactide microparticles using a spray-drying techniques. J Pharm Pharmacol 1988; 40: 754-57.

[65]   Wise DL, McCormick GJ, Willet GP, Anderson LC. Sustained release of an antimalarial drug using a co-polymer of glycolic/lactic acid. Life Sci 1976; 19: 867-74.

[66]   Gibbs BF, Kermasha S, Alli I, Mulligan CN. Encapsulation in the food industry:A review. Int J Food Sci Nutr 1999; 50: 213-24.

[67]   Min KH, Park K, Kim YS, et al. Hydrophobically modified glycol chitosan nanoparticles-encapsulated camptothecin enhance the drug stability and tumor targeting in cancer therapy. J Control release 2008; 127: 208-18.

[68]   Brigger I, Dubernet C, Couvreur P. Nanoparticles in cancer therapy and diagnosis. Adv Drug Deliv Rev 2002; 54: 631-51.

[69]   Kingsley JD, Dou H, Morehead J, Rabinow B, Gendelman HE, Destache CJ. Nanotechnology: a focus on nanoparticles as a drug delivery system. J Neuroimmune Pharmacol 2006; 1: 340-50.

[70]   Liu Z, Jiao Y, Wang Y, Zhou C, Zhang Z. Polysaccharides-based nanoparticles as drug delivery systems. Adv Drug Deliv Rev 2008; 60: 1650-62.

[71]   Park SJ, Kim KS. Influence of hydrophobe on the release behavior of vinyl acetate miniemulsion polymerization. Colloids Surf B: Biointerfaces, 2005; 46: 52-6.

[72]   Sahoo SK, Labhasetwar V. Nanotech approaches to drug delivery and imaging. Drug Discov Today 2003; 8: 1112-20.

[73]   Huang CY, Chen CM, Lee YD. Synthesis of high loading and encapsulation efficient paclitaxel-loaded poly(n-butyl cyanoacrylate) nanoparticles via miniemulsion. Int J Pharm 2007; 338: 267-75.

[74]   Moghimi SM, Hunter AC, Murray JC. Long-circulating and target specific nanoparticles: theory to practice. Pharmacol Rev 2001; 53: 283-318.

[75]   Vinagradov SV, Bronich TK, Kabanov AV. Nanosized cationic hydrogels for drug delivery: preparation, properties and interactions with cells. Adv Drug Deliv Rev 2002; 54: 135-47.

[76]   Husseini GA, Pitt WG. Micelles and nanoparticles for ultrasonic drug and gene delivery. Adv Drug Deliv Rev 2008; 60: 1137-52.

[77]   Illum L. Nanoparticulate systems for nasal delivery of drugs: a real improvement over simple systems? J Pharm Sci 2007; 96: 473-83.

[78]   Pinto Reis C, Neufeld RJ, Ribeiro AJ, Veiga F. Nanoencapsulation I. Methods for preparation of drug-loaded polymeric nanoparticles. Nanomedicine 2006; 2: 8-21.

[79]   Xiaoling L, Bhaskara RJ. Design of controlled release drug delivery systems. MaGraw-Hill, New York, USA 2006.

[80]   Boer GJ, Kmisbrink J. A polymeric controlled drug delivery device for peptides based on a surface desorption/diffusion mechanism. Biomaterials 1987; 8: 265-74.

[81]   Chamarthy SP, Pinal R. Plasticizer concentration and the performance of a diffusion-controlled polymeric drug delivery system. Colloids Surfaces A: Physicochem Eng Aspects 2008; 331: 25-30.

[82]   Benita S. Microencapsulation: Methods and industrial applications. Marcel Dekker, New York Basel Hong Kong 1996.

[83]   Siepmanna J, Göpferich A. Mathematical modeling of bioerodible, polymeric drug delivery Systems. Adv Drug Delive Rev 2001; 48: 229-47.

[84]   Devine DM, Devery SM, Lyons JG, Geever LM, Kennedy JE, Higginbotham CL. Multifunctional polyvinylpyrrolidinone-polyacrylic acid copolymer hydrogels for biomedical applications. Int J Pharm 2006; 326: 50-9.

[85]   Kashyap N, Kumar N, Kumar M. Hydrogels for pharmaceutical and biomedical applications. Crit Rev Ther Drug Carr Syst 2005; 22: 107-49.

[86]   Venkatesh S, Byrne ME, Peppas NA, Hilt JZ. Applications of biomimetic systems in drug delivery. Expert Opin Drug Deliv 2005; 2: 1085-96.

[87]   Peppas NA, Khare AR. Preparation, structure and diffusional behavior of hydrogels in controlled release. Adv Drug Delive Rev 1993; 11: 1-35.

[88]   Peppas NA. Physiologically responsive gels. J Bioact Compat Polym 1991; 6: 241-6.

[89]   Kono K, Henmi A,Yamashita H, Hayashi H., Takagishi T. Improvement of temperature-sensitivity of poly(N-isopropylacrylamide)-modified liposomes. J Control Release 1999; 59: 63-75.

[90]   Rodríguez R, Alvarez-Lorenzo F, Concheiro CA. Cationic cellulose hydrogels: kinetics of the cross-linking process and characterization as pH-/ion-sensitive drug delivery systems. J Control Release 2003; 86: 253-65.

[91]   Alvarez-Lorenzo C, Concheiro A. Reversible adsorption by a pH- and temperature-sensitive acrylic hydrogel. J Control Release 2002; 80: 247-57.

[92]   Coviello T, Matricardi P, Marianecci C, Alhaique F. Polysaccharide hydrogels for modified release formulations. J Control Release 2007; 119: 5-24.

[93]   Peppas NA, Mongia NK. Ultrapure poly(vinyl alcohol) hydro gels with mucoadhesive drug delivery characteristics. Eur J Pharm Biopharm 1997; 43: 51-8.

[94]   Davis KA, Anseth KS. Controlled release from crosslinked degradable networks. Crit Rev Ther Drug Carr Syst 2002; 19: 385-423.

[95]   Lin CC, Metters AT. Hydrogels in controlled release formulations: network design and mathematical modeling. Adv Drug Delive Rev 2006; 58: 1379-408.

[96]   Hamidi M, Azadi A, Rafiei P. Hydrogel nanoparticles in drug delivery. Adv Drug Delive Rev 2008; 60: 1638-49.

[97]   Amsden B. Solute diffusion within hydrogels. Mechanisms and models. Macromolecules 1998; 31: 8382-95.

[98]   Canal T, Peppas NA. Correlation between mesh size and equilibrium degree of swelling of polymeric networks. J Biomed Mater Res 1989; 23: 1183-93.
[99]   Siepmann J, Peppas NA. Modeling of drug release from delivery systems based on hydroxypropyl methylcellulose (HPMC). Adv Drug Deliv Rev 2001; 48: 139-57.
[100]   Bettini R, Colombo P, Massimo G, Catellani PL, Vitali T. Swelling and drugrelease in hydrogel matrices- polymer viscosity and matrix porosity effects. Eur J Pharm Sci 1994; 2: 213-9.
[101]   Kanjickal DG, Lopina ST. Modeling of drug release from polymeric delivery systems-a review. Crit Rev Ther Drug Carrier Syst 2004; 21: 345-86.
[102]   Rabenstein DL. NMR Spectroscopy: past and present. Anal Chem 2001; 73: 214A-23A.
[103]   Moniot KI, Kuchel PW. Pulsed field gradient nuclear magnetic resonance as a tool for studying drug delivery systems. Concepts Magn Reson Part A 2003; 19A: 51-64.
[104]   Price WS, Elwinger F, Vigouroux C, Stilbs P. PGSE-WATERGATE, a new tool for NMR diffusion-based studies of ligand-macromolecule binding. Magn Reson Chem 2002; 40: 391-5.
[105]   Kwak S, Lafleur M. NMR self-diffusion of molecular and macromolecular species in dextran solutions and gels. Macromolecules 2003; 36: 3189-95.
[106]   Lafitte G, Thuresson K, Jarwoll P, Nyden M, Transport properties and aggregation phenomena of polyoxyethylene sorbitane monooleate (polysorbate 80) in pig gastrointestinal mucin and mucus. Langmuir 2007; 23: 10933-9.
[107]   Trampel R, Schiller J, Naji L, Stallmach F, Karger J, Arnold K. Self-diffusion of polymers in cartilage as studied by pulsed field gradient NMR. Biophys Chem 2002; 97: 251-60.
[108]   Kwak S, Lafleur M. Self-diffusion of macromolecules and macroassemblies in curdlan gels as examined by PFG-SE NMR technique. Colloids Surfaces A, Physicochem Eng Aspects 2003; 221: 231-42.
[109]   Colsenet R, Soderman O, Mariette F. Effect of casein concentration in suspensions and gels on poly(ethylene glycol) NMR self-diffusion measurements. Macromolecules 2005; 38: 9171-9.
[110]   Lafitte G, Soderman O, Thuresson K, Davies J. PFG-NMR diffusometry: a tool for investigating the structure and dynamics of noncommercial purified pig gastric mucin in a wide range of concentrations. Biopolymers 2007; 86: 165-75.
[111]   Assifaoui A, Champion D, Chiotelli E, Verel A. Characterization of water mobility in biscuit dough using a low-field ^{1}H NMR technique. Carbohydr Polym 2006; 64: 197-204.
[112]   Hua L, Weisan P, Jiayu L, Ying Z. Preparation, evaluation, and NMR characterization of vinpocetine microemulsion for transdermal delivery. Drug Dev Ind Pharm 2004; 30: 657-66.
[113]   Regan DG, Momot KI, Martens PJ, Kuchel PW, Poole-Warren LA. NMR measurement of small-molecule diffusion in PVA hydrogels: a comparison of CONVEX and standard PGSE methods. Diffus Fundam 2006; 4: 11-8
[114]   Okamura E, Kakitsubo R, Nakahara M. NMR determination of the delivery site of bisphenol A in phospholipid bilayer membranes. Langmuir 1999; 15: 8332-5.
[115]   Lubach JW, Padden BE, Winslow SL, et al. Solid-state NMR studies of pharmaceutical solids in polymer matrices. Anal Bioanal Chem 2004; 378: 1504-10.
[116]   Thomasa CP, Platts J, Tatchell T, Heard CM. Probing the skin permeation of fish oil/EPA and ketoprofen 1. NMR spectroscopy and molecular modeling. Int J Pharm 2007; 338: 207-12.
[117]   Ferrero C, Massuelle D, Jeannerat D, Doelker E. Towards elucidation of the drug release mechanism from compressed hydrophilic matrices made of cellulose ethers. I. Pulse-field-gradient spin-echo NMR study of sodium salicylate diffusivity in swollen hydrogels with respect to polymer matrix physical structure. J Control Release 2008; 128: 71-9.
[118]   Leal C, Rognvaldsson S, Fossheim S, Nilssen EA, Topgaard D. Dynamic and structural aspects of PEGylated liposomes monitored by NMR. J Colloid Interface Sci 2008; 325: 485-93.
[119]   Ek R, Gren T, Henriksson U, Nyqvist H, Nyström, Ödberg L. Prediction of drug release by characterization of the tortuosity in porous cellulose beads using a spin echo NMR technique. Int J Pharm 1995; 124: 9-18.
[120]   Kreilgaard M, Pedersen EJ, Jaroszewski JW. NMR characterisation and transdermal drug delivery potential of microemulsion systems. J Control Release 2000; 69: 421-33.
[121]   Collins JHP, Gladden LF, Hardy IJ, Mantle MD, Characterization the evolution of porosity during controlled drug release. Appl Magn Reson 2007; 32: 185-204.
[122]   Melia CD, Rajabi-Siahboomi AR, Bowtell RW. Magnetic resonance imaging of controlled release pharmaceutical dosage forms. Pharm Sci Technol Today 1998; 1: 32-9.
[123]   Rajabi-Siahboomi AR, Bowtell RW, Mansfield P, Henderson A, Davies MC, Melia CD. Structure and behavior in hydrophilic matrix sustained release dosage forms: 2. NMR imaging studies of dimensional changes in the gel layer and core of HPMC tablets undergoing hydration. J Control Release 1994; 31: 121-8.
[124]   Rajabi-Siahboomi AR, Bowtell RW, Mansfield P, Davies MC, Melia CD. Structure and behavior in hydrophilic matrix sustained release dosage forms: 4. Studies of water mobility and diffusion coefficients in the gel layer of HPMC tablets using NMR imaging. Pharm Res 1996; 13: 376-80.
[125]   Madhu B, Hjärtstam J, Soussi B. Studies of the internal flow process in polymers by H-1 NMR microscopy at 500 MHz. J Control Release 1998; 56: 95-104.
[126]   Kojima M, Nakagami H. Investigation of water mobility and diffusivity in hydrating micronized low-substituted hydroxypropyl cellulose, hydroxypropylmethyl cellulose, and hydroxypropyl cellulose matrix tablets by magnetic resonance imaging (MRI). Chem Pharm Bull 2002; 50: 1621-4.
[127]   Fyfe CA, Blazek AI. Investigation of hydrogel formation from hydroxypropylmethylcellulose (HPMC) by NMR spectroscopy and NMR imaging techniques. Macromolecules 1997; 30: 6230-7.
[128]   Fyfe CA, Blazek-Welsh AI. Quantitative NMR imaging study of the mechanism of drug release from swelling hydroxypropylmethylcellulose tablets. J Control Release 2000; 68: 313-33.
[129]   Dahlberg C, Fureby A, Schuleit M , Dvinskikh SV, Furó I. Polymer mobilization and drug release during tablet swelling. A 1H NMR and NMR microimaging study. J Control Rel 2007; 122: 199-205.
[130]   Hyde TM, Gladden LF. Simultaneous measurement of water and polymer concentration profiles during swelling of poly (ethylene oxide) using magnetic resonance imaging. Polymer 1998; 39: 811-9.
[131]   Messaritaki A, Black SJ , van der Walle CF, Rigby SP. NMR and confocal microscopy studies of the mechanisms of burst drug release from PLGA microspheres. J Control Release 2005; 108: 271-81.

[132]   Sutcha JCD, Rossb AC., Köckenbergerc W, Bowtellc RW, MacRaed RJ, Stevensb HNE, Melia CD. Investigating the coating-dependent release mechanism of a pulsatile capsule using NMR microscopy. J Control Release 2003; 92: 341-7.

[133]   Fyfe CA, Grondey H, Blazek-Welsh AI, Chopra SK, Fahie BJ. NMR imaging investigations of drug delivery devices using a flow-through USP dissolution apparatus. J Control Release 2000; 68: 73-83.

[134]   Katzhendler I, Mäder K, Friedman M. Structure and hydration properties of hydroxypropyl methylcellulose matrices containing naproxen and naproxen sodium. Int J Pharm 2000; 200: 161-79.

[135]   Masaro L, Ousalem M, Baille WE, Lessard D, Zhu XX. Self-diffusion studies of water and poly(ethylene glycol) in solutions and gels of selected hydrophilic polymers. Macromolecules 1999; 32: 4375-82.

[136]   Baille WE, Malveau C, Zhu XX, Marchessault RH. NMR imaging of high-amylose starch tablets. 1. Swelling and water uptake. Biomacromolecules 2002; 3: 214-8.

[137]   Tritt-Goc J, Piślewski N. Magnetic resonance imaging study of the swelling kinetics of hydroxypropylmethylcellulose (HPMC) in water. J Control Release 2002; 80: 79-86.

[138]   Fahie BJ, Nangia A, Chopra SK, Fyfe CA, Grondey H, Blazek A. Use of NMR imaging in the optimization of a compression-coated regulated release system. J Control Release 1998; 51: 179-84.

[139]   Tritt-Goc J, Kowalczuk J. *In situ*, real time observation of the disintegration of paracetamol tablets in aqueous solution by magnetic resonance imaging. Eur J Pharm Sci 2002; 15: 341-6.

[140]   Malaterre V, Metz H, Ogorka J, Gurny R, Loggia N, Mäder K. Benchtop-magnetic resonance imaging (BT-MRI) characterization of push–pull osmotic controlled release systems. J Control Release 2009; 133: 31-6.

[141]   Hyde TM, Gladden LF, Payne R. A nuclear magnetic resonance imaging study of the effect of incorporating a macromolecular drug in poly (glycolic acid-co-DL-lactic acid). J Control Release 1995; 36: 261-75.

[142]   Kojima M, Ando S, Kataoka K, Hirota T, Aoyagi K, Nakagami H. Magnetic resonance imaging (MRI) study of swelling and water mobility in micronized low-substituted hydroxypropylcellulose matrix tablets. Chem Pharm Bull 1998; 46: 324-8.

[143]   Ghi PY, Hill DJT, Whittaker AK. NMR Imaging of Water Sorption into Poly (hydroxyethylmethacrylate-co-tetrahydrofurfurylmethacrylate).Biomacromolecules 2001; 2: 504-10.

[144]   Narasimhan B, Snaar JEM, Bowtell RW, Morgan S, Melia CD, Peppas NA. Magnetic resonance imaging analysis of molecular mobility during dissolution of poly(vinyl alcohol) in water. Macromolecules 1999; 32: 704-10.

[145]   Davies NM. Biopharmaceutical considerations in topical ocular drug delivery. Clin Exp Pharmacol Physiol 2000; 27: 558-62.

[146]   Mainardes RM, Urban MC, Cinto PO, Khalil NM, Chaud MV, Evangelista RC, Gremiao MP. Colloidal carriers for ophthalmic drug delivery. Curr Drug Targets 2005; 6: 363-71.

[147]   Taravella MJ, Balentine J, Young DA, Stepp P. Collagen shield delivery of ofloxacin to the human eye. J Cataract Refract Surg 1999; 25: 562-5.

[148]   Li SK, Lizak MJ, Jeong EK. MRI in ocular drug delivery. NMR Biomed 2008; 21: 941-56.

[149]   Berkowitz BA, Roberts R, Luan H, Peysakhov J, Mao X, Thomas KA. Dynamic contrast-enhanced MRI measurements of passive permeability through blood retinal barrier in diabetic rats. Invest Ophthalmol Vis Sci 2004; 45: 2391-8.

[150]   Alikacem N, Yoshizawa T, Nelson KD, Wilson CA. Quantitative MR imaging study of intravitreal sustained release of VEGF in rabbits. Invest Ophthalmol Vis Sci 2000; 41: 1561-9.

[151]   Kim HC, Robinson MR, Lizak MJ, Tansey G, Lutz RJ, Yuan P, Wang NS, Csaky KG. Controlled drug release from an ocular implant: An evaluation using dynamic three-dimensional magnetic resonance imaging. Invest Ophthalmol Vis Sci 2004; 45: 2722-31.

[152]   Metais A, Cambert M, Riaublanc A, Mariette F. Influence of fat globule membrane composition on water holding capacity and water mobility in casein rennet gel: a nuclear magnetic resonance self-diffusion and relaxation study. Int Dairy J 2006; 16: 344-53.

[153]   Schachter DM, Xiong J, Tirol GC. Solid state NMR perspective of drug–polymer solid solutions: a model system based on poly(ethylene oxide). Int J Pharma 2004; 281: 89-101.

[154]   Alfrey TJ, Gurnee EF, Lloyd WG. Diffusion in glassy polymers. J Polym Sci Part C: Polym Sympo 1966; 12: 249-61.

[155]   Okamura E, Nakahara M. NMR study directly determining drug delivery sites in phospholipid bilayer membranes. J Phys Chem B 1999; 103: 3505-9.

[156]   Hanoulle X, Wieruszeski JM, Rousselot-Pailley P, Landrieu I, Baulard AR, Lippens G. Monitoring of the ethionamide pro-drug activation in mycobacteria by ^{1}H high resolution magic angle spinning NMR. Biochem Biophys Res Commun 2005; 331: 452-8.

[157]   Joquevielc C, Martino R, Gilard V, Malet-Martino M, Canal P, Niemeyer U. Urinary excretion of cyclophosphamide in humans, determined by phosphorus-31 nuclear magnetic resonance spectroscopy, Drug Metab Dispos 1998; 26: 418-28.

[158]   Peters GJ, Ackland SP. New antimetabolites in preclinical and clinical development. Exp Opin Invest Drugs 1996; 5: 637-79.

NMR Screening Methods in Fragment-Based Drug Discovery

Consonni Roberto[*,1] and Veronesi Marina[2]

[1]*Institute for the Study of Macromolecules (ISMAC), NMR Laboratory,
National Council of Research, Milan, Italy*

[2]*Italian Institute of Technology, Genoa, Italy*

Abstract: In the late 90's, NMR based screening emerged as a powerful technique in the identification of new targeted molecule in drug discovery, at both industrial and academic levels. The capacity of finding ligands with low affinity has proved NMR to be a leader technique for Fragmented Based Drug Discovery (FBDD). This approach, complementary to HTS (High Throughput Screening), is based on the idea that it is easier to develop small and low affinity molecules endowed with BEI (Binding Efficiency Index) comparable to potent drug molecules with respect to the HTS identified molecules. Several NMR screening methods have been developed in the last 20 years to identify these fragments, and they were generally based on the observation of protein signals of interest (e.g. SAR by NMR) or on ligand signals (e.g. waterLOGSY). In this paper, different NMR techniques and their pharmaceutical applications will be summarized and discussed.

Keywords: NMR screening, fragment based drug discovery, ^{19}F-NMR, ^{1}H-NMR, STD, Water-LOGSY, 3-FABS, FAXS, SAR by NMR, ligand efficiency index (BEI), fragment based approach.

INTRODUCTION

The NMR Potentiality in the Drug Discovery Process

Drug discovery process consists of several steps linked together and requiring a huge amount of time, money and energy in both human and experimental resources. In the past years hits identification in the early stage of drug discovery was based on High Throughput Screening (HTS) [1]; afterwards, Fragment Based Drug Discovery (FBDD) [2,3] was proposed for overcoming the HTS limitations. Different biophysical techniques have been applied to FBDD investigations: Nuclear Magnetic Resonance (NMR), Surface Plasmon Resonance (SPR), Fluorescence Intensity (FI), Fluorescence Anisotropy (FA) and Fluorescence Lifetime (FL), Mass Spectrometry (MS), Thermal Shift (TS), Differential Scanning Calorimetry (DSC), Isothermal Titration Calorimetry (ITC) and Waveguide Grating Based Technique (WBT), X-Ray. Among the possible listed approaches, NMR is playing an ever growing role in drug discovery process and these perspectives were recently reviewed [4]; this success was mainly due to the improvements experienced either in terms of stronger magnetic fields development or in more sensitive probes aimed to improve both resolution and sensitivity. Concerning this latter issue, the combined use of micro-coil and cryogenic coil cooling system, in which the probe coil is cooled down at almost liquid Helium temperature, increased dramatically the sensitivity of the measurement, assuring a theoretical 200-fold reduction in experimental time with respect to normal systems. Over the past years, NMR has been extensively used for structure determination of ligand molecules in combination with molecular dynamic simulations; subsequently biological targets have also been investigated. In drug discovery general form, NMR can be largely employed to follow the entire process (Fig. **1**). NMR screening methods, recently reviewed in Klages paper [5], are based on ligand or target characterization, and the ligand-based approach is largely preferred. These two approaches present their own advantages and drawbacks. Briefly, ligand-based approaches use mono-dimensional NMR experiments, small amounts of both ligand and target and are not limited by the target molecular mass. The target-based approach instead is typically applicable to small biological target (MW $\leq$ 40 kDa), requires large amounts of soluble (^{15}N and/or ^{13}C) isotope enriched protein and acquisition of multidimensional NMR experiments. Usually this approach involves longer acquisition time. The initial step of the target-based approach is the target protein over-expression preferably with isotope labeling, for solving the three-dimensional structure via NMR experiments.

* Corresponding author: Tel: +39-02-23699578; E-mail: roberto.consonni@ismac.cnr.it

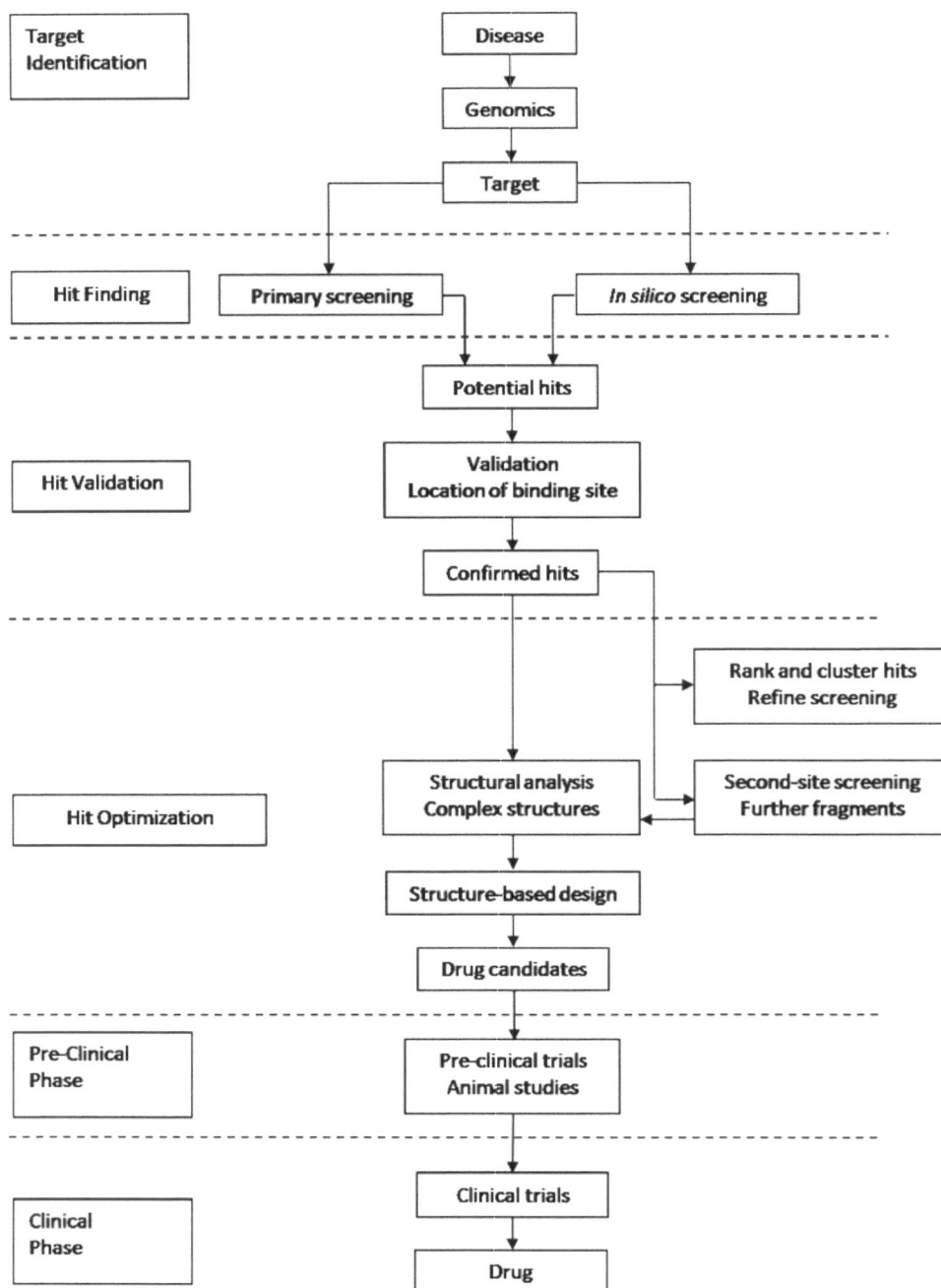

Fig. (1). A flow chart outlining the drug discovery process; steps allowing the application of NMR techniques are shaded. Reproduced with permission from Klages J, Coles M, Kessler H. NMR-based screening: a powerful tool in fragment-based drug discovery. Mol BioSyst 2006; 2: 318-23.

Then compounds libraries are screened, as single compounds or as mixtures, to find initial hits. Upon binding with the macromolecule, both global and local effects can be observed: global effects are size dependent and therefore evaluable by looking at the ligand, while local effects can be monitored on both target and ligand (Fig. **2**) [6]. In the ligand-based approach, some disadvantages are still present: the dissociation constant (K_d) must be in the order of milli/micro molar, large differences between molecular weight of ligand and protein, difficulties in distinguishing between specific and unspecific binding. Ligands with high affinity, $K_d < \mu M$, gives false negatives because of their slow dissociation rates: as a matter of fact strong binders are usually not detected by simple ligand-based techniques, and competition-based screening [7,8] could be the alternative approach. Target-based methods present some disadvantages as well: molecular weight is a limiting factor, because relaxation rates are very fast and spectra suffer from poor resolution and signal intensity.

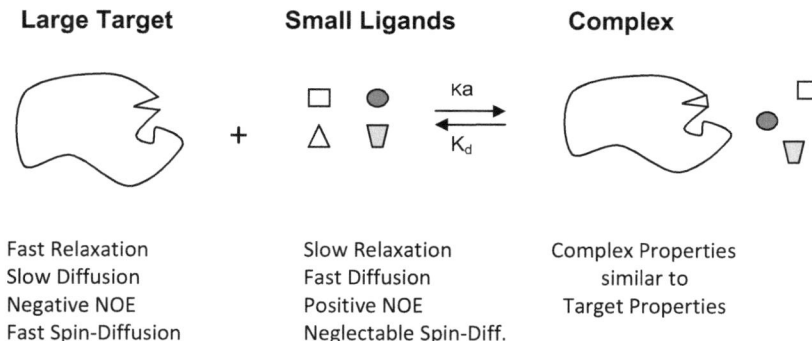

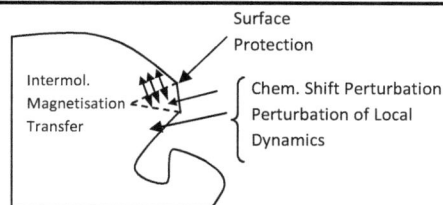

Fig. (2). The effects that arise from binding of a ligand to a target protein. The upper part shows the global effects that are advantageously monitored on the ligand. Local effects (lower part) can be detected on both the ligand and target. Adapted with permission from Heller M, Kessler H. NMR spectroscopy in drug design. Pure Appl Chem 2001; 73: 1429-36.

High Throughput Screening

The advent of molecular biology, combinatorial chemistry and automated technologies have led to the concept of "magic large number" in drug discovery [9]: large number of hypothetical targets are tested *in vitro* and *in vivo* against a large number of compounds (from 10^5 to 10^6) to rapidly identifying active antibodies, genes and hits, which modulate a particular biomolecular pathway and to profile cellular phenotypes. This approach goes under the name of High-Throughput Screening (HTS). The assays are performed with a single point format in a micro plate (from 96 to 1536 wells) using a very small amount of reagent in a fully automated liquid handling and signal detection assembly; this enables analysis of thousands of compounds a day, with adequate sensitivity, reproducibility and accuracy the main requirements. In the last 20 years HTS has been successfully applied to many screening projects in pharmaceutical companies. Among the different biophysical techniques developed for HTS screening, most widely used are the traditional SPA (Scintillation Proximity Assay) [10] and the homogeneous fluorescence methods [11]: Fluorescence Polarization (FP); Fluorescence Lifetime (FL); FRET (Fluorescence Resonance Energy Transfer). Fluorescence methods have very versatile application; they can be used in direct and competition-binding studies, in different types of enzymatic assays using both fluorescent labeled substrate (protease, kinase, phosphatase etc.) and in direct/competitive immunodetection. The high sensitivity of these methods makes them very suitable in the automation processes, for both *in vitro* and *in vivo* experiments. Nevertheless these assays are subjected to many false positive or negative hits due for example to the compound interferences with the detection method. Therefore a combination of different biophysical approaches could be used [12]. Multiple independent HTS screening for the same target maximizes the assay's high-throughput, reducing the number of false negative and false positives. The presence of the same classes of hits in different screening protocols, confirm them as robust, "real" hits and help the researcher to select the candidate for the lead optimization phase. The rapid increase in genomic and proteomic discovery in the last years produced a strong improvement in developing HTS technologies such as (cABPP) [13-15] to assign functional and interactive relationships to the proteins encoded by genomes and the possibility in finding their inhibitors. Despite the numerous progresses in HTS techniques, the drawback still remains the high cost that makes this kind of research unaffordable even to large companies, and the huge number of false positives and negatives, that slow-down the drugs development.

Fragment Based Drug Discovery

Despite the quantitative advantages, the very high costs of HTS, in terms of highly specialized personnel and expensive screening laboratories, limit the use of this approach to large companies or large research institutions. These drawbacks, have led to the search for new approaches, which are less expensive and more reliable, like the Fragment Based Approach (FBA). This method is based on the idea that it is easier to improve a weak affinity small molecule (fragment), than a strong affinity complex compound. Early work suggesting the use of small molecules in drug discovery processes appeared in the eighties. For example, Jencks [16] proposed that the affinity of the whole molecule could be considered as the sum of its separated parts. Nakamura and Abeles [17] further demonstrated that small molecule inhibitors binding to HMG-CoA reductase, could be considered as linkage of two fragments, each of them binding to distinct sites of the target protein. However the application of this approach encountered many difficulties; a) low affinity ligands searches produced a large number of false positive, due to the large intrinsic error of the applied technique, b) the sensitivity of these methods were not suitable for weak binder identification and c) even if weak binders were identified, the approach required two compounds binding the target in close proximity, in order to link them together. These limitations diminished the impact of FBA on drug discovery until 1996 when Shuker and co-workers [18] of Abbott proposed an NMR-based method to identify, optimize and link together small molecules that bind to proximal sites to obtain high-affinity inhibitors: the so called "SAR by NMR" (Structure Activity Relationship by Nuclear Magnetic Resonance). In 2001 Hann and co-workers [19] introduced the concept of molecular complexity and based on these findings they demonstrated that the probability to find randomly lead compounds strongly decreased with the increasing of the molecule complexity: small fragments can better fit the binding pocket than a big molecule. Also, by increasing molecular complexity, the diversity and "drug space" increased exponentially as well. In fact it has been calculated that more than 10^{63} drug like compounds [20] are present, whereas the number of compounds with molecular weight ≤ 160 Da [21] is limited to 10^7. Due to

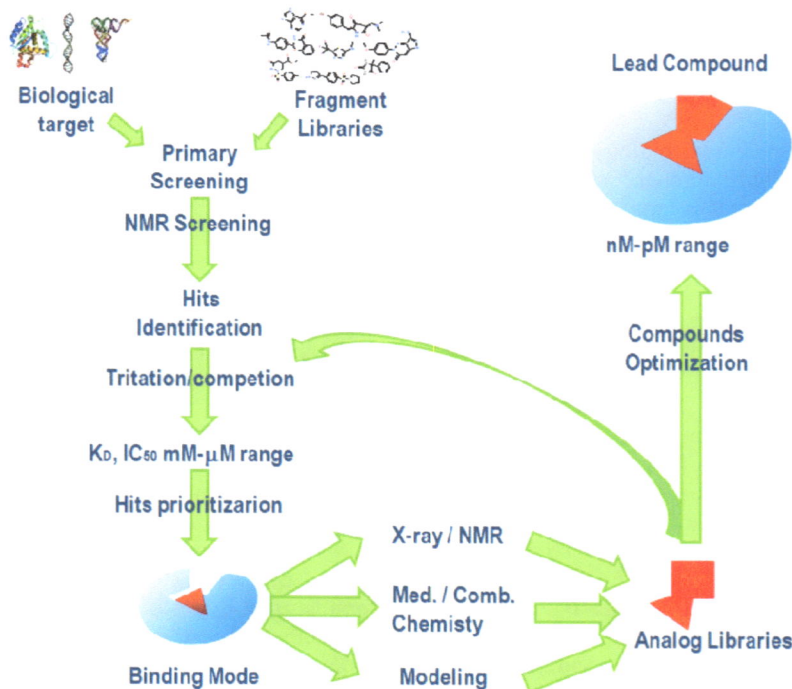

Fig. (3). The general flow-through of fragment based drug discovery from the primary screening to Lead Identification. The simultaneously application of different biophysical and structural methods is essential for the generation of the lead compounds.

their chemical simplicity and reduced functionality, the resulting hits from fragment based screening (FBS) show a weak affinity for the target (µM-mM range). Therefore sensitive detection methods, structural information and computational index are required to prioritize and consequently generate lead candidates. Integrated collaboration among biologists, biophysicists, crystallographers, computational and medicinal chemists is required for the success of this approach. The general flow-through of fragment based drug discovery (FBDD) is schematized in Fig. (**3**). A plethora of NMR screening methods have been developed in the last thirteen years in both academic and industry and will be explored in detail in the next chapters, highlighting the strength, sensitivity, reliability and versatility of NMR techniques. Several reviews and books have been published focusing on the applications of FBA [2, 22-28], including a recent review [29]. In Table **1** compounds currently in clinical development derived from FBA are shown. Interestingly, several compounds originated mainly from small biotech companies such as Astex, Plexxikon,Vernalis and SGX pharmaceuticals, that made FBA their religion.

Table 1. Fragment Derived Compounds and Furthest Stage of Clinical Development. *Reproduced with Permission from Chessari G. and Woodhead AJ. Drug Discov Today 2009; 14: 668-75*

Compound	Company	Target	Progress	Detection Method
ABT-263	Abbott	Bcl-XL	Phase 2	NMR
AT9283	Astex	Aurora	Phase 2	X-ray
LY-517717	Lilly/Protherics	Fxa	Phase 2	Computational/X-ray
NVP-AUY-922	Novartis/Vernalis	Hsp90	Phase 2	NMR
Indeglitazar	Plexxikon	PPAR agonist	Phase 2	HCS/X-ray
ABT-518	Abbott	MMP-2 & 9	Phase 1	NMR
AT7519	Astex	CDK2	Phase 1	X-ray
AT13387	Astex	Hsp90	Phase 1	NMR/X-ray
IC-776	Lilly/ICOS	LFA-1	Phase 1	NMR
PLX-4032	Plexxikon	B-Raf[V600E]	Phase 1	HCS/X-ray
PLX-5568	Plexxikon	Kinase Inhibitor	Phase 1	HCS/X-ray
SGX-523	SGX Pharmaceauticals	Met	Phase 1	X-ray/HCS
SNS-314	Sunesis	Aurora	Phase 1	MS

Before looking further at the different fragment-based screening methods we wish to first give an overview on two fundamental steps of the FBDD process: the fragment library construction and prioritizing indexes.

Fragment Library Construction

Historically Lipinskis "rule of five" [30] was widely used and lead compounds must limit the MW to 500 Da, the LogP <5, the number of hydrogen bond donors ≤ 5 and numbers of hydrogen bond acceptors ≤ 10 in order to obtain oral drugs. Consequently other physicochemical limits are added [31-33] such as the number of heavy atom to 10-70, the number rotatable bonds (NROT) to 2-8, number of rings 1-7 which aromatic < 3. Even though these characteristics lead to identification of drug-like molecules, this process is contradictory to the FBA approach, which is focused on obtaining a good starting point (lead). The fragment library should be "lead-like" rather than "drug-like" and small polar compounds with high solubility and no aggregation properties should be included. The solubility and aggregation state are important features since fragments must be tested at high concentration. Unfortunately present prediction methods still lack accuracy and therefore experimental compounds characterization is requested during library construction. Congreve and co-workers [34] from Astex, analyzing the fragment hits on several different targets, proposed the "rule of three" for building fragment libraries able to identify efficient lead compounds. Therefore a good fragment should have MW ≤ 300 Da, with the number of hydrogen bond donors ≤ 3, the number of hydrogen bond acceptors ≤ 3 and ClogP ≤ 3 and suggested NROT (≤3) and Polar Surface Area (PSA ≤60). Currently the rule of three is well accepted and used [2, 24, 25, 35, 36] for fragment libraries construction, but other features need to be considered developing a screening detection system, especially in NMR screening. For instance the SHAPES library [37] was built with the aim of reducing the costs, improving the synthetic accessibility, diversity, solubility and simplification of NMR spectra. Therefore a commercial database [38] of commercially available drugs was first examined for common chemical structures (shapes). This resulted in the observation that 50% of all drugs were represented by 32 different shapes. This framework number increased when atom type, bond order and common drugs side chains were incorporated into the analysis. The most common shapes and drug side chains were used to computationally select compounds among different available compounds databases (MCM, ACD etc). The 132 resulting compounds constituted the SHAPES library; these compounds had a MW comprised between 68 and 341 Da, they were pure and soluble compounds and did not aggregate up to the concentration of mM in water. In addition they did not contain reactive moiety, were commercially available, and showed a simple 1D ^{1}H NMR spectrum, with at least two protons within 0.5 nm apart. The authors reported several successes screening the SHAPES library by NMR experiments on p38 MAP kinase and 5'-monophosphate dehydrogenase, and showed that the hit rate of compounds containing SHAPES scaffolds was 8-10 fold higher then HTS compounds looking at the 50% of inhibition. A similar library was proposed [35] in order to find inhibitors of the rabbit muscle creatine kinase (RMCK) [39], an enzyme which catalyzes the reversible transfer of a phosphate group between ATP and creatine. RMCK was considered an interesting model system for studying pharmaceutical targets such as brain creatine kinase and arginine kinase. Therefore 61 commercially available SHAPES compounds were selected. Seven were immediately discarded on the basis of their low water solubility or chemical instability or their unsuitable NMR spectral properties The remaining 53 small molecules were tested against the RMCK by NMR experiments, identifying 19 possible binders. Interestingly, a phenylfuroic acid, showing a strong NMR binding effect, was not displaced by increasing and saturating amounts of MgADP, thus suggesting a different binding site. The docking of this molecule to the crystal structures of creatine kinase indicated that it could bind in a pocket close by the nucleotide binding site. Using this fragment as an anchor together with *in-silico* studies, four analogues that partially accessed the nucleotide binding site were synthesized, providing novel micromolar reversible inhibitors with a mixed ATP competitive/non-competitive mechanism. The limited number of tested compounds and the combination of different experimental and computational methods is further proof of the potential of the fragment libraries and the FBDD. Hit fragments are typically discovered by screening generic libraries, but in some cases, knowledge about initial ligands or the protein binding site can be used to build focused libraries [40]. Recently Abbot researchers described [41] a systematic approach to rapidly design libraries allowing the discovery of novel, potent and selective kinase inhibitors with good new intellectual property (IP) space. The method implemented a scaffold-based strategy and utilises the experience of Abbott in synthetic chemistry, fragment-based screening, computational techniques and X-ray crystallography. Investigating the hinge region [42] in the ATP binding site of several different kinases by three parallel approaches (one and two dimensional NMR screening, biochemical assay at high compounds concentration and virtual screening) and combining "in house" and commercial compound collections, they initially selected about 500 fragments with a potency less than 100mM. Unfortunately the provenance of the numerous active compounds from large already used compound collections resulted in a very low IP novelty. Therefore the authors decided to build ex novo new scaffolds, maintaining only some common structural motives for binding at the hinge region, such as hydrogen bond donor–acceptor–donor atoms at specific distances from each other, and donor–acceptor / acceptor–donor pairs, by adding, subtracting, expanding and contracting the rings. This synthetic elaboration generated more than 5000 compounds in 50 series, with sub-millimolar activity on several kinases. The resulting hinge-core elements were then confirmed by functional assays and X-ray structural analysis. The structural studies showed that in some cases the fragment modification induces a change of its binding mode to the same kinase, and moreover that several fragments showed different binding mode in different kinases, highlighting the possibility to differentiate multiple pathways to improve potency and selectivity of those scaffolds. An example of their results is reported in the Fig. (**4**).

Several advantages of this approach have been reported: a) potency and specificity for a large number of kinases can be achieved even by fragment scaffolds b) the possibility to improve the quality, the drug-like and the biological relevancy of the compound c) the possibility to increase the hit rates of corporate compound collections. Another important aspect is that by looking at the ATP binding site, the resulting functionalized adenine mimetics can be active not only on kinases but also on all those proteins that use or bind adenosine, largely expanding the hit screening rates of the collections. Libraries are generally based on the lead-like and drug-like concepts, but recently others libraries based on the "Metabolite-likeness" criterion [43] have been proposed. In this library the metabolites are the endogenous molecules of primary

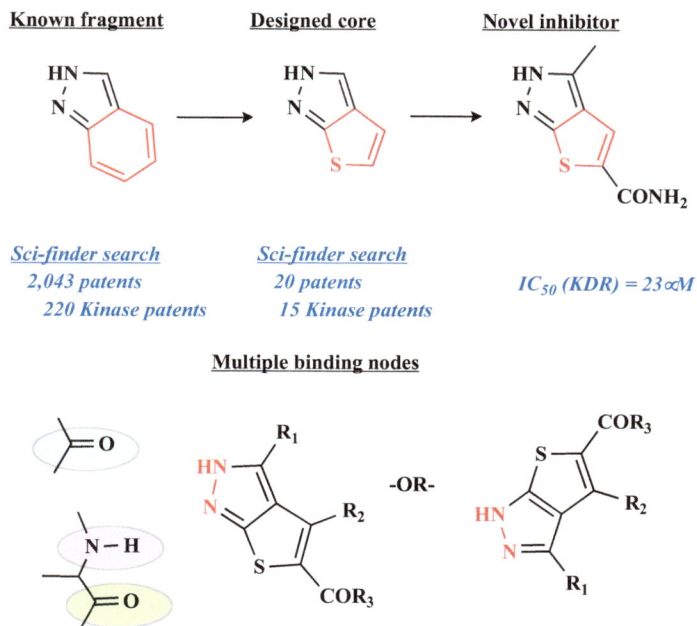

Fig. (4). *De novo* design. An example of the *de novo* design of a kinase core by starting with the well known kinase binder indazole and synthetically pursuing the, at that time, unknown hinge region element thienopyrazole analog. In the right is shown a representative molecule synthesized to evaluate the utility of this core, from wich multiple compound libraries were derived. At the bottom is schematized the multiple binding mode experimentally observed for this chemotype, based on tautomerization of the pyrazole ring. *Reprinted with permission from Akritopoulou-Zanze I, Hajduk PJ. Kinase-targeted libraries: The design and synthesis of novel, potent, and selective kinase inhibitors. Drug Discov Today 2009; 14: 291-297.*

metabolism and not the reaction drug products resulting from the action of drug metabolism enzymes. By comparing human metabolites with commercial drugs and library compounds with several molecular descriptors, it was found that the commercial drugs were more metabolite-like, than the compounds forming the common screening libraries. Therefore the metabolite-likeness can be used as a filter for the construction of pharmaceutical screening libraries that cover the total diversity of human metabolites [43]. With the advent of new screening methods, such as ^{19}F and ^{31}P NMR spectroscopy, specific libraries have been developed [44, 45]. Finally we wish to draw the attention to a new type of compound library built with the intention of assigning the function to new proteins coming from different genomic programs: the NMR functional chemical library [46]. This library is based on the idea that proteins sharing the same biological function should have similar binding sites and therefore similar ligands binding. Analyzing many databases for known compound interactions with known protein systems, a library was built containing small ligands, such as substrates, metabolites, known drugs, co-factors and inhibitors, with known biological activity. The functional compounds cover as much as possible the different biological functions and obey the "rule of five". Moreover NMR functional libraries have to be modelled to the NMR requirements and therefore show a good water solubility, no aggregation, precipitation, reactivity in mixtures and have a unique NMR resonances to avoid long deconvolution procedures. The importance of building a good fragment library and the increasing popularity of FBA have induced industries to develop and to sell fragment libraries and follow-up libraries, as summarized in Table **2** [29]. This allows even small biotech companies and academic groups to use these libraries without losing time and resource to create new ones.

Prioritizing Indexes

Due to the large number of weak-affinity binders coming from FBS it becomes necessary to identify criteria for validating and selecting the lead compound. Based on Kuntz and coworkers [47] paper, that demonstrated that the maximum affinity per atom for organic compounds is -1.5 Kcal mol^{-1}, Hopkins and co-workers [48] introduced the concept of ligand efficiency (LE), a way for normalizing the potency and the dimension to provide a useful comparison among different compounds. LE is calculated dividing the

Table 2. Chemical Suppliers that Provide Screening Libraries. *Reproduced with Permission from Chessari G. and Woodhead AJ. Drug Discov Today 2009; 14: 668-675*

Supplier	Library Name	Number of Fragments
ACB Block	Fragment library	1280
ACB Block	F-Library for NMR	760
ASDI	Fragment screening collection	nd
Asinex	Building blocks	6200
BIONET	Fragment library	3228
Chembridge	Fragment library	3800
Enamine	Fragment library	11717
Maybridge	Br-fragment collection for X-ray	1500
Maybridge	F-fragment collection for NMR	5300
IOTA Pharmaceuticals	Fragment library	5500
Zenobia Therapeutics	Commercial fragment library	352

free energy of binding (ΔG) of a ligand for a specific protein, by the number of non-hydrogen atom, or heavy atom count (HAC), according with the eq. 1:

$$LE = -\frac{\Delta G}{HAC} \approx -\frac{RTln(IC_{50})}{HAC} \tag{1}$$

The LE lower limit is around 0.3 Kcal mol^{-1}, that correspond to the LE of "acceptable" drugs with a 500 MW compound and a binding constant of 10nM, the maximum limit is 1.5 Kcal mol^{-1} therefore LE value between 0.3 Kcal mol^{-1} and 1.5 Kcal mol^{-1} are desirable to obtain oral drug candidates [49]. Based on LE, Abad-Zapatero and Mets [50] introduced three indexes, two of them based on molecular weight (MW in kDa) scale and percentage of efficiency (PEI) and the ligand efficiency index (BEI) respectively. The former is calculated using the single spot percentage of inhibition at a specific concentration, whereas the latter is calculated using the more accurate dissociation constant K_D value or inhibition constant K_I or IC_{50} values according to the equations 2 and 3:

$$PEI = \frac{\%\ of\ inhibition\ at\ a\ specific\ concentration}{MW} \tag{2}$$

$$BEI = \frac{-LogK_D,\ LogK_I,\ -LogIC_{50}}{MW} \tag{3}$$

The third index is the surface-binding efficiency index (SEI), function of the polar surface area (PSA) as important parameter for permeability and oral bioavailability, given by the equation 4:

$$SEI = \frac{-LogK_D,\ LogK_I,\ -LogIC_{50}}{PSA} \tag{4}$$

PEI is a relative index and usually is used to evaluate HTS results, because in HTS screening the molecules activity is determined by measuring the percentage of inhibition at a determined concentration. It is important when comparing different PEI to ensure that the percentages of inhibition have been calculated at the same concentration. Since BEI and SEI follow the potency per additional kDa and the potency per exposed PSA respectively, they can be even used simultaneously to monitor the ligand optimization: compounds with high SEI-BEI pair-wise-value, have highest probability of having favourable PK and become drugs [50]. In his recent review Zhao [51], looking at the nature of the binding pocket, drawn the attention to lipophilic and hydrophilic molecules properties. He noticed that in a target endowed with preferential lipophilic compounds binding, hydrophobic interactions could be promptly identified;

conversely molecules with more polar interactions prefer hydrophilic pockets. Isothermal Tritation Calorimetric (ITC) experiments allow the entropy-driven and enthalpy-driven compounds selection and let easier the optimization for hydrophilic and lipophilic binding pockets respectively.

Other two indexes recently have been proposed and have been reviewed by Congreve and co-workers [52]: the Group Efficiency (GE) and the Ligand Lipophilicity Efficiency (LLE) [53]. The former enables the evaluation of a single group contribution (e.g. Methyl, phenyl ring, amide etc.) with respect to the free energy binding wall, and it is calculated similarly to LE being $GE = -\Delta G/HAC$, where HAC is the number of heavy atoms of the single group. GE evaluates if there is part of the molecule that should be substituted ($GE < 0.3$) or if the substitutions are properly made to maintain drug-like proprieties ($GE > 0.3$). The LLE index, allows the correlation between potency and lipophilicity; high lipophilicity is typical of a promiscuous inhibitor and could increase the risk of non-specific toxicity [53]; in such a way the goal is to increase the potency by maintaining constant the lipophilicity. This can be achieved by looking at LEE index, defined as $LLE = pIC_{50}$ (or PK_I) – cLogP (or LogD): oral drugs show a mean cLogP value of 2.5 and potency of 1-10nM, so compounds with $LLE \geq 5\text{-}7$ could be considered as good starting point for preclinical candidates. As highlighted by Congreve and co-workers this analysis is a further advantage of FBA. According to Astex experience, fragments improvement lead to compounds with lower molecular weight and higher polarity. Finally recently the Golden Ratio ($\Phi = 1.6180339887...$) has been proposed as a tool for prioritizing the fragment hits [54]. For instance studying the Kuntz [47] results and plotting the LE value of strongest compounds as function of the HAC (heavy atom), it has been observed that when HA halves itself the maximal LE increases of 1.6 fold. Based on this observation, the percentage of LE (%LE) has been introduced [54]. %LE is defined as $\%LE = (LE/maxLE) \times 100$, where $maxLE = 1.614^{(10/HA)} \cong \Phi^{(10/HA)}$. The %LE relates the LE of a compound with the strongest binding hit with the same number of HA, and can be useful both for analyzing HTS results and prioritizing the hits of the FBA.

METHODS DIFFERENT FROM NMR IN FRAGMENT-BASED APPROACH

As previously mentioned, the main problem either with HTS or FBDD protocols is the approved and definitive demonstration of a direct binding between hits and target molecules. The further hits compound optimization is based on this required result. The role of HTS has been critically pointed out recently by Macarron [55] suggesting a critical evaluation of successful drug discovery engines. It was already pointed out by Gribbon [56] that HTS is just a tool to explore the chemical resource presented in corporate collections. In the last few years several new technologies have been developed to improve the quality and the efficiency of the drug discovery process. The new tools for the identification of new targets were essentially based on genetics and gene expression profiling, on the other side, different physical methods were applied to screening compounds libraries for direct target binding process. Recently some NMR methods and few other most popular biophysical techniques were reviewed by Dalvit [57] and Gòmez-Hens [58]: we have summarized most of them here with some important applications.

Label Free Biosensor Techniques: Optical Methods

Label free biosensors are largely used for detection of biological agents [59-61] but are also particularly valuable in the drug discovery process because they provide fast results on biomolecular interactions and for the money saving labeling absence. Currently they have gained a higher level of automation and their combined use with other techniques, including NMR, X-ray crystallography and mass spectrometry, have made them a very valuable tool. Advantages especially in fragment-based screening were reviewed recently [62, 63] with special attention to very recent models available on the market. Direct optical methods can be differentiated into three possible detection methods: Surface Plasmon Resonance (SPR), grating couplers (Waveguide Based graTing couplers, WBT) and Reflectometric Interference Spectroscopy (RIfS).

The first two are essentially detector systems based on a biosensor, able to detect changes of refractive index as a function of sample concentration or surface density, thus interacting with the evanescent field. In biological applications, the immobilization of a protein onto the surface, usually a chip, allows the screening of several hundreds of molecules. When protein-protein or protein-ligand interactions are investigated, both affinity and kinetics can be explored, thus providing the binding constant determination. SPR-based biosensors are the most common used systems in label free biomolecular interaction analysis and Biacore system is probably the largest used in academic and industrial research. Few papers are present in the literature concerning Biacore based fragment screening applications: Myszka group [64] joined together several research groups on binding constant measurements of 10 selected compounds and carbonic anydrase, performed with different instrumentations. The measured dissociation constants obtained by

BIACORE were compared to those obtained by ITC determinations, showing excellent agreement. The kinetic distribution plots of the compounds analyzed, provided a clear insight into the possible variation that could be expected for this type of analysis. Differences in results were due to poor data quality for different instruments characteristics. The Wright group [65] approached the discovery of a non nucleoside inhibitor of hepatitis C virus (HCV) NS5b RNA polymerase, starting from fragments with millimolar binding affinity discovered by X-ray crystallographic screening down to nanomolar affinity using structure-based design. SPR analysis using BIACORE technology was used in the early stages of the work. SPR analysis also helped to diagnose biochemical false positives during fragment optimization. Finally, Karlsson group [66] described in their paper a general methodology that uses biosensor technology to approach the study of kinases, using complementary methods of screening and enzyme activity measurement. Two important factors were indicated for successful binding determination: 1) immobilize sufficient levels of kinase while maintaining binding capacity, that could be achieved in most cases by avoiding the pH lowering below 6.0 and 2) using an improved buffer containing essential cofactors (i.e. bivalent metal ions). SPR are considered as "informative rich" methods: k_{on}, k_{off}, binding stechiometry, and K_D are available within the measurement.

Few examples are present in the literature of SPR analysis in fragment-based screening: quite recently Nordström [67] group identified small inhibitors of matrix metalloproteinase 12 (MMP-12) approached by using a specifically designed fragment library. This experimental design was successfully in overcoming the instability of MMP-12 and allowed the identification of fragments that interacted specifically with the active-site of MMP-12 but not with the reference protein. Two effective inhibitors were rapidly identified, and the highest affinity compound was confirmed to be a competitive inhibitor with an IC_{50} of 290 nM and a ligand efficiency of 0.7 kcal/mol heavy atom. RIfS spectroscopy is based on the interference of light beams, which are reflected at interfaces with different refractive indices [68]. It is used to monitor biomolecular interactions but also for volatile organic compounds (VOCs) and other environmental relevant analytes. The first application appeared back in 1996 concerning the characterization of an optical immunoassay where Fab-fragments reacted with analyte molecules [69]. RIfS in contrast to SPR can detect optical thickness without an evanescent field, thus resulting in a more stable determination against instabilities due to thermal variations.

Labeled Fluorescence-Based Methods

Fluorescence-based techniques have proved to be one of the most well suited detection methods because of their sensitivity and small sample quantity requirements. They have emerged as alternative to the radio-labeled based assays and in general conventional fluorimetry is quite sensitive to background effects. However, FA and FL are usually applied in binding or biochemical assays due to their high absolute sensitivity. FL refers to the average time the fluorophore remains in its excited state before emitting a photon, while FA gives indications on the molecule tumbling, in other words it could be related to the molecular shape/size.

This technique is based on the emission of polarized light from the excited fluorophore: the polarization of the emitted light will increase with the slowing of the molecular tumbling (the ideal case is a small fluorophore fused to a larger molecule). In this case, K_D can be determined. FA is used for screening purpose in competition assays with a fluorophore-labeled reporter. The weakness of the fluorescence-based methods is that in case of protein-fragment interaction studies, identification of binders can be obtained only when fragment concentration is comparable to the K_D of the fragment. Furthermore, optical interferences can generate false responses. As already pointed out [57], FL and FA show the same properties and requirements with the exception of the protein size sensitivity.

Mass Spectrometry

In MS spectrometry, molecules are investigated and quantified on the basis of their molecular masses, with high sensitivity. Samples need to be ionized (and this is achieved by using different methods) then the ions are separated on the basis of their masses within an electro-magnetic field and then detected by different detection methods relayed on the high or low molecular weights. Samples ionization can be obtained by different processes: electronically, chemically, by electrospray or by laser (matrix assisted laser desorption ionization, MALDI). Before mass detection, a separation process need to be performed and usually liquid chromatography (LC) is the most common used technique. Mass techniques for HTS in drug discovery have been reviewed [70]. Recent applications of non-covalent mass spectrometry were reported by Moore [71] and Ritschel [72] in determining the ligand binding process. This technique enables accurate mass measurements of intact non-covalent complexes, thus providing reliable information on fragment binding and identification of promiscuous multiple binders.

Thermal Methods

Thermal methods are essentially based on the protein unfolding process monitored by measurements of the difference in the melting point between free and ligand bound state of the protein, which is related to the affinity binding of the ligand. One method, in particular the use of environmentally sensitive dyes, was first reported in 1997 by Poklar [73] and adapted to enable HTP screening in 2001 by Salemme group [74]. The measurement of the different fluorescence intensity during the temperature rising of a fluorescence dye hydrophobically bound to the protein (differential scanning fluorimetry, DSF) is a screening method that allows the identification of ligands bound to a protein. This "thermal shift" is best monitored by using conventional real-time PCR, where ligand solutions from a storage plate are added to a solution of protein and dye distributed into the wells of the PCR plate. It has been demonstrated [75] that the increase in melting temperature for a protein upon ligand interaction can be proportional to the concentration and the affinity of the ligand. This correlation is not true for ligands with different enthalpic and entropic components of target interaction. A second method is based on the use of static light scattering (differential static light scattering, DSLS) to monitor protein thermal denaturation and subsequent aggregation. The use of light scattering was reported originally in 2002 by Kurganov [76] and then patented in 2004 by Sinisterra [77]. In a relatively recent paper [78] these two screening platforms were compared to measure the increase in protein thermal stability upon binding of a ligand without the need to monitor the enzyme activity. In this paper 221 proteins from humans and human parasites were successfully screened against small molecule libraries, leading to identification of new ligands for these proteins. However these methods are not believed to be properly reliable for significant numbers of cases and are subject to interferences.

X-Ray Crystallography

A key aspect of FBDD involves the use of X-ray crystal structure of the protein target, thus relating to a specific chemical strategy, called Structure-Based Drug Discovery (SBDD). This approach would be difficult to apply in case of membrane proteins coupled to receptors or ion channels. In any case X-ray crystallography remains extremely valuable in FBBD, due to the high content of structural information available. Typically mixtures of ligands are soaked in a preformed crystal of the target macromolecule to determine the structural modification upon binding. Several examples appeared in the literature concerning the use of template screening and hit validation through protein ligand X-ray crystallography validation. Particular attention was focused on kinase proteins in these last years, because of their general involvement in cellular signalling and their amenable inhibition by small organic molecules. All kinase proteins contain a sequence of amino acids making a sort of a hinge connecting two lobes of the catalytic domain [42]; these amino acids constitute the binding site for ATP and allow the phosphorylation. Thus the blocking of this binding is translated into a direct protein kinase inhibition. In this respect, the Abbot approach to enable rapid discoveries of kinase targeted libraries has been recently reviewed [41]. Just to mention few examples, in the recent paper [79] a weakly active purine template hit was found, progressively iterated to submicromolar pyrazolopyridine derivative ligand with good efficiency of inhibition against Checkpoint kinase 1 (CHK1) protein, an oncology target. This result is a good example of sequential combination of in silico low molecular weight template selection, high concentration biochemical assays and hit validation via X-ray crystallography. In another recent paper, the combined use of X-ray crystallography and virtual screening was applied for hit identification of inhibitors for the cyclin dependent kinase 2 (CDK2) [80]. In this work, apo-crystals of CDK2 were soaked with cocktails of targeted fragments. For some of the interesting fragment hits, detailed structure information were investigated and the optimization of the indazole hit led to the discovery of the potent, ligand efficient inhibitor of CDK2, with good activity against a range of human tumor cell lines. (Fig. (**5**) the discovery scheme for the final compound).

Similarly selective inhibitors of another kinase, involved in the control of cell proliferation and survival, in particular protein kinase B (PKB), were found [81] in a FBS with azoindole moiety, by using iterative X-ray crystallography of chimera complexes inhibitor-PKA-PKB. Structure-based design of chemical series was driven by the crystallographic data from PKA and PKA-PKB chimeras. The structure determination was followed by assays on compound activity in enzyme assays and cellular mechanistic, to drive efficiently potency and selectivity of compounds. Furthermore, selectivity for PKB *vs.* PKA was observed for selected inhibitors, 4-aminopiperidine derivatives, essentially due to the basic amine and lipophilic binding elements, as well as the flexibility of piperidine ring and the variable hybridization of the piperidine nitrogen. The most selective PKB inhibitor, showed significant differences in binding mode between PKA and PKA-PKB chimera. Other examples of X-ray crystallography applied to hit generation are summarized in reference [82].

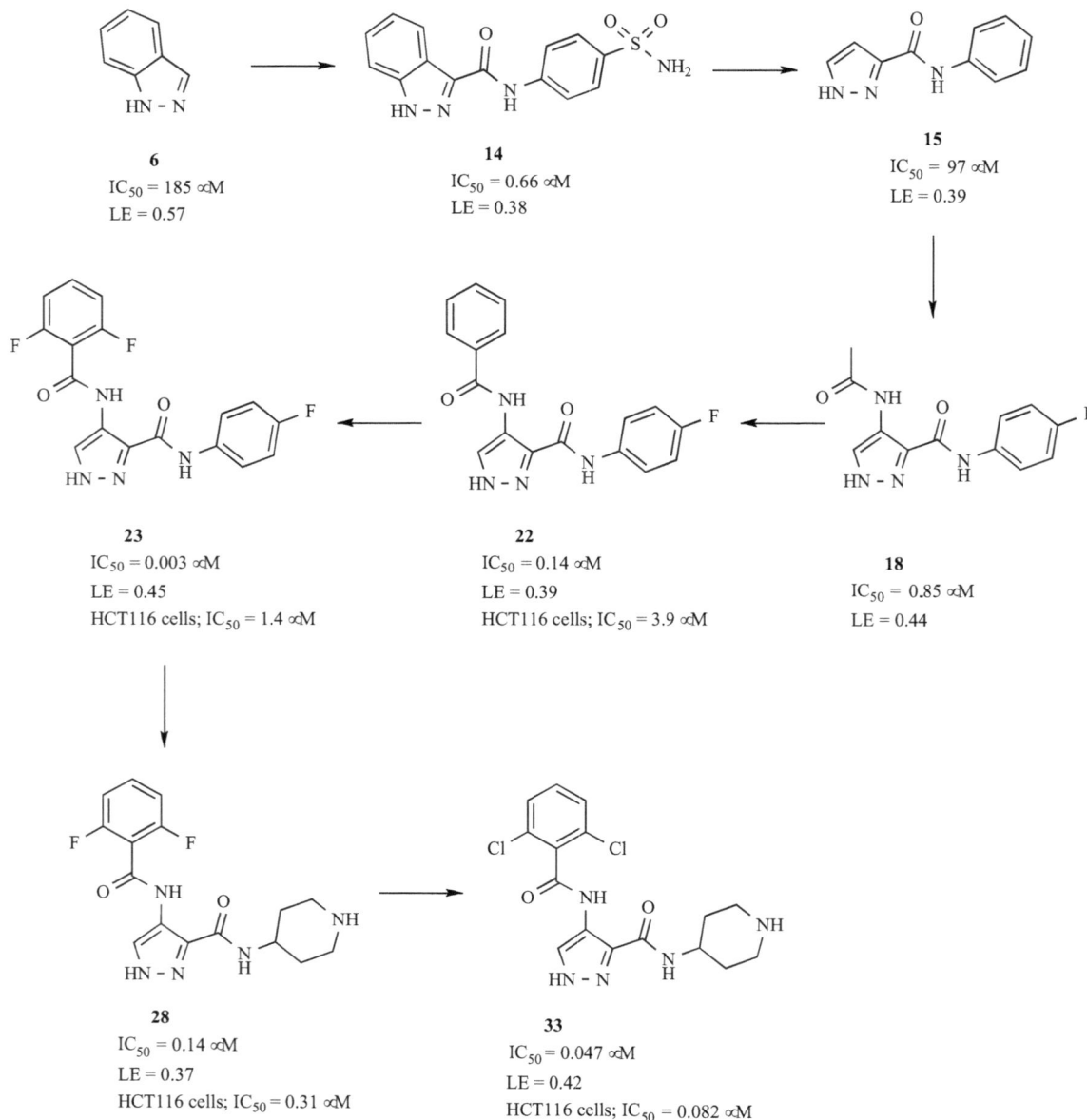

Fig. (5). Key compounds in the optimization of fragment 6 to the clinical candidate 33. *Reprinted with permission from Wyatt PG, Woodhead AJ, Berdini V, et al. Identification of N-(4-Piperidinyl)-4-(2,6-dichloro,benzoylamino)-1H-pyrazole-3-carboxamide (AT7519), a Novel Cyclin Dependent Kinase Inhibitor Using Fragment-Based X-Ray Crystallography and Structure Based Drug Design. J Med Chem **2008**; 51: 4986-4999.*

NMR METHODS FOR FRAGMENT-BASED APPROACH

Drug discovery process takes advantage of several NMR methods to gain insight into the recognition process, as well as lead optimization. Fragmentation of the drug into small pieces or down to functional groups has been used to simplify the computational analysis of ligand binding. This approach makes the assumption that binding affinity results from the sum of individual interactions effects. Development and evolution have been recently illustrated in the Hajduk review [26].

Looking at the Protein: Protein Chemical Shift Perturbation

NMR is a very useful technique for the FBDD process due to its ability in detecting very weak molecular interactions at atomic level. The high sensitivity of NMR chemical shift to the atomic environment yields information about the binding of a small molecule to the biological target (protein, nucleic acid, etc). Chemical shift perturbation can be used to map the ligand binding sites and to screen libraries for identifying active compounds. For this purpose mono e bi-dimensional NMR experiments can be run on labeled and unlabeled protein.

Unlabeled Protein

Interaction studies of small molecules and organic solvents to unlabeled proteins have been carried out by NMR since the end of 90's. In contrast, X-ray diffraction studies have shown that the organic solvents are able, not only to bind in the substrate binding sites, thus inhibiting the enzyme catalytic activity, but also in secondary pockets [83]. Molecular information about active sites and those alternative pockets can help to increase the specificity and potency of lead compounds. Liepinsh and Otting [84] applied an NMR approach to detect the molecular interactions of hen egg-white lysozyme (HEWL) in the presence of different organic solvents and to calculate the corresponding binding constant. The dipole-dipole interaction (nuclear Overhauser effect, NOE) between protein and ligand protons can be detected only if the two protons are less than 0.5 nm distance apart. Therefore structural information can be achieved observing NOEs between different protons. Applying a filtering scheme in two-dimensional NOESY they made it possible that the intermolecular NOEs visible in the spectrum were only those of the aminoacids interacting with the organic solvent. Intermolecular ligand-HEWL NOEs have made it possible to determine both the binding site and mode of interaction for the ligand, whereas the binding constants have been calculated following the ^{1}H chemical shift of HEWL in presence of increasing concentration of organic solvent, taking into account the volume changes upon titration. Dalvit and co-workers [85] by applying an efficient multiple solvent suppression technique were able to detect and quantify very weak intermolecular interactions between organic solvents and unlabeled proteins in solution. The use of multiple solvents suppression increases the detection dynamic range especially in the spectra containing multiple large peaks of the solvent. Based on previously in house and literature data, that confirmed the binding of DMSO to the FK506 binding protein (FKBP12), they decided to characterize the interaction between the organic solvent and the protein. Determination of binding site and binding mode of DMSO to the protein was obtained by recording one-dimensional ePHOGSY spectra with multiple solvents suppression for the complex FKBP12/DMSO. In this spectrum only NOEs between specific aminoacids and DMSO were visible. It is important to note that the sensitivity of the experiment allowed the use of only 0.3 mM of protein whereas usually at least 1-10 mM of protein was required. Protein chemical shift changes in presence of increasing concentration of DMSO, allowed the calculation of a K_D of 275 mM for the DMSO. The detection of several NOEs between the protein and such very weak-affinity compounds demonstrates the sensitivity of the approach and indicate that this could be a good method to screen fragment compounds and obtaining structural and affinity data. Recently a paper has been published [86] reporting the use of tryptophan-mutated proteins for NMR screening by 1D proton NMR version of the antagonist induced dissociation assay-NMR (AIDA-NMR) [87]. The chemical shift of the NH side chain ($^NH^\epsilon$) of tryptophan (Trp), usually it is around 10 ppm at the physiological pH, far away from the bulk amide protons region. This makes it a good reporter in the NMR screening experiments. Trp it not so frequent in the protein sequences, but is often present in the hot spot regions of the protein that are the major contributors to the binding energy [88, 89]. Protein lacking in Trp residues can be engineerized in the region of interest (binding site, protein-protein interaction regions); this was successfully applied for studying and characterizing the p53-Mdm2 complex, the p53-Mdm2 complex antagonist and cyclin-dependent kinase 2 (CDK2) antagonists by using only 50-100 µM of protein. Since AIDA is a competition binding experiment K_D of the protein-protein complex, protein inhibitors, and protein-protein antagonist can be obtained by a single NMR experiment [87, 88]. The combination of 1D Trp screening with a selective NMR pulse sequence SEI-AIDA (Selective Excitation Inversion) [90] has permitted to decrease the acquisition time of the experiments. The drawback of this interesting and useful approach still remains the necessity to introduce Trp in the protein sequence and the following time consuming processes of mutants analysis and characterization. Furthermore the observed Trp signal is due only to one proton; this limits the sensitivity of the method even in presence of the experimental filter introduced. Observation of endogenous aminoacids methyl groups can result in a high sensitivity, since three protons contributes to the NMR signal intensity allowing a minor protein and ligand consumption. Moreover, by looking at the methyl group it is possible to quickly monitor the whole protein without the necessity to label and mutate the biological target, saving a lot of time, and reducing the costs.

Labeled Protein

In the last 23 years many chemical shift-based binding studies by selectively labeled protein have been proposed. The common methods consist in the observation of protein resonance shift in 2D [^{15}N,^{1}H]-HSQC spectra after compounds addition. Many are the reasons of the success of these approaches such as the simplicity of the assay system, which involves only the target and test compounds, the complete absence of background signals via spectral editing, the possibility to discriminate between different binding sites of the protein and the achievement of structural and affinity information. The dominant role in these approaches is played by SAR by NMR (Structure Activity Relationship) [18]. In this approach, after the identification of a weak-affinity binding compound, a second compound that binds to the target close in proximity to the first site, is searched using a chemical shift mapping based approach. Structural characterization of the ternary complex by NMR will give the potential scaffold for a bi-dentate binding compound with higher affinity. Alternatively specific aminoacid can be labeled. Hajduk *et al.* [91] proposed to monitor the ^{13}C/^{1}H chemical shift perturbation of valine, leucine, and isoleucine (δ1 only) methyl group resonances by 2D [^{13}C,^{1}H]-HSQC experiments. They demonstrate that this approach increases the sensitivity by nearly 3-fold compared with ^{1}H/^{15}N chemical shifts NMR-based screening as show in Fig. (**6**).

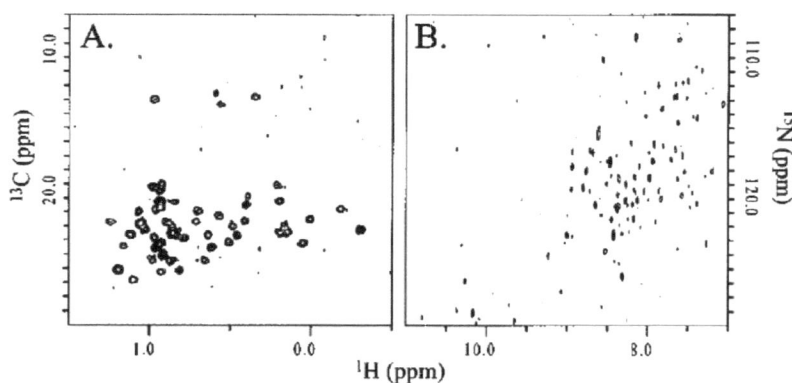

Fig. (6). (**A**) ^{1}H/^{13}C-HSQC and (**B**) ^{1}H/^{15}N-HSQC spectra acquired at 500 MHz in 10 min on a 50 mM sample of ^{13}C(methyl)/U-^{15}N-labeled Bxl-xL. The ^{1}H/^{13}C-HSQC spectrum shown in (**A**) is suitable for NMR based screening, while the corresponding) ^{1}H/^{15}N-HSQC spectrum shown in (**B**) is not. *Reprinted with permission from Hajduk PJ, Augeri DJ, Mack J, et al. NMR-Based Screening of Proteins containing ^{13}C-Labeled Methyl Groups. J Am Chem Soc 2000; 122: 7898-7904.*

Combination 2D [^{13}C,^{1}H]-HSQC with cryogenic probe allows the decrease in protein concentrations as low as 15 ∝M. Moreover they developed an efficient synthetic way, alternative to the classical expensive ^{13}C-glucose, to obtain ^{13}C labeled aminoacids. The increased sensitivity and the new synthetic selective labeling way allow rapid acquisition of ^{1}H/^{13}C-HSQC spectra with low protein consumption and decrease of the screening costs. By combining the methyl group labeling with the protein-perdeuteration the authors were able to overcome the protein dimension limit and detect ligand-binding to proteins with a molecular weight higher than 100 kDa. An interesting paper on large protein, that avoid the protein-perdeuteration, has been recently published by Tarragó and co-workers [92]. They report a cost-effective ^{15}N labeling strategy, where the ^{15}N isotope is incorporated into the tryptophan side chain of the Prolyl Oligopeptidase (POP) a serine protease of 80 kDa. ^{13}C selective Trp labeling for protein-protein interaction and protein-ligand interaction studies was previously proposed and applied to different biological targets [93, 94]. Based on these works Tarragó *et al.* used ^{15}N-labeled indole instead of ^{13}C-labeled indole to selectively ^{15}N label the side chain of Trp. They compared the TROSY spectra of three different labeled POP, the perdeuterated and uniformly ^{15}N-labeld POP, the Trp [^{13}C-indole]-POP and Trp [^{15}N-indole]-POP, proving that the last labeling is more convenient for large proteins. Indeed the TROSY spectra of the Trp [^{15}N-indole]-POP showed a satisfactory signal-to-noise and a well NMR signal dispersion for 11 of the 12 tryptophans. The authors ascribed this result principally to two motives. Firstly the ^{15}N CSA tensor deviates less from axial symmetry than chemical shift tensor for aromatic ^{13}C [95] and secondly CSA value for amide ^{1}H spins is much higher than aromatic ^{1}H spins [96]. The resulting more efficient TROSY effect for ^{1}H-^{15}N moieties leads to an increase of the signal-to-noise ratio. Titration experiments in presence of increasing concentrations of the known POP inhibitor ZZP showed the shifting of the Trp [^{15}N-indole] cross peak in the enzyme active site, suggesting the possibility to use this approach for screening compounds libraries against large proteins without the need of perdeuteration. In addition to the classical ^{15}N, ^{13}C, and ^2H labels, ^{19}F was proposed to selective labeling protein aminoacids in order to study the protein ligand interactions. Leone *et al.* [97] published in 2003 a communication where they presented a new method for the simultaneous incorporation of ^{19}F into all protein Trp side chains. In this way selectively 5-F-Trp labeling of the third Baculovirus inhibitor of apoptosis repeat domain (BIR3) of the inhibitor of apoptosis XIAP [98] was achieved. 1D-^{19}F NMR spectra permit following the binding events and calculating the K_D values. The approach can be used to study binding events in screening campaigns. It strongly decreases the labeling cost, is applicable even to large proteins and provides some structural information on the binding site. All these advantages make it a good method for the FBDD process.

Looking at the Ligand

The NMR methods that use the ligand as observable are largely used with respect to the counterpart's methods that use protein as observable. This essentially is because of the simpler feasibility of the methods; they do not require isotopic labeled samples and are not restricted by the size of the protein. The ligand-based methods perform best, in principle, in the case of large difference in molecular weight between protein and ligand: the large proteins "tumble" much slower than the ligand and therefore increase the nuclear spin relaxation of the bound molecules. The principal methods based on detecting ligand binding have been reviewed very recently [99] and are listed below.

STD and Tr-NOESY

Among the NMR methods that use the ligand as observable, the saturation transfer difference experiment (STD) is largely used for screening, due to its high sensitivity and robustness, coupled with very short experimental time. The experiment is based on the saturation transfer from selectively irradiated signals of the protein to the ligand, which is in exchange between bound and Free State (Fig. **7**). Spectral difference between irradiated and non irradiated protein signals results in residual signal for the ligand signals due to the bound atoms. Typically protein–ligand ratio around 1:100 is used in these experiments. The basic sequence can be incorporated into other NMR sequences, like homo- (STD-TOCSY) [100] or hetero-nuclear (STD-HSQC/HMBC) [101] multidimensional experiments, in order to better characterize the binding epitope. The original application of this technique was introduced in the late 1999, by Mayer [100] and applied to a mixture of seven saccharides and wheat germ agglutinin. This new experiment, inserted within the bioaffinity NMR methods, allowed the detection and identification of binding molecules directly from mixtures. Only two years later, in 2001 the same group proposed the GEM protocol [102] (Group Epitope Mapping), based on STD experiments, studying the interactions between two saccharides and the lectin RCA_{120} using large excess of ligand over the protein, and showing with competitive studies that the two saccharides bind at the same receptor site.

In the screening version of this experiment, a pool of ligands, or compound library, is used: the bound ligand can be identified among the others non interacting molecules. Several papers are present in the literature describing the use of STD experiment in elucidating structural details of complexes between large protein and small molecules. In 2005 Krishnan [103] reviewed the use of STD experiment in ligand screening, in combination with other NMR methods as a practical guide in their use and designing. In a relatively recent paper, Poppe group [104] used a pegylated form of Nurr1 protein, which resulted in a 2-fold increase in the strength of the saturation transfer experiment. This effect was interestingly supported by the measured hydrodynamic radii of the two protein forms: 6 nm against 3.4 nm for the pegylated Nurr1 with respect to the unpegylated form. In this work, the authors used a second site screening to further refine the lead optimizing process, originally proposed by Jahnke [105]. Gomez-Paloma group [106] reported a novel STD-based approach to explore DNA binding ligands. The proposed method, named DF-STD (Differential Frequency STD) consists of two parallel STD experiments in which protons from different regions of the DNA helix (aromatic, sugar, backbone) are saturated by irradiation: then STD effects are obtained by subtraction of off resonance irradiation experiments, as usual. This approach relied on the fact that a ligand in close contact with aromatic base protons will receive more saturation upon irradiation of DNA aromatic protons rather than irradiation of DNA proton from backbone or sugar protons. They tested this approach on poly(dG-dC)poly(dG-dC) copolymer as receptor and six small molecules that bind in different DNA pockets. Another modification of the STD experiment was proposed by Meyer group [107], called STDD (Saturation Transfer Double Difference). It was successfully applied to observe ligands bound to membrane bound proteins, usually not easily investigated. Intact human platelets mixed with the cyclo(-RGDfV) inhibitor, a potent ligand for integrin $\alpha_V\beta_3$ with nanomolar binding affinity were investigated. Experimentally, the STD spectrum of cell suspension added with the ligand was acquired and from this, the STD spectrum of sample containing only cell suspension was subtracted resulting only the STD response for the ligand.

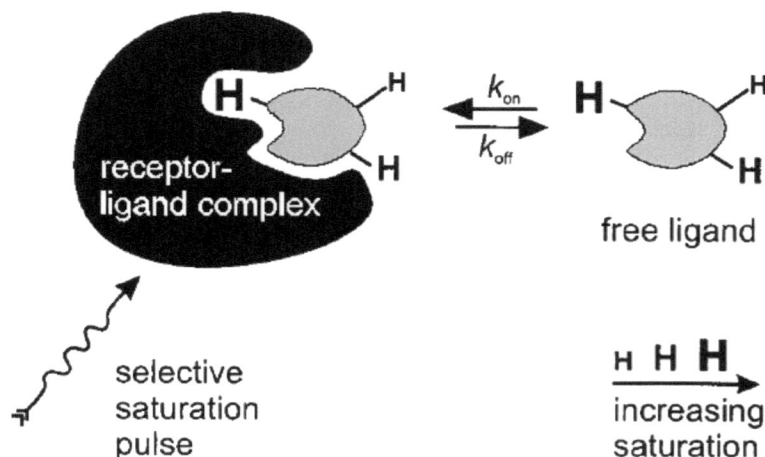

Fig. (7). Irradiation of the protein at a resonance where no ligand signals are present leads to a selective and very efficient saturation of the entire protein by spin diffusion. Saturation is transferred to the binding parts of the ligand by intermolecular saturation transfer. Here, groups represented by the large proton are in close contact with the protein, the medium-sized proton symbolizes a group with less interaction. The smallest proton represents a group with almost no contact with the protein, thus receiving minimal saturation. Therefore, the degree of saturation of the individual protons of a small ligand molecule reflects the proximity of these to the protein surface. *Reprinted with permission from Mayer M, Meyer B. Group epitope mapping by saturation transfer difference NMR to identify segments of ligand in direct contact with a protein receptor. J Am Chem Soc 2001; 123: 6108-6117.*

Cutting and co-workers [108] presented a comparison of saturation schemes with the aim of improving the sensitivity of the STD experiments. Their studies performed on HSA (human serum albumin) and tryptophan complex, suggested that the use of E-burp 1 selective pulse in substitution of the normal Gaussian pulse, and decreases selectivity of the pulse train and leads to a larger extent of protein saturation, thus enhancing the STD signal. Signal overlapping is frequently found, typically when working with carbohydrates. In this case the extension of the STD standard experiment into 2D version is sufficient to resolve the signals. Fehér [109] proposed a combined approach using isotope editing and filtering schemes in STD experiments. By using a labeled reference ligand, the filtering scheme allowed detection of unlabeled ligand while the editing scheme produces only signals from the labeled ligand: this result could be very useful especially in competition experiments, to resolve signal overlapping.

In a recent paper, Mercier [110] showed a multi-step NMR screening process for evaluating chemical leads for drug discovery. It consists of combined use of multidimensional NMR experiments to differentiate between non specific and stoichiometric binders, determine a semi-quantitative dissociation constant, identify the ligand binding site and rapidly determine the protein-ligand co-structure, by using HSQC and STD-NMR experiments. For quantitative interpretation of STD-NMR peak intensities a CORCEMA [111] theory and associated algorithms have been developed.

The transferred NOE (Tr-NOESY) originally proposed by Bothner-By [112] in late 1972, was reviewed in 1997 by Peters [113] for screening applications. Tr-NOESY is the NOE between chemical exchanging molecules, widely used in large proteins interacting with small ligands. As a matter of fact ligands in fast exchange between bound and free states experience negative NOE from the protein complex to the free molecules population. So in 2D NOESY spectra, cross peaks of protein ligands are of the same sign of the diagonal, while free compounds cross peaks will disappear or in the reverse sign. In large overlapping spectra, like mixtures of compounds, 3D experiments could be executed by combining Tr-NOESY and other sequences (TOCSY-Tr-NOESY) [114].

Magnetic Transfer with Water

Another experiment based on saturation transfer was proposed by Dalvit group in 2000 [115] and relied on the general use of bulk water to detect molecules interacting with a protein. This experiment utilizes the water molecules involved in protein-ligand interactions and other water molecules located in cavities close to the binding site for transferring the magnetization of the bulk water to the ligand (Fig. **8**). This process can be performed in general, with two different types of NOE-based NMR experiments: in one case, steady state NOE with on resonance saturation applied on the water signal is performed and in the other case a selective inversion of water signal followed by long mixing time for NOE build up is performed. The use of gradients in these experiments avoids largely the presence of artefacts due to radiation damping and subtraction. The experiments are called Water-LOGSY (water-ligand observation with gradient spectroscopy). In 2001, the same group published a guide on the use of Water-LOGSY for determining the binding constant [116]. A variant of the original experiment is the competition Water-LOGSY (c-Water-LOGSY [7]) experiment that enables the detection of strong binders with the use of medium and low affinity reporter compounds. In addition the binding constant values for the identified molecules can be extracted from these experiments.

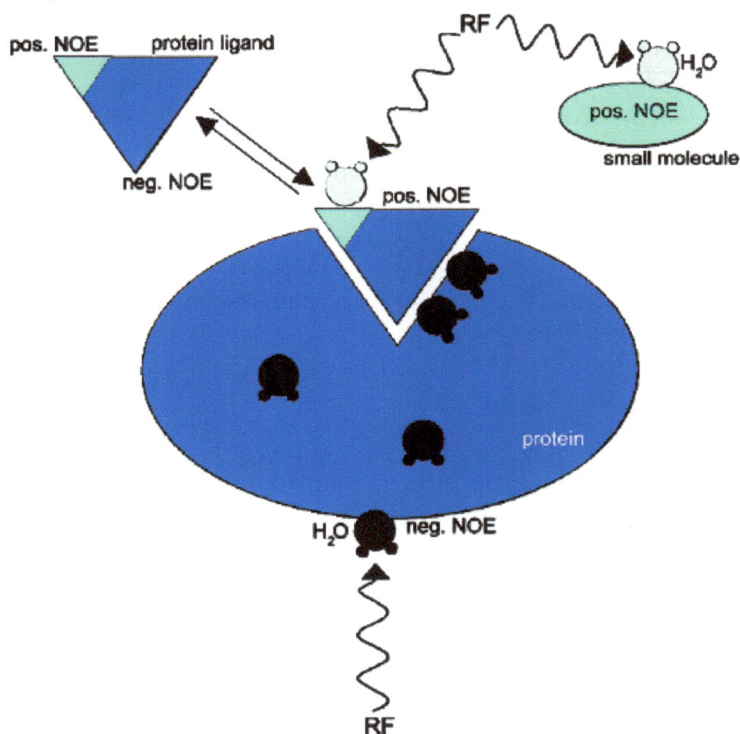

Fig. (8). Schematic representation of the ePHOGSY-NOE principle. Selective radiofrequency irradiation of the solvent resonance followed by a NOESY transfer leads to signals with negative NOEs for ligands (protein ligand) that bind to the protein (blue triangles). For molecules that do not bind to the protein (small molecules), opposite sign signals arise from a positive NOE polarization transfer (cyan). Parts of the protein-bound ligand may retain a positive NOE (cyan triangle in protein ligand) for solvent protons that remain solvent accessible. The polarization transfer may start from water molecules in the active site or on the protein surface or water exchanging with protons on the surface of the protein. *Reprinted with permission from Ludwig C, Michiels PJA, Wu X, et al. SALMON: Solvent accessibility, ligand binding, and mapping of ligand orientation by NMR Spectroscopy. J Med Chem 2008; 51: 1-3.*

Recently the Water-LOGSY experiment was successfully applied for deriving solvent accessible epitopes for ligands [117, 118]. This approach gives information similar to those available from the STD experiment, where the protons bound to the protein but not solvent accessible are identified. The authors investigated the binding for ligands to two hydrogenases (AKR1C3 and HSD17β1) showing very different binding pockets. In the first case, binding between the anti-inflammatory indomethacin to aldochetoreductase 1C type 3(AKR1C3) was investigated by using Water-LOGSY and STD-NMR experiments, resulting in an interaction involving the aromatic protons of the indomethacin molecule, which resulted in good agreement with solvent accessibility data obtained from the X-ray crystallography of the complex. In the case of complex between 17β-hydroxysteroid dehydrogenase type 1 (HSD17β1) and estradiol, the binding pocket is deeply buried inside the protein and not solvent accessible. In this case SALMON and STD results showed different information due to the scarcely solvent accessible binding pocket.

Transverse Relaxation

Transverse relaxation experiments have been largely used in molecular recognition studies [119]. This approach is based on the observation of transverse relaxation rate changes of the small molecules upon binding with the protein. A mixture of nine aromatic compounds against FK506 binding protein was screened using this method.

3.2.4 Hit/Lead Optimization NMR Method

The dominant role in this process is played by SAR, already discussed in previous section. Alternative approaches for the design of high affinity bi-dentate compounds rely on the use of paramagnetic labeling of the first ligand. This method, called SLAPSTIC (Spin Label Attached to Protein Side chain as a Tool to identify Interacting Compounds) has been proposed to alleviate the demanding requirement in terms of large soluble protein concentration in the NMR screening stage [105]. It relies on the use of a paramagnetic moiety such as TEMPO (2,2,6,6-tetramethyl1-piperidine-1-oxyl) covalently bound to one ligand. When a second ligand binds close to the spin label moiety its protons experience paramagnetic relaxation enhancement caused by the spin label, thus resulting in broadening of its resonances. Unfortunately these methods are usually applicable only to small proteins (MW max 30-40 kDa) and show a low sensitivity which limits the screening rate even by using cryogenic probe.

Other approaches make use of the protein mediated ligand-ligand magnetization transfer, know as SAR by ILOEs [120]. Negative InterLigand nuclear Overhauser Effect (ILOE) are detected for ligands that bind simultaneously the target protein and are close in space. A complete approach was recently presented [121], in which a combined trNOEs, ILOEs screening, chemical shift mapping and virtual docking was used for designing a series of bi-dentate compounds targeting the hydrophobic groove near BH3 region of protein Bid, a key molecule in the control of apoptosis.

The magnetization transfer between two different ligands that bind in the same pocket can be observed by using the INPHARMA approach [122]. This method requires two ligands at reasonable concentration to be present and if the structure is known for one of them, the orientation of the other ligand can be derived from additional NOEs observed in the NOESY spectrum. In this paper the authors observed intermolecular NOEs among epothilone A and baccatin III in the presence of tubulin.

^{31}P NMR Spectroscopy

In last century the ^{31}P NMR spectroscopy has been largely used both *in vivo* and *in vitro* to profile neoplastic events, comparing the different amount of phosphorylated metabolites in normal and tumoral tissues and study the *in vivo* drug effects [123-126]. With the arrival of NMR screening in drug discovery process, it had been proposed to use ^{31}P NMR spectroscopy for identifying active compounds against pharmaceutical biological targets, both *in vivo* and *in vitro*. ^{31}P has a spin-$^1/_2$ nucleus with gyromagnetic ratio of 17.25 MHz T^{-1}, a natural abundance of 100%, a large chemical shift anisotropy (CSA) and wide chemical shift dispersion (about 100 ppm) that make quite difficult the NMR signals overlapping; all these characteristic make the ^{31}P a good tool for detecting low affinity binders and for monitoring enzymatic reactions. Both functional and binding NMR experiments have been proposed.

^{31}P Functional Screening

A simple ^{31}P functional NMR method [127] was proposed to observe the activity of bacterial response regulators and to identify low affinity leads in a structure activity relationship (SAR) for developing new antibacterial drugs [128]. Only in bacteria fungi, slim molds and plants, in response to environmental changes, exists a sensing phosphotransfer system (TCST), where a histidine kinase transfers a phosphate from ATP to a response regulator protein. Following the ^{31}P NMR signal of the inorganic Pi in the presence of chemical compounds it is possible to study the turnover inhibition of one of the isolated components of TCST. This approach allowed the successful screening of a number of small molecules that could not be previously tested in HTS approach for their low solubility or their fluorescence properties. Among the advantages of ^{31}P NMR spectroscopy there are the absence of interference with the detection system and the possibility to use high concentration of non deuterated detergents: those increasing the solubility of the hydrophobic compounds. As suggested in the mentioned paper, this biological NMR based assay can be applied to enzymes such as phosphatase, kinase, dehydrogenases and so on, that use phosphate compounds as substrate or cofactors. For example applications of this approach have been shown in *Yersinia* outer protein H (YopH), a phosphatase effector protein from *Yersinia pestis* and in conversion of pyruvate to lactate by lactate dehydrogenase [129]. An interesting use of ^{31}P NMR-functional assay for characterizing the radial spoke protein-2 (RSP2), involved in the flagella formation of bacterium *Chlamydomonas reinhardtii*, has been reported [130]. In this work the selective hydrolysis of ATP versus GTP of RSP2 demonstrated the previously hypothesis that this enzyme was a kinase [131]. All kinase enzymes show a slow ATPase activity in absence of substrate. For several days the conversion of both ATP in ADP and GTP in GDP in the presence of RSP2 by ^{31}P NMR spectroscopy was monitored. Only ATP-γ-phosphate and ADP-β-phosphate decrease and increase were respectively observed whereas the GTP signal kept constant (Fig. **9**), confirming that RSP2 is a kinase.

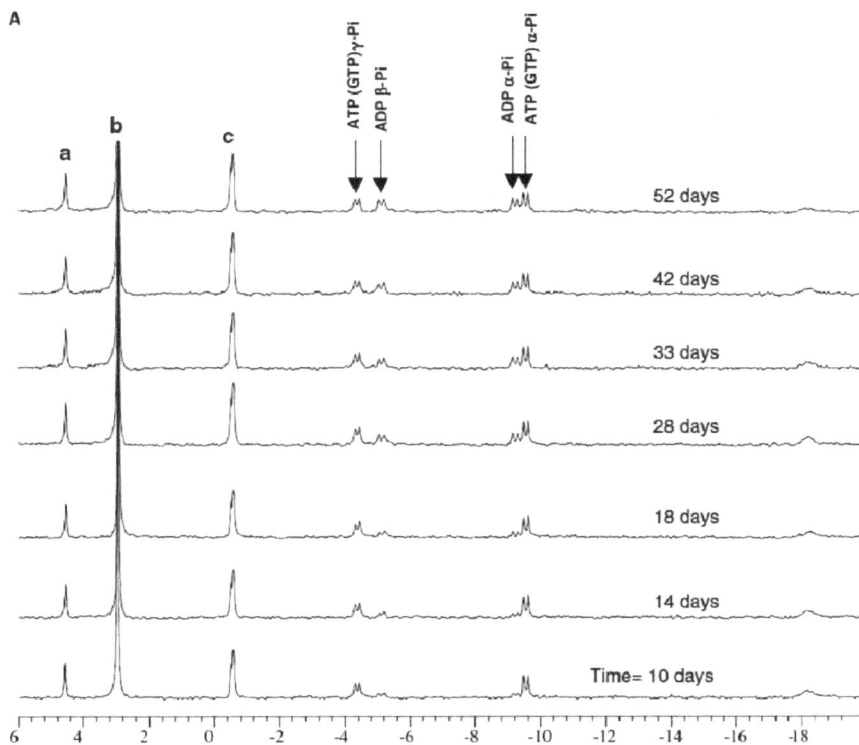

Fig. (9). 1D^{31}P spectra for a mixture of 6 cofactors (ATP, GTP, cAMP, cGMP, cCMP and 50AMP, each 1 mM) in presence of RSP2, recorded at different days. The ^{31}P spectrum at the bottom (10 days) was the first spectrum in which the ADP phosphate resonances were observed. The α and γ phosphate signals of ATP and the α and β phosphate signals of ADP are marked. The letters **a**, **b**, **c** indicate the 5'AMP, the inorganic phosphate and the three cAMP, cGMP and cCMP ^{31}P NMR signals respectively. Control sample with no RSP2 showed no ADP-phosphate signals after 28 days. *Reprinted with permission from Yao H; Sem DS. Cofactor fingerprinting with STD NMR to characterize proteins of unknown function: identification of a rare cCMP cofactor preference. FEBS Lett 2005; 579: 661-666.*

The sensitivity of this method in discriminating different substrates suggested the use ^{31}P NMR spectroscopy even in functional genomics studies.

^{31}P Binding Screening

Very recently Manzenriede and co-workers [45] proposed, the use of ^{31}P NMR ligand-based screening aimed at detecting weak binding events by screening phosphorus-containing compound libraries against biological targets. The direct binding of small phosphate compounds to the target was monitored by observing the line broadening of the NMR signals at increasing protein concentration by proton decoupling ^{31}P-1D-NMR experiments. The screening of a mixture of six phospho-natural compounds against the well characterized Thermolysin protein [132], identified the Hepsera drug, normally used for hepatitis B treatment, as new inhibitor of Thermolysin. Successively competition binding experiments in the presence of a strong known inhibitor confirmed the specificity of the Hepsera drug in binding the active site of Thermolysin. In fact the Hepsera ^{31}P NMR signal disappears completely in presence of protein, while the signal is restored after addition of the known inhibitor CbzL^P(O)LA [133] whereas the intensity of the other signals remain constant, as shown in Fig. (**10**).

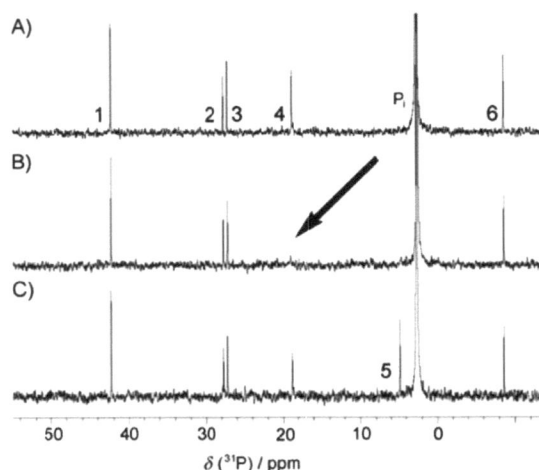

Fig. (10). ^{31}P NMR spectra of a small ligand library in phosphate solution, **A**) in absence and **B**) and presence of thermolysin. The signal of 4 disappears in the presence of protein (see the arrow). **C**) The recovery of the vanished signal upon addition of the tight binder phosphoramidon, validates the binding site of compounds **4**. *Reprinted with permission from Manzenrieder F, Frank AO, Kessler H. Phosphorus NMR Spectroscopy as a Versatile Tool for Compound Library Screening. Ang Chem Int Ed 2008; 47: 2608-2611.*

This screening presents several advantages: a) the large CSA of the ^{31}P allows the identification of weak binders making it suitable for fragment-base drug discovery; b) the large chemical shift dispersion allows the construction and the screening of large phosphorylated compound mixtures and decreases the problem of signals overlapping; 2D-^{1}H-^{31}P-COLOC experiments can be recorded to resolve possible signal overlapping; d) large non phosphated libraries can be screened as mixtures by competition binding experiments looking at the displacement of a phospho-molecule that binds with low-medium affinity to the target. Competition binding experiments also avoid the identification of unspecific binders. Finally, since there are many phosphorylated substances mimicking the tetrahedral intermediate of peptide bond hydrolysis (transition-state analogs) [134] such as phospho-amide, phosphonic acid and other moieties, the authors suggested the use of ^{31}P for screening large mixtures of molecules against proteases.

Despite the numerous advantages of ^{31}P spectroscopy, the ^{31}P nucleus shows a lower sensitivity when compared to ^{19}F nucleus.

^{19}F NMR Spectroscopy

The ^{19}F isotope atom has a natural abundance of 100%, a sensitivity comparable with that of ^{1}H, large chemical shift dispersion (> 200 ppm in organofluorine compounds) and its chemical shift is very sensitive to the environment changes occurring upon binding. In addition 15% of all commercial molecules contain at least one fluorine atom. In the last ten years ^{19}F NMR spectroscopy obtained a footing in the hit identification and optimization phases especially in the Fragment Based Drug Discovery. A more detailed overview of the most used ^{19}F NMR techniques and their practical applications is presented in this section.

^{19}F Binding Screening

^{19}F NMR spectroscopy has been used to study the effect of ligand binding on the protein structure and dynamics [135-137], the protein folding/unfolding events [138] and to discriminate between free and bound compound fraction in a lipid membrane system [139]. Direct binding ^{19}F NMR experiments have been proposed in the literature for screening of small fluorinated compounds libraries [44, 104, 140]. For example Tengel [140] applied this method for identifying small molecules able to bind the chaperones PapD and FimC. These chaperons are involved in the assembly of virulence-associated, hair-like protein structures termed "pili" on the surface of uropathogenic *E. coli* [141]. The combination of ^{19}F NMR titration experiments and other NMR techniques (such as dSTD), successfully conducted found novel binders of the Nurr1 receptor [104], a possible target for Parkinson's diseases, as it plays an important role in the development of dopaminergic neurons in the midbrain [142]. Interestingly the HTS campaign of 4.000.000 compounds on the same target Nurr1 was able to identify only a couple of compounds with a K_D of about 5 µM and no one with a K_D < 1 µM. Direct binding observation shows some limits such as inability to detect strong ligands or low soluble compounds and false positive detection. Therefore, to overcome these limitations, Dalvit and co-workers [143] proposed a fluorine ligand-based competition binding screening named FAXS (Fluorine chemical shift Anisotropy and exchange for Screening). The aim of the approach was to identify compounds of pharmaceutical interest and to calculate in a reliable way their K_D values. In their work, the authors showed that transverse relaxation rate R_2 is a very sensitive parameter for these studies, due to the large Chemical Shift Anisotropy (CSA) of ^{19}F nucleus and to the large exchange contribution. A simulation reported in this work showed the difference in ^{19}F resonance line width, due to CSA contribution, of a small molecule free in solution and when bound to a large macromolecule (Fig. **11**). The effect increases proportionally with the square of the strength of the magnetic field and therefore at the strong magnetic field used today, the FAXS experiments represent one of the most sensitive biophysical techniques for the detection of small molecules interacting with the receptor.

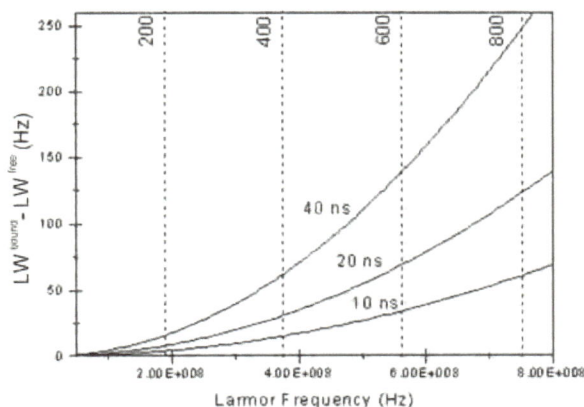

Fig. (11). Difference in linewidth due to CSA interaction of the ^{19}F signal of a small molecule free in solution and when bound to a large macromolecule as a function of the [19]F Larmor frequency. This simulation was performed with a ^{19}F CSA of 100 ppm and a correlation time τ_c of 200 ps for the small molecule when free in solution. Different correlation times for the macromolecule corresponding to different sizes of the macromolecule were considered (values indicated with the curves). The dashed vertical lines indicate some of the commercially available spectrometers. The value corresponding to the [1]H Larmor frequency of these spectrometers are indicated with the vertical lines. *Reprinted with permission from Dalvit C, Flocco M, Veronesi M, Stockman BJ. Fluorine-NMR Competition Binding Experiments for High-Throughput Screening of Large Compound Mixtures. Comb Chem & HTS 2002; 5: 605-611.*

The idea is schematically shown in the Fig. (**12**).

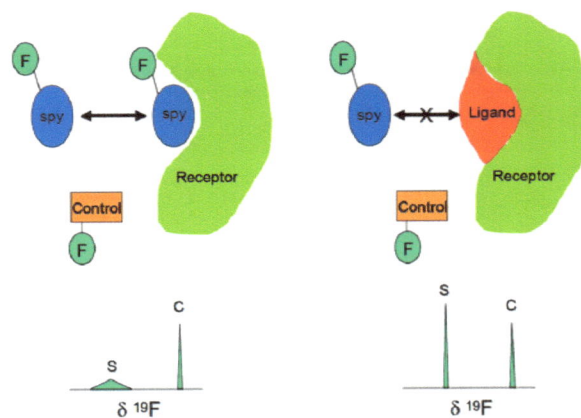

Fig. (12). Schematic diagram of the FAXS experiment. The biological target, a weak-to medium affinity ligand of known binding constant as the SPY (S) and in presence of an additional molecule that does not interact with the receptor as the CONTROL (C) are shown. Changes in the transverse relaxation rates of the Spy resonances are monitored. The broad signal of the spy molecule when binds to the target becomes sharp in the presence of a competitive ligand due to the displacement of the spy molecule from the receptor. The signals from the molecules screened are not utilized. *Reprinted with permission from Dalvit C. Ligand and substrate-based ^{19}F screening: Principles and application to Drug Discovery. Prog Nucl Magn Res Spect 2007; 51: 243-271.*

In the FAXS approach a library of fluorinated molecules is initially screened with ^{19}F-R_2 filter NMR experiments in presence/absence of the biological target for the identification of compounds that interact with the receptor. A candidate compound (spy) is selected, among the identified hits, based on strong signal attenuation after protein binding and on the presence of a single ^{19}F signal; the specific binding of the spy in the active pocket of the protein is confirmed by its complete displacement by the natural substrate or by a known strong inhibitor. The K_D of the spy molecule is determined by ITC, Fluorescence Spectroscopy or Plasmon Surface Resonance; titration experiments of the spy molecule in presence of increasing concentration of protein are then performed by R_2 filtered experiments with CMPG (Carr-Purcell-Meiboom-Gill) [144] spin-echo schemes in order to set-up the screening conditions (protein and spy molecule concentration) and plot observable parameter against the bound ligand fraction. This plot is necessary for the calculation of the K_I [8] of the identified hits. The identification of the strong inhibitor SU-13901 against the Serine/Threonine p21-activated kinase PAK4 among a mixture of 7 compounds is shown in Fig. (**13**): the use of a fluorinated small molecule that do not bind to the receptor (control) permits the estimation of the signal intensity change of the spy molecule with a single experiment. As internal control they used the TFA and recorded single 1D ^{19}F experiments of both spy (PHA-739917) and control (TFA) in the absence of protein (a), in the presence of PAK4 (b), of a mixture of 6 compounds (c) and in the presence of SU-13901: the total displacement of PHA-739917 in the presence of SU-13901 confirmed this compound as inhibitor.

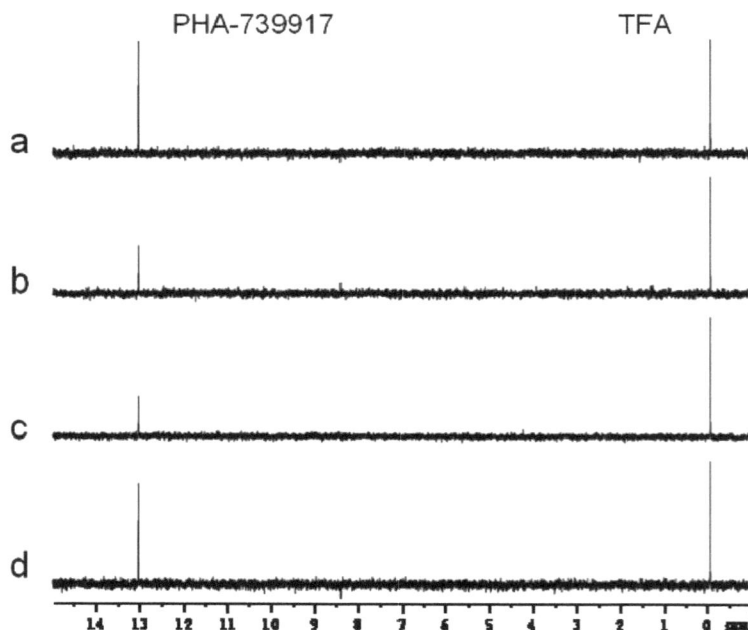

Fig. (13). Screening and deconvolution performed with one-dimensional ^{19}F spectra recorded in the presence of the weak-affinity PAK4 ligand PHA-739917, TFA, and PAK4. The chemical shifts are referenced to TFA. (**a**) Spectrum recorded in the absence of the protein. (**b**) Spectrum recorded in the presence of PAK4, (**c**) Spectrum recorded in the presence of PAK4 and a six compounds mixture (**d**) Spectrum recorded in the presence of PAK4 and the same six compounds mixture plus the molecule SU-13901. The presence of the competing molecule SU-13901 results in almost complete displacement of the reference compound from the protein (d) and the spectrum is similar to the spectrum for the two molecules in PBS (a). *Reproduced with permission from Dalvit C, Flocco M, Veronesi M, Stockman BJ. Fluorine-NMR Competition Binding Experiments for High-Throughput Screening of Large Compound Mixtures. Comb Chem HTS 2002; 5: 605-611.*

Moreover they suggested the possibility of using alternatively to the control molecule an ERETIC signal [145,146], which is based on an electronically generated signal with a defined frequency, linewidth and amplitude. Several recent articles and reviews analyzed the theoretical aspects and the practical application of the FAXS approach [143,147-149], underlining its strength when compared with other competition binding experiments or screening techniques. The most important characteristic of FAXS is its sensitivity to the binding events that allow the use of very low fractions of bound spy molecule and permits to detection of both very strong (nM-low µM range) and very weak (high µM-mM range) affinity binders as show in the simulation in Fig. (**14**). The percentage of reduction of fraction of bound spy molecule as function of the K_I of the competing molecule and concentration of spy molecule is plotted. Lower is the spy molecule concentration and weaker are the binders detected.

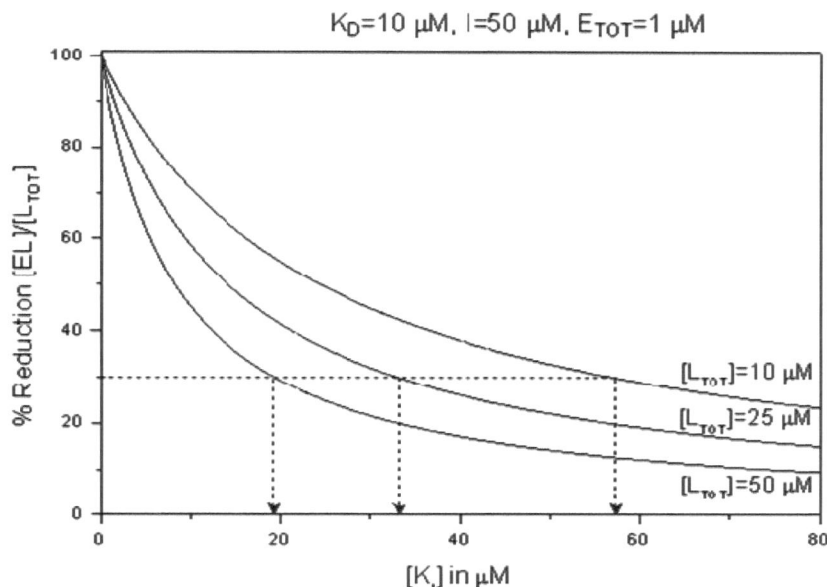

Fig. (14). Simulation showing the percentage reduction of the fraction of bound spy molecule as a function of the binding constant K_I of the competing molecule for different concentrations of the spy molecule. The K_D of the spy molecule was 10 µM, and the concentration of the competing molecule and protein was 50 and 1 µM, respectively. The horizontal line drawn at 30% reduction in the fraction of spy molecule bound to the receptor $[EL]/[L_{TOT}]$ (where $[EL]$ and $[L_{TOT}]$ are the bound and total spy molecule concentration respectively) represents a conservative detection limit of a FAXS run. For this value, the detection limit of K_I is 57, 33, and 19 µM for $[L_{TOT}]$ of 10, 25, and 50 µM, respectively. *Reprinted with permission from Dalvit C, Mongelli N, Papeo G, et al. Sensitivity Improvement in ^{19}F NMR-Based Screening Experiments: Theoretical Considerations and Experimental Applications. J Am Chem Soc 2005; 127: 13380-13385.*

This behaviour makes FAXS one of the most useful methods for the FBDD, where the focus is not the throughput, but the capacity of identifying fragments with weak affinity for the target of interest. In 2007 Taylor and co-workers [150] proposed a fragment identification protocol based on a virtual fragment screening applied to a large compound database, followed by ^{19}F NMR screening and ITC analysis against the phosphotyrosine binding site of the SH2 domain of the tyrosine kinase v-Src to identify novel phosphotyrosine (pY) mimetic small molecule to attach to a suitable scaffold and generate a high-affinity specific ligand. Starting from a subset of 12969 small molecules from the Available Chemical Database (ACD) and after a sequential molecular docking process, Chemical Similarity Analysis and prompt availability from suppliers they selected a final group of nine diverse fragments for the NMR screening that were tested with FAXS. This allowed the discovery of fragments interacting with the receptor. The presence in chemical commercial libraries of a large number of fluorinated compounds permits the selection of several low-medium affinity binders to use as spy molecules and allows the easy set-up of FAXS. In the last 50 years an increasing number of commercial drugs were fluorinated [151]. The introduction of ^{19}F by chemists is mainly due to improve the absorption, distribution, metabolism and excretion (ADME) proprieties of the lead compounds. The substitution of hydrogen by fluorine may block the metabolism of the compound or increase its membrane permeability. The presence of ^{19}F in drug molecules can have also a positive large contribution in drug selectivity and efficacy as accurately reviewed by Muller and co-workers [152]. They underlined the ^{19}F preference to electropositive area of the binding site, due to its high electronegativity of Fluorine; the presence of different fluorophilic regions, including the ubiquitous peptide bonds which undergo multipolar C-F···HN, C-F···C=O and C-F···H-Cα interaction, as well as side chain amides residue of Asn and Gln and positively side chain of Arg. By considering the overall positive effect and the propensity of fluoroalkyl groups to occupy the apolar aromatic regions, they suggested to decor the ligand with fluoroalkyl residues even if this substitution does not seem to increase the affinity for the lipophilic pockets compared with the alkyl residues. In conclusion they suggested systematic fluorine scans of ligand in lead optimization to strengthen protein ligand binding interaction as well as ADME proprieties. The finding of "fluorophilic protein environment" on the protein surface and the experimental observation of the influence of local environment around the ^{19}F atom in presenting the fluorine in the right orientation to the fluorophilic binding site has resulted in a very important breakthrough for the drug discovery process. On the basis of this observation Vulpetti and co-workers [153] recently proposed to use the "fluorine finger print" to build a new library of fluorinated fragments LEF (Local Environment of Fluorine) to combine with the FAXS screening for the identification and optimization of novel fragment scaffolds of pharmaceutical interest. They clustered the "in house" libraries by adapting the topological torsion descriptors proposed by Nilakantan and co-workers [154] and other standard algorithms and thresholds to build an interpretable fingerprint which starts from the fluorine atom or CF$_3$ moiety, as shown in Fig. (**15**). For the other details we refer to the article.

Finally they selected 1400 fluorinated fragments representing the different fluorine chemical environments and screened them in large mixtures against Bovine Trypsin, a Serine protease, by ^{19}F R$_2$ filter CPMG experiments in absence and presence of protein. The ^{19}F NMR signal intensity reduction in the presence of protein quickly allowed the identification of binders in the mixture. Displacement of putative binders by the addition of a known strong inhibitor confirms the competition events and validates the identified hits. The selected "LEF" binders are then used as spy molecules in FAXS experiments for

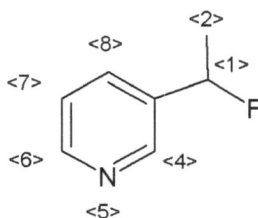

Length	Path	Atom Codes	Bit Id
1	F-C1	(F-0-0,C-0-2)	b1
2	F-C1-C2	(F-0-0,C-0-1,C-0-0)	b2
2	F-C1-C3	(F-0-0,C-0-1,C-1-2)	b3
3	F-C1-C3-C4	(F-0-0,C-0-1,C-1-1, C-1-1)	b4
3	F-C1-C3-C8	(F-0-0,C-0-1,C-1-1, C-1-1)	b4
4	F-C1-C3-C4-N5	(F-0-0,C-0-1,C-1-1, C-1-0, N-1-1)	b5
4	F-C1-C3-C8-C7	(F-0-0,C-0-1,C-1-1, C-1-0, C-1-1)	b6
5	F-C1-C3-C4-N5-C6	(F-0-0,C-0-1,C-1-1, C-1-0, N-1-0, C-1-1)	b7
5	F-C1-C3-C8-C7-C6	(F-0-0,C-0-1,C-1-1, C-1-0, C-1-0, C-1-1)	b8

Fig. (15). (Top) Sample molecule with atoms numbered using angle brackets. (Bottom) Set of all paths of length one to five bonds rooted at the fluorine atom. Atom types are generated for the atoms in the path, and these atom-typed paths are hashed to generate integer bit id's. *Adapted with permission from Vulpetti A, Hommel U, Landrum G, et al. Design and NMR Based Screening of LEF, a Library of Chemical Fragment with Different Local Environment of Fluorine. J Am Chem Soc 2009; 131: 12949-12959.*

a) optimization phase, b) screening analogue compounds, c) new chemical scaffolds identification and d) screening different compounds libraries. The identification of molecules belonging to the same F-cluster in the optimization phase suggests the presence of a fluorophilic protein environment. In conclusion, this approach presents several advantages: a) the possibility to screen large compounds mixtures, that strongly reduces the acquisition time, as well as protein and compounds consumption, b) the possibility to screen molecules with low solubility, c) the possibility of simultaneous detection of multiple ligands belonging to the same mixture, due to the increased relative sensitivity of the NMR assay by working at low concentration and d) the identification of very-high affinity and/or covalent binders. For the theoretical aspects of these advantages we defer to the original article. Dalvit [57] has recently published a very interesting and accurate comparison and theoretical evaluation of NMR screening experiments applied to the FBA. In this paper, NMR techniques were compared with the most largely used Fluorescence Anisotropy (FA), Fluorescence Life Time (FL) and Surface Plasmon Resonance (SPR). The simulation and data analysis demonstrated that, even if NMR screening is the methodology with the lowest intrinsic sensitivity, it shows the largest dynamic range and allows the detection of very weak binding events (in particular the ^{19}F NMR based experiments), by testing the molecules at concentration orders of magnitude lower of their K_D. On the contrary FA, FL and SPR suffered from a limited dynamic range and required the molecules to be tested at concentration comparable to their K_D, thus limiting the screening to those compounds with a sufficient solubility. Simultaneous application of different biophysical and structural methods is in any case necessary to obtain information required by the FBDD process and to generate new scaffolds with high-potential therapeutic agents.

^{19}F NMR Functional Screening

A ^{13}C NMR biochemical screening [155] was proposed at the end of the last century for measuring the enzyme inhibition and for identifying new lead compounds. Unfortunately the low sensitivity of the approach required high substrate concentration, and strongly confined its application to the identification of only strong inhibitors. Therefore this method was not suitable for the initial phase of FBDD, where the essential requirement is the detection of small fragments with low-medium affinity (μM to mM range) for the target of interest. Afterwards, Dalvit and co-workers [156] developed a very simple biochemical method that overcame this limitation and was suitable for the fragment base approach. The method, called 3-FABS (Three Fluorine Atoms for Biochemical Screening), is based on the ^{19}F NMR spectroscopy. The assay requires the substrate labeling of an enzymatic reaction, even far from the modification site, with a CF_3 moiety and then the use of ^{19}F- NMR spectroscopy to detect and quantify the substrate and the enzymatic reaction products signals. The possibility of monitoring both signals in a non-destructive way allows measurement of many properties of complex enzymatic reactions and mechanisms of inhibition. A schematic exemplification of the 3-FABS [157] is reported in Fig. (**16**).

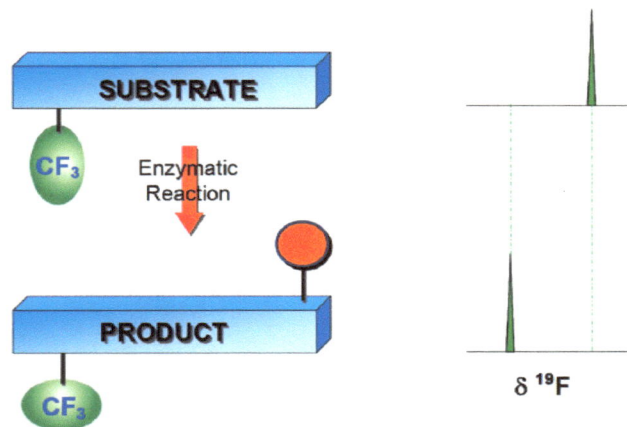

Fig. (16). Schematic diagram of the 3-FABS method. The enzymatic modification of the substrate changes the electronic cloud of the CF_3 moiety, resulting in different isotropic chemical shifts for the product and substrate fluorine signals. The high sensitivity of fluorine to the chemical environment allows the insertion of the CF_3 moiety even far from the reaction site. *Reprinted with permission from Dalvit C. Ligand and substrate-based ^{19}F screening: Principles and application to Drug Discovery. Prog Nucl Magn Res Spect 2007; 51: 243-271.*

The 3-FABS process has been extensively reviewed [148] and it includes: 1) determination of the linear range of the enzymatic reaction, necessary to select the delay after which to quench the reaction and for reliable and comparable IC_{50} (percentage of inhibitor at which the 50% inhibition of the reaction is achieved) determination. This is achieved by reaction progression in the NMR tube, by recording a ^{19}F spectra at different intervals and by plotting product or substrate signal integrals as function of the reaction evolution time. 2) Determination of the K_M of the substrate and cosubstrate, achieved by performing the reaction at different substrate (cosubstrate) concentrations and by plotting the reaction velocity (determined as the ratio between ^{19}F signal intensity of products and the incubation time) against the substrate (cosubstrate) concentration. This is necessary for the substrate concentration set-up and for calculating the inhibitor binding constant K_I from the IC_{50} value, according to the equation 5:

$$K_I = \frac{IC_{50}}{1 + \frac{[S]}{[K_M]}} \tag{5}$$

Where [S] is the concentration of substrate and K_M the Henry-Michaelis-Menten constant.

3) Screening of molecules in small or large mixtures in an end point format at a substrate concentration in the range 2–5 folds of K_M. In every 3-FABS run, a sample without compounds is recorded. This sample represents the reference with 0% inhibition. Deconvolution of the active mixtures is then performed for the identification of the inhibitors (hits). 4) IC_{50} determination at different inhibitor concentrations are performed for the inhibitor and then the corresponding ^{19}F signal intensity of product (or substrate) is plotted against the inhibitor concentration as illustrated in Fig. (**17**).

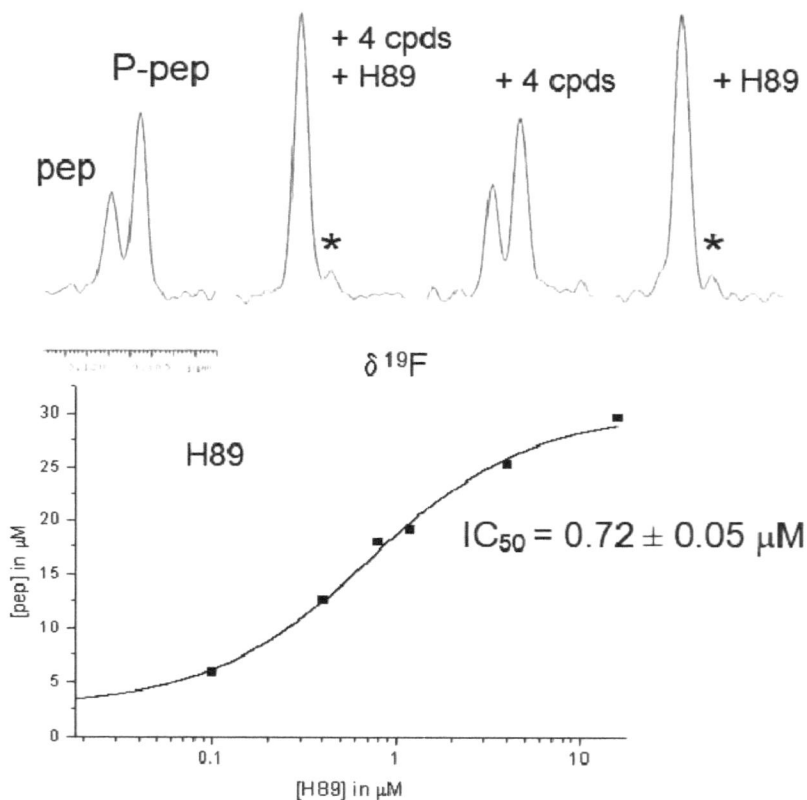

Fig. (17). Screening and deconvolution (top) and IC50 measurement (bottom). Top: ^{19}F NMR spectra for 1 recorded from left to right in the absence of compound, in the presence of a mixture of 5 compounds containing H89, in the presence of the same mixture without H89, and in the presence of only H89. The asterisks indicate the tiny amount of phosphorylated peptide. Bottom: Unphosphorylated peptide concentration (measured with integration of the ^{19}F NMR signal) as a function of the inhibitor concentration and the best fit of the experimental data with the derived IC50 for H89 of 0.72 ± 0.05 μM (compared to the value of 0.5 μM determined with enzymology). The same IC50 value was obtained using he integral of the ^{19}F signal of the phosphorylated peptide. *Reprinted with permission from Dalvit C, Ardini E, Flocco M, Fogliatto GP, Mongelli N, Veronesi M. A general NMR method for rapid, efficient, and reliable biochemical screening. J Am Chem Soc 2003; 125: 14620-14625.*

The authors pointed out that with this approach it is possible to obtain a meaningful IC_{50} value with a single point measurement in absence of allosteric effect, since the value for the two plateaus are known. For instance, looking at the product signals the two plateaus result in respectively the ^{19}F product signal intensity in the absence of inhibitor (0 % inhibition) and the 0 value (100% inhibition). The method is versatile, easy to set-up and properly suited for studying the inhibition of almost all kinds of enzymes, such as kinase, methylase, protease, phosphatase. 3-FABS can also be used to identify the best substrate for an enzymatic reaction and to develop inhibitors with a reduced affinity for the HSA (human serum albumin). In the hit to lead optimization phase 3-FABS performed in absence/ presence of HSA can provide early on in the project important structural information for designing analogues with reduced affinity for HSA. The application of 3-FABS to functional genomics studies have been proposed by Dalvit [157]; this could be very important especially nowadays, when the sequencing of the human genome has resulted in the identification of thousands of new proteins with unknown function. The idea consists firstly in the construction of a library of CF_3-labeled substrates of enzymes of known function, where ^{19}F chemical shifts of the starting and modified substrate are recorded: in some cases the same molecule can act as substrate for different enzymes. In this latter case unambiguous and distinct ^{19}F chemical shifts depending of the enzyme function are observed. The appearance of the corresponding ^{19}F product signal in the presence of the unknown enzyme allows the inference of the enzyme function. An example is shown in the test case of Fig. (**18**) where the enzyme PAK4 has been identified as kinase by comparing the three spectra of a substrate in the presence of kinase (up) in the presence of a protease (middle) and in the presence of PAK4 (down).

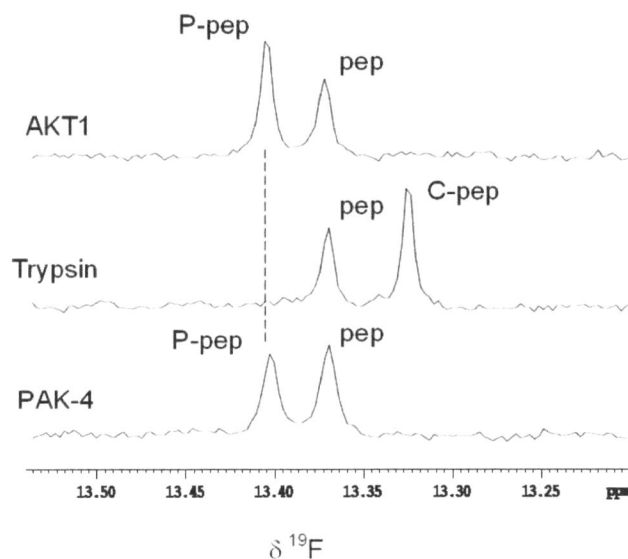

Fig. (18). Application of 3-FABS to functional genomics with ^{19}F spectra of a ^{19}F labeled peptide in the presence of three different proteins, AKT1 (top), trypsin (middle), and PAK4 (bottom). The peptide is a good substrate for AKT1 and for trypsin. The chemical shift of the cleaved peptide is quite different when compared to the chemical shift of the phosphorylated peptide. The reaction performed in the presence of PAK4 results in the appearance of a signal at the chemical shift of the phosphorylated peptide. *Reproduced with permission from Dalvit C, Ardini E, Flocco M, et al. A general NMR method for rapid, efficient, and reliable biochemical screening. J Am Chem Soc 2003; 125: 14620-14625.*

Several applications of 3-FABS to different targets of pharmaceutical interest have recently been reported, most of all in the hit identification phase. For istance Pellecchia and co-workers [129] reported the use of 3-FABS to monitor the kinetics of a phosphatase effector from *Yersinia pestis*: the *Yersinia* outer protein H (YopH) following hydrolysis of the 4-threefluorophenylphosphate (4-CF_3PP) by the enzyme. In 2005 Forino and co-workers [158] used the 3-FABS as detection method in the Fragment Based Approach applied to the anthrax lethal factor (LF), a metalloproteinase with an important role in the pathogenesis of the disease. The necessity to screen the fragment library at high concentration, in order to find weak inhibitors induced the authors to prefer the ^{19}F-NMR functional assay over the classical fluorescence-based assay. This is due to the large number of false positive and/or false negatives in the latter principally due to fluorescence interference introduced by high concentration of the test compounds. A specific peptide labeled at the C-term with a CF_3 moiety was used as substrate for the enzymatic reaction; this approach reached the goal to find a new small scaffold with a IC_{50} of 140 μM. The iterative optimization of this fragment resulted in a series of selective derivatives with nanomolar inhibitory activity against LF. The data were confirmed with *in vitro* tests and cell-based assays. Moreover these novel compounds were demonstrated not affecting the prototype human metalloproteinases, that are structurally similar to LF, and were able to protect macrophages from LF-induced cytotoxicity at concentrations below those needed with a nonselective hydroxamate-based protease inhibitor and finally showed synergistic protection with ciprofloxacin *in vivo*. This work shows the ability of FBA and 3-FABS in identifying inhibitors that are optimal starting points for the development of new, safe, and effective drugs. In the same year a new strategy [159] was proposed, combining NMR functional screening and computational docking methods to identify non-peptide inhibitors of caspase, an enzyme responsible of the cell death [160]. The method was called SUBITO (Structure based Iterative Optimization approach), to stress its rapidity and low costs. The idea is to use a ^{19}F-NMR functional assay to screen a small fragment library, consisting of the most common scaffolds frequently found in drugs and displaying hundreds of commercially available analogues, and identifying compounds with very weak affinity. So, instead of screening all available commercial analogues, they restricted the analogues numbers to test, on the basis of Docking analysis results. In this way not only the number of screened compounds is drastically reduced (few dozen at least) but also, and more relevant, intrinsic structural information can be obtained. The possibility to easily screen extracts of natural product with 3-FABS has been pointed out by Dalvit *et al.* [156] and has been successfully applied by the group of Giralt [161,162] for the identification of inhibitors of HIV-1 protease and of prolyl oligopeptidase (POP) from the Traditional Chinese Medicinal (TCM) plant extracts. Aqueous extracts solution of the TCM natural products were screened against the two targets and inhibitors originating from *Cortex cinnamon* and *Medula junci extracts* were identified for the POP and HIV-1 protease respectively. The detection of the active principles of the identified extracts is then achieved by 3-FABS screening of the different fractions separated by HPLC. The lack of colorimetric interference with the NMR system avoids the possibility of false negative and positive results and makes the combination of 3-FABS and HPLC technologies one of the most helpful methods for the study of the inhibitory activity of natural compounds.

Methods for Identifying False Positive and Negative from High Throughput Screening

Despite the multiple success and the large investments of pharmaceutical companies and the improvements in both technology and knowledge, it was shown that HTS has failed in reaching the expectations. The number of new drugs per year introduced into the market constantly dropped in the last ten years [163]. This induced many researchers to analyze the HTS process and to propose new approaches to overcome the failure. There are several reasons for this failure and one of these is certainly the presence of promiscuous compounds in the screened libraries. These molecules result in a large number of false positive. Many physical characteristics of the chemical compounds can interfere with the detection system (compound absorbance or fluorescence) or with the target (covalent compound binding or denaturing compounds) [164]. During the last years the aggregates formation of chemical compounds has resulted as one of the most frequent cause of false positives in HTS campaigns [165]. Therefore numerous biochemical and biophysical assays have been set-up to early identify aggregating compounds in the HTS libraries [166-169].

NMR spectroscopy has also been proposed for identification of false positive in HTS results. The observation of the remarkable stability of NMR spectra of the La protein in the presence and absence of reducing agents, even after several days at room temperature and the presence in the protein of a cysteine very sensitive to modification by electrophilic compounds prompted to Huth and co-workers [170] to monitor the DTT-dependent ^{13}C chemical shift changes of LA protein in presence/absence of single compounds for identifying thiol reactive false positives. The method named ALARM NMR (A La Assay to detect Reactive Molecules by Nuclear Magnetic Resonance) has been validated by the authors testing hundreds of negative control compounds as well as thousands of hits from actual drug screening. The conclusion reported by the authors, was firstly the ability and sensitivity of ALARM NMR assay in identifying reactive compounds, not captured by *in silico* filter programs [171] or not resulting from active routine biochemical assay. Secondly there is a highly characteristic susceptibility of different targets to reactive compounds, and therefore the likelihood of give false positive. Indeed whereas an average of 12% of hits are reactive false positive, several targets have exhibited up-to 60% of high percentage of reactive compounds. Moreover they implemented computational tools introducing thiol reactivity index (TRI) to predict the probability of a compound to be a thiol-reactive therefore helping to increase the quality of the compounds to add to the screening library. This results in an improvement of the quality of the HTS hits and in time reduction for hit validation. ALARM by NMR has the drawbacks that cannot be performed at the same time of the screening and require different experiments and reagents than those used for the screening. Moreover they are able to detect false positive but not false negative. Another NMR binding method for quality control of the hits has been proposed by Dalvit and co-workers [172]: three ^{1}H NMR spectra (1D reference spectrum, a Water-LOGSY [115] spectrum and a selective longitudinal relaxation filter or a transverse relaxation filter spectrum [173,174]) are recorded after the screening NMR-experiments in the presence of a reference molecule for directly characterizing the identity, purity, solubility, stability and aggregation state of the active compounds in water solution. So only the compounds that pass this so named SPAM (Solubility, Purity and Aggregation of the Molecule) filter approach can be considered as true inhibitors and starting points for lead optimization. From the 1D NMR spectrum, looking at the NMR signals of the compounds, it is possible to obtain precise information about their identity and purity, and, comparing the integrals of their NMR signals to those of the reference, to calculate their solubility and concentration. These data certainly allow deducing reliable and robust IC_{50} and/or K_D values, but they even permit to identify false negatives. The low affinity of a compound in a HTS screening can result from its low solubility and/or purity. Therefore the real concentration is lower than the nominal concentration resulting in a large underestimation of the inhibitor potency. The importance of this analysis is well exemplified in the article. They show that a molecule exhibiting only 12% of inhibition at a nominal concentration of 20 µM ($IC_{50} \sim 113$ µM in their experimental condition) in a NMR screening against the kinase human protein AKT1 [175] resulted completely absent in the 1D NMR spectrum of Fig. (**19**) (lower), but after purification it was detectable in 1D NMR spectrum at the expected concentration (upper), and it showed a strong affinity for the biological target ($IC_{50} \sim 1.08$ µM). Without the SPAM filter experiment the compound would have been discarded!

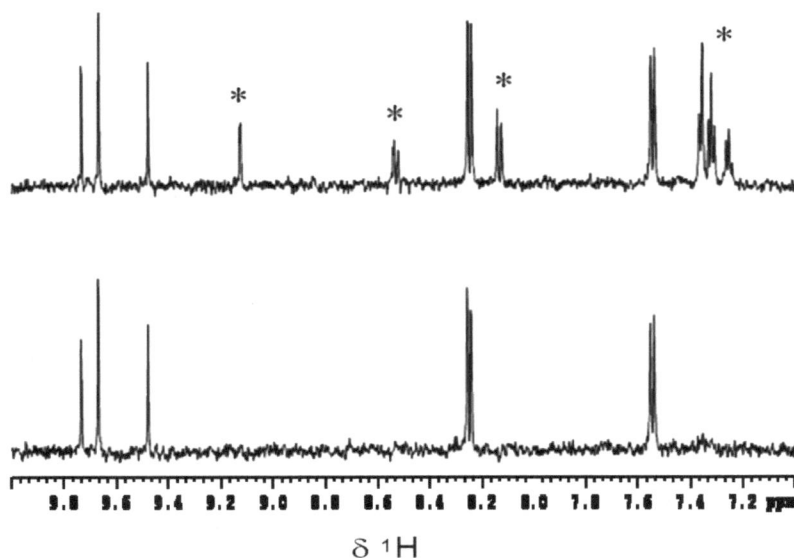

Fig. (19). Lower trace for 25 µM of the reference compound (**C**) and nominal 25 µM of the compound (**X**) showing a 12% of inhibition against AKT1. Upper trace same expanded region of the 1D ^1H spectrum recorded for 25 µM the **C** and 25 µM **X** after purification. The sample are in PBS solution containing DMSO. The signals originating from the control molecule are the only visible signals in the lower spectrum. The signals of the tested molecule visible in the upper spectrum are indicated with an asterisk. The concentration of the tested molecule in the upper spectrum calculated using the signals of the reference molecule corresponds to 25 µM. *Reproduced with permission from Dalvit C, Caronni D, Mongelli N, et al. NMR-Based Quality Control Approach for the Identification of False Positives and False Negatives in High Throughput Screening. Curr Drug Discov Tech 2006; 3: 115-124.*

Water-LOGSY and/or selective R1 and R2 filter experiments in the presence of the reference compound are instead used to study the aggregation state of the molecules. Since the reference interacts in a nonspecific way with the surface of aggregates formations (often not detectable in the NMR spectrum for their dimension), its NMR signal in absence/presence of the aggregating molecule will be different in Water-LOGSY and R1/R2 filter experiments. For example in Water-LOGSY experiments, in the presence of aggregation the selective magnetization of the bulk water is transferred to the aggregates and consequently to the reference bound to the aggregates. This magnetization transfer is then detected as the chemical shifts of the free molecule as positive NMR signals. On the contrary in the absence of aggregates the Water-LOGSY signal is negative.

In Fig. (**20**) the authors detected the aggregation of the known multi target inhibitor Quercitin observing that in its presence the NMR signal of the reference changes from negative (upper spectrum) to positive (middle spectrum) whereas the NMR signal of the DMSO that does not interact with the aggregate still remains negative; in the figure only the aliphatic region of the NMR spectra is shown and the bottom spectrum is the 1D NMR of the reference in the presence of the Quercitin.

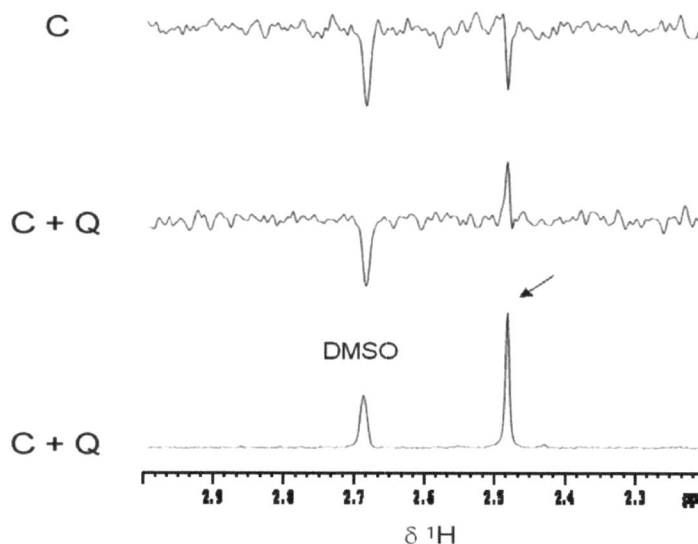

Fig. (20). Lower trace: expanded region of the 1D ^{1}H spectrum for 50 µM of the reference compound (**C**) and 25 µM of quercetin (**Q**) in PBS containing the DMSO and the CH_3 signal (arrow) of compound **C**. Upper trace: WaterLOGSY spectrum recorded for 50 µM **C** in the absence of **Q**. Middle trace: WaterLOGSY spectrum recorded for 50 µM **C** in the presence of 25 µM of **Q**. In the waterLOGSY experiment in the presence of quercetin the CH_3 signal of compound **C** changes sign and becomes positive indicating the presence of aggregates. *Reproduced with permission from Dalvit C, Caronni D, Mongelli N, et al. NMR-Based Quality Control Approach for the Identification of False Positives and False Negatives in High Throughput Screening. Curr Drug Discov Tech 2006; 3: 115-124.*

The advantage of SPAM-filter compared to the others methods, is the possibility to check the characteristic of the compounds just after the screening and most of all, in the same experimental condition of the screening campaign; moreover this approach not only is very robust and trustworthy in detecting false positive, but is able to identify also false negatives, which are lost with the other screening methodologies developed for detecting promiscuous hits. SPAM-filter results are very used to build good new library screening compounds and speed-up the hit-to-lead and lead optimization phase. Finally it is important to highlight that false negatives can even be due to the binding of the compounds to one or more of the numerous binding-sites of the bovine serum albumin (BSA) a protein that is generally added at high concentration to coat the wall of the wells in the plates and avoid the binding of the studied protein to the plate, as demonstrated by Dalvit and co-workers [156]. The binding of the tested molecules to BSA results in a minor amount of free compound available for the pharmaceutical target. Therefore so the different affinity for the BSA of the different compounds can result in a deceitful SAR. To avoid this inconvenient and obtain a reliable SAR it is better use detergents, like the Triton X-100, at a concentration lower than their CMC (critical micelles concentration) rather than using high BSA concentration.

CONCLUSIONS AND REMARKS

FBDD has proven to be a valid alternative to HTS since the first theoretical suggestion in late 1981 [16], and it has now been largely accepted for its important contributions in the drug discovery process. The idea of FBDD relies on a better explored drug space with smaller libraries of lighter molecules with respect to larger compounds, because the number of fragments with 12 heavy atoms was evaluated to be only 10^7 molecules in the Flink study [21]. In such a way a larger part of the drug space could be explored by fragment-based screening with respect to HTS. Since fragment based screening hits shows typically a low medium affinity (micro/milli-molar range) for the target of interest, the development and the combination of biophysical and structural methods able to detect very weak binders is essential for generating new scaffolds with high-potentially therapeutic agents generation capability. In this respect the NMR–based screening method resulted the better suited technique allowing detection of very weak interactions, thanks to its large dynamic range, particularly in the ^{19}F monitored NMR experiments [57]. Furthermore, several advantages like the non invasive characteristic of NMR, the high sensitivity in case of weak binder events, the progressive increase of the sensitivity due to technology of cryoprobes, and finally the high speed in generating data on protein–ligand interactions, enforced the choice of NMR as the leader technique in the fragment-based drug discovery.

5. ACKNOWLEDGEMENTS

The authors wish sincerely thanks Dr. Claudio Dalvit for his generous scientific contribution and for very helpful discussions. Dr. Bains Sandip and Dr. Laura Ruth Cagliani are kindly acknowledged for their technical assistance.

REFERENCES

[1]   Bleicher KH, Bohm HJ, Muller K, Alanine AI. Hit and lead generation: beyond high-throughput screening. Nat Rev Drug Discov 2003; 2: 369-78.
[2]   Jahnke W, Erlanson DA. Fragment based approaches in drug discovery. Wiley-Vch Verlag GmbH & Co. KGaA: Weinheim 2006.
[3]   Zartler ER, Shapiro MJ. Fragment based drug discovery. John Wiley & Sons Ltd: Chichester 2008.
[4]   Pellecchia M, Bertini I, Cowburn D, et al. Perspectives on NMR in drug discovery: a technique comes of age. Nat Rev Drug Discov 2008; 7: 738-45.
[5]   Klages J, Coles M, Kessler H. NMR-based screening: a powerful tool in fragment-based drug discovery. Mol BioSyst 2006; 2: 318-23.
[6]   Heller M, Kessler H. NMR spectroscopy in drug design. Pure Appl Chem 2001; 73: 1429-36.
[7]   Dalvit C, Fasolini M, Flocco M, et al. NMR-based screening with competition water-ligand observed via gradient spectroscopy experiments: Detection of high-affinity ligands. J Med Chem 2002; 45: 2610-4.
[8]   Dalvit C, Flocco M, Knapp S, et al. High-Throughput NMR-Based Screening with Competition Binding Experiments. J Am Chem Soc 2002; 124: 7702-9.
[9]   Drews J. Drug discovery: a historical perspective. Science 2000; 287: 1960-4.
[10]   Cook ND. Scintillation proximity assay: a versatile high-throughput screening technology. Drug Disc Today 1996; 1: 287-94.
[11]   Owicki JC. Fluorescence polarizarion and anisotropy in high throughput screening: perspectives and primer. J Biomol Screen 2000; 5: 297-306.
[12]   Qian J, Voorbach MJ, Huth JR, et al. Discovery of Novel inhibitors of Bcl-xL using multiple high-throughput screening platform. Anal Bioch 2004; 328: 131-8.
[13]   Cravatt BF, Wright AT, Kozarich JW. Activity-based protein profiling: from enzyme chemistry to proteomic chemistry. Annu Rev Biochem 2008; 77: 383-414.
[14]   Bachovchin DA, Brown SJ, Rosen H, Cravatt BJ. Identification of selective inhibitors of uncharacterized enzymes by high-throughput screening with fluorescent activity-based probes. Nat Biotech 2009; 27: 387-94.
[15]   Jessani N, Niessen S, Wei BQ, et al. A streamlined platform for high-content functional proteomics of primary human specimens. Nat Methods 2005; 2: 691-7.
[16]   Jencks WP. On the attribution and additivity of binding energies. Proc Natl Acad Sci USA 1981; 78: 4046-50.
[17]   Nakamura CE, Abeles RH. Mode of interaction of beta-hydroxy-betamethylglutaryl coenzyme A reductase with strong binding inhibitors: compactin and related compounds. Biochemistry 1985; 24: 1364-76.
[18]   Shuker SB, Hajduk PJ, Meadows RP, Fesik SW. Discovering high-affinity ligands for protein: SAR-by-NMR. Science 1996; 274: 1531-4.
[19]   Hann MM, Leach AR, Harper G. Molecular complexity and its impact on the probability of finding leads for drug discovery. J Chem Inf Comput Sci 2001; 41: 856-64.
[20]   Bohacek RS, McMartin C, Guida WC. The art and practice of structure based drug design: a molecular modelling perspective. Med Res Rev 1996; 16: 3-50.
[21]   Fink T, Bruggesser H, Reymond JL. Virtual exploration of the small-molecule chemical universe below 160 Daltons. Angew Chem Int Ed Engl 2005; 44: 1504-8.
[22]   Rees DC, Congreve MS, Murray CW, Carr R. Fragment-based lead discovery. Nat Rev Drug Discov 2004; 3: 660-72.
[23]   Carra HJ, McHugh CA, Mulligan S, Machiesky LM, Soares AS, Millard CB. Fragment-based identification of determinants of conformational and spectroscopic change at the ricin active site. BMC Struct Biol 2007; 7: 72.
[24]   Bartoli S, Fincham CI, Fattori D. The fragment-approach: An update. Drug Discov Today Tech 2006; 3: 425-31.
[25]   Ciulli A, Abell C. Fragment-based approaches to enzyme inhibition. Curr Opin Biotechnol 2007; 18: 489-96.
[26]   Hajduk PJ, Greer J. A decade of fragment-based drug design: strategic advances and lessons learned. Nat Rev Drug Discov 2007; 6: 211-9.
[27]   Murray WC, Rees DC. The rise of fragment-based drug discovery. Nat Chem 2009; 1: 187-92.
[28]   De Kloe GE, Bailey D, Leurs R, De Esch IJP. Transforming fragments into candidates: small becomes big in medicinal chemistry. Drug Discov Today 2009; 14: 630-45.
[29]   Chessari G, Woodhead AJ. From fragment to clinical candidate a historical perspective. Drug Discov Today 2009; 14: 668-75.
[30]   Lipinski CA, Lombardo F, Dominy BW, Feeney PJ. Experimental and computational approaches to estimate solubility and permeability in drug discovery and development settings. Adv Drug Deliv Rev 1997; 23: 3-25.
[31]   Ghose AK, Viswanadhan VN, Wendoloski JJ. A knowledge-based approach in designing combinatorial or medicinal chemistry libraries for drug discovery. 1. A qualitative and quantitative characterization of known drug databases. J Comb Chem 1999; 1: 55-68.
[32]   Xu J, Stevenson J. Drug-like index: a new approach to measure drug-like compounds and their diversity. J Chem Inf Comput Sci 2000; 40: 1177-87.

[33]   Oprea TI. Property distribution of drug-related chemical databases. J Comput -Aided Mol Des 2000; 14: 251-64.
[34]   Congreve M, Carr R, Murray C, Jhoti H. A "rule of three" for fragment-based lead discovery? Drug Discov Today 2003; 8: 876-7.
[35]   Bretonnet AS, Jochum A, Walker O, et al. NMR screening applied to the fragment-based generation of inhibitors of creatine kinase exploiting a new interaction proximate to the ATP binding site. J Med Chem 2007; 50: 1865-75.
[36]   Siegal G, Ab E, Schultz J. Integration of fragment screening and library design. Drug Discov Today 2007; 12: 1032-9.
[37]   Fejzo J, Lepre CA, Peng JW, et al. The SHAPES strategy: an NMR-based approach for lead generation in drug discovery. Chem Biol 1999; 6: 755-69.
[38]   Bemis GW, Murcko MA. The properties of known drugs.1. Molecular frameworks. J Med Chem 1996; 39: 2887-93.
[39]   McLeish MJ, Kenyon GL. Relating structure to mechanism in creatine kinase. Crit Rev Biochem Mol Biol 2005; 40: 1-20.
[40]   Villar HO, Hansen CK. Computational techniques in fragment based drug discovery. Curr Top Med Chem 2007; 7: 1-5.
[41]   Akritopoulou-Zanze I, Hajduk PJ. Kinase-targeted libraries: The design and synthesis of novel, potent, and selective kinase inhibitors. Drug Discov Today 2009; 14: 291-7.
[42]   Noble MEM, Endicott JA, Johnson LN. Protein kinase inhibitors: insights into drug design from structure. Science 2004; 303: 1800-5.
[43]   Dobson PD, Patel Y, Kell DB. 'Metabolite-likeness' as a criterion in the design and selection of pharmaceutical drug libraries. Drug Discov Today 2009; 14: 31-40.
[44]   Dalvit C, Fagerness PE, Hadden DTA, Sarver RW, Stockman BJ. Fluorine-NMR experiments for high-throughput screening: theoretical aspect, practical consideration and range of applicability. J Am Chem Soc 2003; 125: 7696-703.
[45]   Manzenrieder F, Frank AO, Kessler H. Phosphorus NMR spectroscopy as a versatile tool for compound library screening. Angew Chem Int Ed 2008; 47: 2608-11.
[46]   Mercier KA, Germer K, Powers R. Design and characterization of a functional library for NMR screening against novel protein targets. Comb Chem High Throughput Screen 2006; 9: 515-53.
[47]   Kuntz ID, Chen K, Sharp KA, Kollman PA. The maximal affinity of ligands. Proc Natl Acad Sci USA 1999; 96: 9997-10002.
[48]   Hopkins AL, Groom CG, Alex A. Ligand efficiency: a useful metric for lead selection. Drug Discov Today 2003; 7: 430-1.
[49]   Hajduk PJ. Fragment-based drug design: how big is too big. J Med Chem 2006; 49: 6972-6.
[50]   Abad-Zapatero C, Metz JT. Ligand efficiency indices as guideposts for drug discovery. Drug Discov Today 2003; 7: 464-9.
[51]   Zhao H. Scaffold selection and scaffold hopping in lead generation: a medicinal chemistry perspective. Drug Discov Today 2007; 12: 149-55.
[52]   Congreve M, Chessari G, Tisi D, Woodhead AJ. Recent developments in fragment-based drug discovery. J Med Chem 2008; 51: 3661-80.
[53]   Leeson PD, Springthorpe B. The influence of drug-like concepts on decision-making in medicinal chemistry. Nat Rev Drug Discov 2007; 6: 881-90.
[54]   Orita M, Ohno K, Niimi T. Two 'Golden Ratio' indices in fragment-based drug discovery. Drug Discov Today 2009; 14: 321-8.
[55]   Macarron R. Critical review of the role of HTS in drug discovery. Drug Discov Today 2006; 11: 277-9.
[56]   Gribbon P, Andreas S. High throughput drug discovery: what can we expect from HTS? Drug Discov Today 2005; 10: 17-22.
[57]   Dalvit C. NMR methods in fragment screening: theory and a comparison with other biophysical techniques. Drug Discov Today 2009; 14: 1051-7.
[58]   Gòmez-Hens A, Aguilar-Caballos MP. Modern analytrical approaches to high throughput drug discovery. Trends Anal Chem 2007; 26: 171-82.
[59]   Feltis BN, Sexton BA, Glenn FL, Best MJ, Wilkins M, Davis TJ. A hand-held surface plasmon resonance biosensor for the detection of ricin and other biological agents. Biosens Bioelect 2008; 23: 1131-6.
[60]   Martin VS, Sullivan BA, Walker K, Hawk H, Sullivan BP, Noe LJ. Surface plasmon resonance investigations of human epidermal growth factor receptor 2. Appl Spec 2006; 60: 994-1003.
[61]   Dunne L, Daly S, Baxter A, Haughey S, O'Kennedy R. Surface plasmon resonance-based inummoassay for the detection of affatoxin B-1 using single-chain antibody fragments. Spec Lett 2005; 38: 229-45.
[62]   Pröll F, Fechner P, Proll G. Direct optical detection in fragment-based screening. Anal Bioanal Chem 2009; 393: 1557-62.
[63]   Neumann T, Junker HD, Schmidt K, Sekul R. SPR-based fragment screening: advantages and applications. Curr Top Med Chem 2007; 7: 1630-42.
[64]   Papalia GA, Leavitt S, Bynum MA, et al. Comparative analysis of 10 small molecules binding to carbonic anhydrase II by different investigators using Biacore technology. Anal Biochem 2009; 359: 94-105.
[65]   Antonysamy SS, Aubol B, Blaney J, et al. Fragment-based discovery of hepatitis C virus NS5b RNA polymerase inhibitors. Biorg Med Chem Lett 2008; 18: 2990-5.
[66]   Nordin H, Jungnelius M, Karlsson R, Karlsson OP. Kinetics studies of small molecules interactions with protein kinases using biosensor technology. Anal Biochem 2005; 340: 359-68.
[67]   Nordström H, Gossas T, Hämäläinen M, et al. Identification of MMP-12 inhibitors by using biosensor-based screening of a fragment library. J Med Chem 2008; 51: 3449-59.
[68]   Lee PH, Gao A, Van Staden C, et al. Evaluation of dynamic mass redistribution technology for pharmacological studies of recombinant and endogenously expressed G protein-coupled receptors. Ass Drug Dev Tech 2007; 4: 83-94.
[69]   Lang T, Brecht A, Gauglitz G. Characterisation and optimisation of an immunoprobe for triazines. Fresenius J Anal Chem 1996; 354: 857-60.
[70]   Roddy TP, Horvath CR, Stout SJ, et al. Mass spectrometric techniques for label-free high throughput screening in drug discovery. Anal Chem 2007; 79: 8207-13.
[71]   Moore CD, Wu H, Bolaños B, et al. Structural and biophysical characterization of XIAP BIR3 G306E mutant: insight in protein dynamics and applicatgion for fragment based drug design. Chem Biol Drug Des 2009; 74: 212-23.
[72]   Ritschel T, Atmanene C, Reuter K, Van Dorsselaer A, Sanglier-Cianferani S, Klebe G. An integrative approach combining noncovalent mass spectrometry, enzyme kinetics and X-ray crystallography to decipher Tgt protein-protein and protein-RNA interaction. J Mol Biol 2009; 393: 8333-47.
[73]   Poklar N, Lah J, Salobir M, Macek P, Vesnaver G. pH and temperature-induced molten globule-like denatured states of equinatoxin II: A study by UV-melting, DSC, far- and near-UV CD spectroscopy, and ANS fluorescence. Biochemistry 1997; 36: 14345-52.

[74]　Pantoliano MW, Petrella EC, Kwasnoski JD, *et al.* High-density miniaturized thermal shift assays as a general strategy for drug discovery. J Biomol Screen 2001; 6: 429-40.

[75]　Matulis D, Kranz JK, Salemme FR, Tod MJ. Thermodynamic stability of carbonic anhydrase: measurements of binding affinity and stoichiometry using ThermoFluor. Biochemistry 2005; 44: 5258-66.

[76]　Kurganov BI. Kinetics of protein aggregation. Quantitative estimation of the chaperone-like activity in test-systems based on suppression of protein aggregation. Biochemistry 2002; 67: 409-22.

[77]　Sinisterra G, Markin E, Yamazaki K, Hui R: US 20040072356 (2004).

[78]　Vedadi M, Niesen FH, Allali-Hassani A, *et al.* Chemical screening methods to identify ligands that promote protein stability, protein crystallization, and structure determination. Proc Natl Acad Sci USA 2006; 103: 15835-40.

[79]　Matthews TP, Klair S, Burns S, *et al.* Identification of inhibitors of checkpoint kinase 1 through template screening. J Med Chem 2009; 52: 4810-9.

[80]　Wyatt PG, Woodhead AJ, Berdini V, *et al.* Identification of N-(4-piperidinyl)-4-(2,6-dichloro, benzoylamino)-1H-pyrazole-3-carboxamide (AT7519), a novel cyclin dependent kinase inhibitor using fragment-based x-ray crystallography and structure based drug design. J Med Chem 2008; 51: 4986-99.

[81]　Caldwell JJ, Davies TG, Donald A, *et al.* Identification of 4-(4-amminopiperidin-1-yl)-7H-pyrrolo[2,3-d]pyrimidines as selective inhibitors of protein kinase B through fragment elaboration. J Med Chem 2008; 51: 2147-57.

[82]　Eitner K, Koch U. From fragment screening to potent binders: strategies for fragment-to-lead evolution. Mini Rev Med Chem 2009; 9: 956-61.

[83]　Mattos C, Ringe D. Proteins in organic solvents. Curr Opin Struct Biol 2001; 1: 761-64.

[84]　Liepinsh E, Otting G. Organic solvents identify specific ligand binding site on protein surfaces. Nat Biotech 1997; 15: 264-5.

[85]　Dalvit C, Floersheim P, Zurini M, Widmer A. Use of organinc solvents and small molecules for locating binding sites on protein in solution. J Biomol NMR 1999; 14: 23-32.

[86]　Rothweiler U, Czarna A, Weber L, *et al.* NMR Screening for lead compounds using tryptophan-mutated proteins. J Med Chem 2008; 51: 5035-42.

[87]　Krajewski M, Rothweiler U, D'Silva L, Majumdar S, Klein C, Holak TA. An NMR-based antagonist induced dissociation assay for targeting the ligand-protein and protein-protein interactions in competition binding experiments. J Med Chem 2007; 50: 4382-7.

[88]　Hajduk PJ, Huth JR, Fesik SW. Druggability indices for protein targets derived from NMR-based screening data. J Med Chem 2005; 48: 2518-25.

[89]　Bogan AA, Thorn KS. Anatomy of hot spots in protein interfaces. J Mol Biol 1998; 280: 1-9.

[90]　Bista M, Kowalska K, Janczyk W, Dömling A, Holak TA. Robust NMR screening for lead compounds using tryptophan-containig proteins. J Am Chem Soc 2009; 131: 7500-1.

[91]　Hajduk PJ, Augeri DJ, Mack J, *et al.* NMR-based screening of proteins containing ^{13}C-labeled methyl Groups. J Am Chem Soc 2000; 122: 7898-904.

[92]　Tarragó T, Clasen B, Kichik N, Rodriguez-Mias RA, Gairí M, Giralt E. A cost-effective labeling strategy for NMR study of large proteins: selective ^{15}N-labeling of the tryptophan side chain of prolyl oligopeptidase. ChemBioChem 2009; 10: 2736-9.

[93]　Rodriguez-Mias RC, Pellecchia M. Use of selective Trp side chain labeling to characterize protein-protein and protein-ligand interaction by NMR spectroscopy. J Am Chem Soc 2003; 125: 2892-3.

[94]　Frutos S, Rodriguez-Mias RC, Madurga S, *et al.* Disruption of HIV-1 protease dimer with interface peptides: structural studies using NMR spectroscopy combined with [2-^{13}C]-Trp selective labeling. Biopolymers (Pept Sci) 2007; 88: 164-73.

[95]　Pervushin K, Ono A, Fernandez C, Szypersky T, Kainosho M, Wütrich K. NMR scalar couplings across Watson–Crick base pair hydrogen bonds in DNA observed by transverse relaxation-optimized spectroscopy. Proc Natl Acad Sci USA 1998; 95: 14147-51.

[96]　Brutscher B, Skrybbikov NR, Bremi T, Bryschweiler R, Ernst RR. Quantitative investigation of dipole–CSA cross-correlated relaxation by ZQ/DQ spectroscopy. J Magn Reson 1998; 130: 346-51.

[97]　Leone M, Rodriguez-Mias RC, Pellecchia M. Selective Incorporation of 19F-labeled Trp side chains for NMR-spectroscopy-based ligand protein interaction studies. ChemBioChem 2003; 4: 649-50.

[98]　Deveraux Q, Reed J. IAP family proteins-suppressors of apoptosis. Genes Dev 1999; 13: 239-52.

[99]　Ludwig C, Guenther UL. Ligand based NMR methods for drug discovery. Front Biosci 2009; 14: 4565-74.

[100]　Mayer M, Meyer B. Characterization of ligand binding by saturation transfer difference NMR spectroscopy. Angew Chem Int Ed 1999; 38: 1784-8.

[101]　Vogtherr M, Peters T. Application of NMR based binding assays to identify key hydroxy groups for intermolecular recognition. J Am Chem Soc 2000; 122: 6093-9.

[102]　Mayer M, Meyer B. Group epitope mapping by saturation transfer difference NMR to identify segments of ligand in direct contact with a protein receptor. J Am Chem Soc 2001; 123: 6108-17.

[103]　Krishnan VV. Ligand screening by saturation transfer difference (STD) NMR spectroscopy. Curr Anal Chem 2005; 1: 307-20.

[104]　Poppe L, Harvey TS, Mohr C, *et al.* Discovery of ligands for Nurr1 by combined use of NMR screening with different isotopic and spin-labeling strategies. J Biomol Screen 2007; 12: 301-11.

[105]　Jahnke W, Rudisser S, Zurini M. Spin label enhanced NMR screening. J Am Chem Soc 2001; 123: 3149-50.

[106]　Di Micco S, Bassarello C, Bifulco G, Riccio R, Gomez-Paloma L. Differential frequency saturation transfer difference NMR spectroscopy allows the detection of different ligand-DNA binding modes. Angew Chem Int Ed 2006; 45: 224-8.

[107]　Claasen B, Axmann M, Meineche R, Meyer B. Direct observation of ligand binding to membrane proteins in living cells by a saturation transfer double difference (STDD) NMR spectroscopy method shows a significantly higher affinity of integrin alpha(IIb)beta(3) in native platelets than in liposomes. J Am Chem Soc 2005; 127: 916-9.

[108]　Cutting B, Shelke SV, Dragic Z, *et al.* Sensitivity enhancement in saturation transfer difference (STD) experiments through optimized excitation scheme. J Am Chem Soc 2007; 45: 720-4.

[109]　Fehér K, Groves P, Batta G, Jiménez-Barbero J, Muhle-Goll C, Kövér KE. Competition saturation transfer difference experiments improved with isotope editing and filtering schemes in NMR based screening. J Am Chem Soc 2008; 130: 17148-53.

[110]　Mercier KA, Shortridge MD, Powers R. A multistep NMR screening for the identification and evaluation of chemical leads for drug discovery. Comb Chem High Through Screen 2009; 12: 285-95.

[111] Jayalakshmi V, Krishna NR. Complete relaxation and conformational exchange matrix (CORCEMA) analysis of intermolecular saturation transfer effects in reversibly forming ligand-receptor complexes. J Mag Res 2002; 155: 106-18.

[112] Balaram P, Bothner-By A, Dadok J. Negative nuclear overhauser effects as probes of macromolecular structures. J Am Chem Soc 1972; 94: 4015-7.

[113] Meyer B, Weimar T, Peters T. Screening mixtures for biological activity by NMR. Eur J Biochem 1997; 246: 705-9.

[114] Herfurth L, Weimar T, Peters T. Application of 3D-TOCSY-tr-NOESY for assignment of bioactive ligands from mixtures. Angew Chem Int Ed Engl 2000; 39: 2097.

[115] Dalvit C, Pevarello P, Tatò M, Veronesi M, Vulpeti A, Sundsrom M. Identification of compounds with biniding affinity to proteins via magnetization transfer from bulk water. J Biomol NMR 2000; 18: 65-8.

[116] Dalvit C, Fogliatto GP, Stewart A, Veronesi, M, Stockman, B. WaterLOGSY as a method for primary NMR screening: practical aspects and range of applicability. J Biomol NMR 2001; 21: 349-59.

[117] Ludwig C, Michiels PJA, Lodi A, Ride J, Bunce C, Günther UL. Evaluation of solvent accessibility epitopes for different dehydrogenase inhibitors. Chem Med Chem 2008; 3: 1371-6.

[118] Ludwig C, Michiels PJA, Wu X, et al. SALMON: Solvent accessibility, ligand binding, and mapping of ligand orientation by NMR Spectroscopy. J Med Chem 2008; 51: 1-3.

[119] Hajduk PJ, Olejniczak ET, Fesik SW. One-dimensional relaxation- and diffusion-edited NMR methods for screening compounds that bind to macromolecules. J Am Chem Soc 1997; 119: 12257-61.

[120] Becattini B, Pellecchia M. SAR by ILOEs: an NMR based approach to reverse chemical genetics. Chemistry (Weinheim an der Bargstrasse, Germany) 2006; 12: 2658-62.

[121] Becattini B, Sareth S, Zhai D, et al. Targeting apoptosis via chemical shift design: inhibition of bid-induced cell death by small organic molecule. Chem Biol 2004; 11: 1107-17.

[122] Sànchez-Pedregal VM, Reese M, Meiler J, Blommers MJJ, Griesinger C, Carlomagno T. The INPHARMA method: protein mediated interligand NOEs for pharmacophore mapping. Angew Chem Int Ed Engl 2005; 44; 4172-5.

[123] Merchant TE, Gierke LW, Meneses P, Glonek T. ^{31}P magnetic resonance spectroscopic profiles of neoplastic breast tiessues. Cancer Res 1988; 48: 5112-8.

[124] Merchant TE, Kasimos JN, Vroom T, et al. Malignant breast tumour phospholipid profiles using ^{31}P magnetic resonance. Cancer Lett 2002; 176: 159-67.

[125] Spasojević I, Zakrzewska J, Bačić G. ^{31}P NMR spectroscopy and polarographic combined study of erythrocytes treated with 5-fluorouracil: cardiotoxicity related changes in ATP, 2,3-BPG and O_2 metabolism. Ann NY Acad Sci 2005; 1048: 311-20.

[126] Spasojević I, Jelić S, Zakrzewska J, Bačić G. Decreased oxygen transfer capacity of erythrocytes as a cause of 5-fluorouracil related ischemia. Molecules 2009; 14: 53-67.

[127] Hubbar JA, MacLachlan LK, Johnson P, et al. A Method for identification of inhibitors of the phosphorylation reactions of bacterial response regulator proteins using ^{31}P nuclear magnetic resonance spectroscopy. Anal Biochem 2001; 299: 31-6.

[128] Barrett JF, Hoch JA. Two-component signal transduction as a target for microbial anti-infective therapy. Antimicrob Agents Chemother 1998; 42: 1529-36.

[129] Pellecchia M, Becattini B, Crowell KJ, et al. NMR-based techniques in the hit identification and optimization processes. Expert Opin Ther Targets 2004; 8: 597-611.

[130] Yao H, Sem DS. Cofactor fingerprinting with STD NMR to characterize proteins of unknown function: identification of a rare cCMP cofactor preference. FEBS Lett 2005; 579: 661-6.

[131] Yang P, Yang C, Sale WS. Flagellar radial spoke protein 2 is a calmodulin binding protein required for motility in Chlamydomonas reinhardtii. Eukaryot Cell 2004; 3: 72-81.

[132] Gettins P. Thermolysin-inhibitor complexes examined by ^{31}P and ^{113}Cd NMR spectroscopy. J Biol Chem 1988; 263: 10208-11.

[133] Bartlett PA, Marlowe CK. Possible role for water dissociation in the slow binding of phosphorus-containing transition-state-analogue inhibitors of thermolysin. Biochemistry 1987; 26: 8553-61.

[134] Doris E, Wagner A, Mioskowski C. α-Aminocyclopropanone hydrates: potential transition-state analog inhibitors of serine proteases. Tetrahedron Lett 2001; 42: 3183-5.

[135] Gerig JT. Fluorine NMR of proteins. Prog Nucl Magn Reson Spectrosc 1994; 26: 293-370.

[136] Lian C, Le H, Montez B, et al. Fluorine-19 nuclear magnetic resonance spectroscopic study of fluorophenylalanine- and fluorotryptophan-labeled avian egg white lysozymes. Biochemistry 1994; 33: 5238-45.

[137] Ahmed AH, Loh AP, Jane DE, Oswald RE. Dynamics of the S1S2 glutamate binding domain of GluR2 measured using 19F NMR spectroscopy. J Biol Chem 2007; 282: 12773-84.

[138] Bann JG, Pinkner J, Hultgren SJ, Frieden C. Real-time and equilibrium 19F-NMR studies reveal the role of domain–domain interactions in the folding of the chaperone PapD. Proc Natl Acad Sci USA 2002; 99: 709-14.

[139] Chekmenev EY, Chow SK, Tofan D, Weitekamp DP, Ross BD, Bhattacharya P. Fluorine-19 NMR chemical shift probes molecular binding to lipid membranes. J Phys Chem B 2008; 112: 6285-7.

[140] Tengel T, Fex T, Emtenäs H, Almqvist F, Sethson I, Kihlberg J. Use of ^{19}F NMR spectroscopy to screen chemical libraries for ligands that bind to proteins. Org Biomol Chem 2004; 2: 725-31.

[141] Hultgren SJ, Jones CH, Normark S. In: Neidhart FC, Ed. Bacterial adhesions and their assembly. Washington DC, ASM Press 1996; pp. 2730-56.

[142] Wallen A, Zetterstrom RH, Solomin L, Arvidsson M, Olson L, Perlmann T. Fate of mesencephalic AHD2-expressing dopamine progenitor cells in NURR1 mutant mice. Exp Cell Res 1999; 253: 737-46.

[143] Dalvit C, Flocco M, Veronesi M, Stockman BJ. Fluorine-NMR competition binding experiments for high-throughput screening of large compound mixtures. Comb Chem HTS 2002; 5: 605-11.

[144] Meiboom S, Gill D. Modified spin-echo method for measuring nuclear relaxation times. Rev Sci Instrum 1958; 29: 688-91.

[145] Akoka S, Barantin L, Trierweiler M. Concentration measurement by proton NMR using the ERETIC method. Anal Chem 1999; 71: 2554-7.

[146] Silvestre V, Goupry S, Trierweiler M, Robins R, Akoka S. Determination of substrate and product concentrations in lactic acid bacterial fermentations by proton NMR using the ERETIC method. Anal Chem 2001; 73: 1862-8.

[147] Dalvit C, Mongelli N, Papeo G, et al. Sensitivity improvement in ^{19}F NMR-based screening experiments: theoretical considerations and experimental applications. J Am Chem Soc 2005; 127: 13380-5.

[148]   Veronesi M, Dalvit C. In: Webb, Graham A, Ed. Modern magnetic resonance. Dordrecht, Springer-Verlag. 2006; pp. 1393-400.

[149]   Dalvit C. Ligand and substrate-based ^{19}F screening: principles and application to drug discovery. Prog Nucl Magn Res Spect 2007; 51: 243-71.

[150]   Taylor JD, Gilbert PJ, Williams MA, Pitt WR, Ladbury JE. Identification of novel fragment compounds targeted against the pY pocket of v-Src SH2 by computational and NMR screening and thermodynamic evaluation. PROTEINS: Struct Funct Bioinformatics 2007; 67: 981-90.

[151]   Purser S, Moore PR, Swallow S, Gouverneur V. Fluorine in medicinal chemistry. Chem Soc Rev 2008; 37: 320-30.

[152]   Muller K, Faeh C, Diederich F. Fluorine in pharmaceuticals: looking beyond intuition. Science 2007; 317: 1881-6.

[153]   Vulpetti A, Hommel U, Landrum G, Lewis R, Dalvit C. Design and NMR based screening of LEF, a library of chemical fragment with different local environment of fluorine. J Am Chem Soc 2009; 131: 12949-59.

[154]   Nilakantan R, Bauman N, Dixon JS, Venkataraghavan R. Topological torsion: A new molecular descriptor for SAR applications. Comparison with other descriptors. J Chem Inf Comp Sci 1987; 27: 82-5.

[155]   Chiyoda T, Iida K, Kzuhiko K, Masahiro K. Screening system for urease inhibitors using ^{13}C-NMR. Chem Pharm Bull 1998; 46: 718-20.

[156]   Dalvit C, Ardini E, Flocco M, Fogliatto GP, Mongelli N, Veronesi M. A general NMR method for rapid, efficient, and reliable biochemical screening. J Am Chem Soc 2003; 125: 14620-5.

[157]   Dalvit C, Ardini E, Fogliatto GP, Mongelli N, Veronesi M. Reliable high-throughput functional screening with 3-FABS. Drug Discov Today 2004; 9: 595-602.

[158]   Forino M, Johnson S, Wong TY, *et al.* Efficient synthetic inhibitors of anthrax lethal factor. Proc Natl Acad Sci USA 2005; 102: 9499-504.

[159]   Fattorusso R, Jung D, Crowell JC, Forino M, Pellecchia M. Discovery of a novel classof reversible non-peptide caspase inhibitors *via* structure-based approach. J Med Chem 2005; 48: 1649-56.

[160]   Philchenkov A, Zavelevich M, Kroczak TJ, Los M. Caspases and cancer: mechanisms of inactivation and new treatment modalities. Exp Oncol 2004; 26: 82-97.

[161]   Frutos S, Tarragò T, Giralt E. A fast and robust ^{19}F NMR-based method for finding new HIV-1 protease inhibitors. Bioorg Med Chem Lett 2006; 16: 2677-81.

[162]   Tarragò T, Frutos S, Rodriguez-Mias RA, Giralt E. Identification by ^{19}F NMR of traditional chinese medicinal plants possessing prolyl oligopeptidase inhibitory activity. ChemBioChem 2006; 7: 827-33.

[163]   Drews J. Strategic trends in the drug industry. Drug Disccov Today 2003; 8: 411-20.

[164]   Inglese J, Johnson RJ, Simeonov A, *et al.* High-throughput screening assays for the identification of chemical probes. Nat Chem Biol 2007; 3: 466-79.

[165]   McGovern SL, Caselli E, Grigorieff N, Shoichet BK. A common mechanism underlying promiscuous inhibitors from virtual and high-throughput screening. J Med Chem 2002; 45: 1712-22.

[166]   Ryan AJ, Gray NM, Lowe PN, Chung CW. Effect of detergent on "Promiscuous" inhibitors. J Med Chem 2003; 46: 3448-51.

[167]   McGovern SL, Shoichet BK. Kinase inhibitors: not just for kinases anymore. J Med Chem 2003; 46: 1478-83.

[168]   Feng BY, Simeonov A, Jadha A, *et al.* A high-throughput screen for aggregation-based inhibition in a large compound library. J Med Chem 2007; 50: 2385-90.

[169]   Inglese J, Auld D, Jadhav A, *et al.* Quantitative high-throughput screening: a titration-based approach that efficiently identifies biological activities in large chemical libraries. Proc Natl Acad Sci USA 2006; 103: 11473-8.

[170]   Huth JR, Mendoza R, Olejniczak ET, *et al.* ALARM NMR: A rapid and robust experimental method to detect reactive false positives in biochemical screens. J Am Chem Soc 2005; 127: 217-24.

[171]   Walters WP, Stahl MT, Murcko MA. Virtual screening -an overview. Drug Disc Today 1998; 3: 160-78.

[172]   Dalvit C, Caronni D, Mongelli N, Veronesi M, Vulpetti A. NMR-based quality control approach for the identification of false positives and false negatives in high throughput screening. Curr Drug Discov Tech 2006; 3: 115-24.

[173]   Freeman R, Hill HDW. High-resolution studies of nuclear spin–lattice relaxation. J Chem Phys 1969; 51: 3140-1.

[174]   Valensin G, Kushnir T, Navon G. Selective and nonselective proton spin–lattice relaxation studies of enzyme–substrate interactions. J Magn Reson 1982; 46: 23-9.

[175]   Hennessy BT, Smith DL, Ram PT, Lu Y, Mills GB. Exploiting the PI3K/AKT pathway for cancer drug discovery. Nat Rev Drug Discov 2005; 4: 988-1004.

Utilization of NMR-Based Techniques in Anticancer Drug Development

V. Raja Solomon[1,2] and Hoyun Lee[*,1,2,3]

[1]*Tumour Biology Group, Northeastern Ontario Regional Cancer Program at the Sudbury Regional Hospital, 41 Ramsey Lake Road, Sudbury, Ontario P3E 5J1, Canada*

[2]*Department of Biology, Laurentian University, 935 Ramsey Lake Road, Sudbury, Ontario P3E 2C6, Canada*

[3]*Department of Medical Sciences, the Northern Ontario School of Medicine, 935 Ramsey Lake Road, Sudbury, Ontario P3E 2C6, Canada*

Abstract: Nuclear magnetic resonance (NMR) utilizes spin changes at the nuclear level when radiofrequency energy is absorbed in the presence of magnetic field. Only nuclei with odd mass numbers (e.g., ^1H, ^{13}C, ^{15}N) give NMR spectra because they have asymmetrical charge distribution and (2I+1) orientations. Since the discovery of NMR phenomenon in 1946 by Purcell and Bloch, it has been used for the study of both synthetic compounds and natural products. NMR has also been used to investigate dynamic molecular properties such as conformational isomerism, molecular asymmetry, hydrogen bonding, and keto-enol tautomerism. Analysis of structure-activity relationship with an NMR technique was introduced in 1996. Using this and other recently advanced NMR techniques, it is now possible to determine the binding site and pattern of a small molecule to its intended molecular target, an extremely powerful tool for the development of effective drugs. One of the most exciting new areas of research is the field of metabonomics, which relies on NMR spectroscopy of biofluids such as urine, plasma, and cerebrospinal fluid. In this review, we will focus on commonly used NMR-based screening approaches for cancer chemotherapeutics development. In doing so, we will first briefly introduce the theoretical aspect of the technique that lays the foundation of NMR-based methods. We will then discuss several different NMR techniques that are commonly used for studying interactions between model ligands and receptors. We will also describe examples of how the NMR spectroscopy has been applied to generate and optimize novel chemical classes of ligands to develop effective anticancer therapeutics. Finally, we will discuss about contributions of the metabonomic NMR spectroscopy to the analysis of drug metabolism and toxicity in the context of apoptosis and tumor control.

Keywords: NMR, spectroscopy, anticancer agent, drug development, structure activity relationship (SAR).

1. INTRODUCTION

Nuclear magnetic resonance (NMR) utilizes spin changes at the nuclear level when radiofrequency energy is absorbed in the presence of magnetic field [1]. In 1950s, NMR spectra have been used for the study of both newly synthesized compounds and natural products isolated from plants, bacteria, and other sources [1]. Generally, organic compounds are composed of hydrogen, carbon, phosphorus, nitrogen and oxygen elements. Additionally, they may also include halogens fluorine, chlorine, bromine, iodine, and sometimes, metal atoms. Each of these elements has an isotopic nucleus which can be detected by NMR. The low natural abundance of ^{15}N and ^{17}O in nature prevents NMR being routinely applied to these elements without the use of labeled substances; however, ^1H, ^{13}C, ^{19}F and ^{31}P based NMR spectroscopy is routinely used [2,3]. Many instruments are equipped with a so-called QNP (quattro nuclei probe) for sequential NMR analysis of ^1H, ^{13}C, ^{31}P and ^{19}F, without the hardware having to be switched [2,3]. Analysis is typically made at proton frequencies between 300 and 500 MHz. Depending on the type of analyses, high field instruments are capable to analyze compounds at the concentration of an mg mL^{-1} range, although a concentration of 1–100 mg mL^{-1} is normal.

A momentous advancement of the NMR technique occurred in 1980's, which was facilitated by the combination of a reliable superconducting magnet with the newly developed and highly sophisticated pulse technique of Fourier transformation [4]. It provided chemists with a method suitable to determine the three-

* Corresponding author: Tel: +1 705 522-6237, Ext. 2703; Fax: +1 705 523 7326; E-mail: hlee@hrsrh.on.ca

Atta-ur-Rahman / M. Iqbal Choudhary (Eds.)

dimensional (3-D) structure of very large molecules such as oligopeptides in solution. Modern NMR spectrometers are available up to field strengths of 21.1Tesla or a proton resonance frequency of 900 MHz.

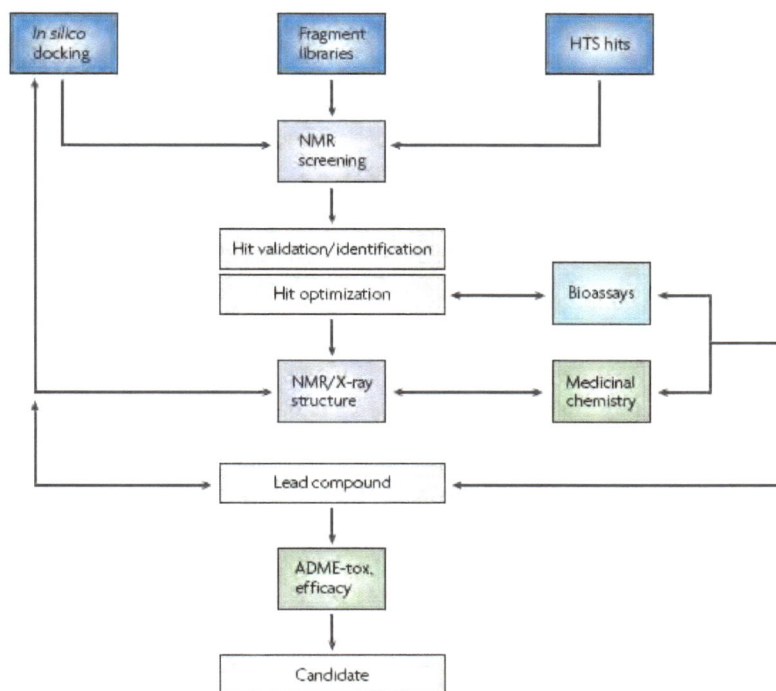

Fig. (1). A schematic diagram of the utilization of NMR in drug discovery. (*Reprinted with permission from Macmillan Publishers Limited [5], copyright 2008*).

In the pharmaceutical industry, NMR spectroscopy is playing an integral and continually expanding role in high-throughput screening (HTS) [6,7] and structure-based drug discovery (SBDD) [8,9]. These approaches have now become the driving force behind the discovery and development of new therapeutics (Fig. **1**) [10]. The NMR based screening process may be incorporated into the following three major steps in drug discovery: the discovery of lead compounds, drug optimization, and clinical validation (Fig. **1**, Table **1**) [11-14]. Given the growing popularity of fragment-based libraries, the NMR-based high-throughput ligand affinity screening is a particularly well-established component of the fragment-based dug discovery (FBDD) and development process [15]. This method can now be used routinely for both validation and identification of novel chemical leads [16,17]. The wide adoption of the FBDD approach enables rapid screening of small chemical libraries by NMR, and the exponential growth of structural space-based HTS methods [18,19]. NMR is also a powerful new tool in the optimization of lead compounds, as it can rapidly determine the complex of the protein–ligand interaction at high resolution [20,21]. In the following sections, we summarize some of the most recent developments in the field, which have significant bearing of cancer chemotherapeutics discovery and development.

2. NMR-BASED SCREENING

NMR-based screening may be defined as the identification of small molecule ligands for macromolecular targets by observation of changes in NMR parameters that occur upon their binding or interaction [22]. Depending on the nature of screening, the NMR-based screening can be divided into two classes: methods that determine target (receptor or macromolecules) resonance, and methods that determine ligand resonance (small molecules). The binding of a ligand to a protein target may be divided into two types: global and local effects (Fig. **2**) [23]. The former is size-dependent and, therefore, suited to the observation on the ligand, while the latter is restricted to the binding region and can be monitored on both the target and the ligand (Fig. **2**). Upon ligand binding to a macromolecule, the molecular weight and

Table 1. NMR-Based Screening Process and Applications

Screening techniques	Types of approach	Description	Applications
Chemical-shift perturbation	Protein or nucleic acids target resonances	Identifies compounds that bind by chemical-shift perturbation of resonances of the target	Primary screening; Hit validation; Site of binding
SAR-by-NMR	Ligand, Target	Design bi-dentate compounds	Structural information; FBDD screening; Compound optimization
SLAPSTIC With first-site spin labeled compound	Ligand	Highly sensitive detection of fragments and weakly interacting second-site compounds	FBDD screening; Compound optimization
SLAPSTIC (spin-labeled protein)	Ligand	Highly sensitive detection of fragments bound	Primary screening
$T_{1\rho}$ and T_2 relaxation; line broadening	Ligand	Binding enhances relaxation; enables affinity estimates; build-up curve identifies interacting functional groups	Primary screening; Hit validation
Diffusion measurements	Ligand	Measures the difference in diffusion rates for ligands in the bound versus free state	Primary screening; Hit validation
Transferred NOEs	Ligand	Provides information about interactions between a ligand and a target; determining the bioactive conformation of flexible ligands such as peptides	Hit validation; Conformation of flexible ligands
STD NMR	Ligand	Identifies compounds that bind weakly; build-up curve identifies interacting functional groups	Primary screening; Hit validation
WaterLOGSY	Ligand	Identifies compounds that bind by water-mediated NOEs	Primary screening
INPHARMA	Ligand-to-ligand	Detects protein mediated ligand-ligand interactions (i.e., competition for the same binding site)	Compound characterization
SAR by ILOEs	Ligand-to-ligand	Detects protein mediated ligand-ligand interactions	FBDD screening; Compound optimization
Pharmacophore by ILOEs	Ligand-to-ligand	Detects protein-mediated ligand-ligand interactions, and uses information for a pharmacophore-based search for bi-dentate compounds	FBDD screening; Compound optimization
FAXS	Reference ligand	Measures the displacement of a fluorinated 'spy' molecule	Primary screening; Hit validation

FAXS, fluorine chemical shift anisotropy and exchange for screening; NOE, nuclear Overhauser effect; SLAPSTIC, spin-labels attached to protein side chains as a tool to identify interacting compounds; STD, saturation transfer difference; TINS, target immobilized NMR screening; $T_{1\rho}$, rotating frame nuclear spin longitudinal relaxation time; T2, transverse nuclear spin relaxation time; WaterLOGSY, water-ligand observed via gradient spectroscopy; FBDD, fragment-based drug design; ILOE, interligand nuclear Overhauser effect; INPHARMA, interligand NOEs for pharmacophore mapping; SAR, structure–activity relationship. (*Reprinted with permission from Macmillan Publishers Limited [5], copyright 2008*).

hydrodynamic radius of a small ligand may be changed by several orders of magnitude. Those changes can be detected with several different NMR methods (Table 1). Although the relaxation filtering and diffusion editing may reveal binding, they cannot provide any structural information [24]. In contrast, NOE-based methods (NOE pumping, reverse NOE pumping) or STD can not only provide information on the interactions, but they can also be used to characterize the binding epitope(s) of the ligand [25]. Some other methods use spin-labeling techniques for second site screening or covalent modifications of a target with a spin-label near the active site, which leads to another dramatic decrease in the amount of protein needed for screening [26,27]. For searching initial hits, ligand based methods can be performed in a high-throughput manner. In medicinal chemistry, possible drug candidates are often stored in libraries. With ligand-based techniques, those libraries can be directly screened as single compounds or in mixtures of 5 to 10 compounds. Binding ligands are directly identified by their NMR spectra [28-30]. Thus, time-consuming deconvolution steps of the mixtures are not necessary.

The NMR-based screening is usually performed in aqueous solution; therefore, the ligands to be examined must have adequate solubility in deuterated water. For ligand-based screening, a concentration of at least 100 μM for each ligand examined should be attained to produce an acceptable signal:noise ratio, using conventional probe head technology. Therefore, the total ligand concentration can be an issue for protein stability if a large mixture is to be examined. The amount of target protein needed varies between 1

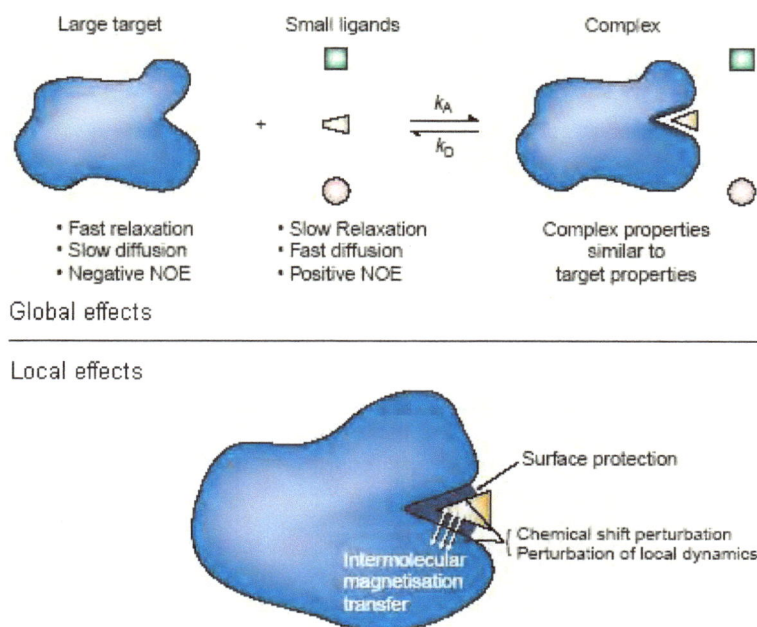

Fig. (2). The Global effects and local effects observed on the ligand and target interaction. (*Reprinted with permission from [23], copyright 2001 Elsevier Ltd.*).

and 50 μM of unlabelled protein per 0.5 mL of screening sample, depending on the technique used. For target-based screening (i.e. chemical shift mapping), 100–300 μM of labeled proteins are required. The use of cryogenic probes can increase probe sensitivity more than three-fold [31,32], with a concomitant lowering the demands on both ligand solubility and the amount of protein required. In most cases, the target molecule is a protein with molecular mass larger than 10 kDa. To overcome signal overlap in the NMR spectra (necessary for both the spectral assignment and target-based screening), stable isotope labeling with ^{15}N or ^{13}C is required. For this reason, an overexpression system with a high cell density and high expression rates of the target protein is essential. For screening proteins up to 45 kDa, global labeling of the target molecule with ^{15}N suffices in most cases. If larger proteins are to be studied, several labeling schemes may be useful. Uniform labeling with ^{15}N, combined with deuteration of all non-exchangeable protons, leads to a dramatic increase in sensitivity using TROSY (transverse relaxaion-optimized spectroscopy)-based experiments [33], which can increase the size limit to >70 kDa. The selective ^{13}C-labeling of valine, leucine and isoleucine methyl groups, which can be achieved by synthesis of the amino-acids from cost effective precursors [34], is also beneficial in target-based screening. The high signal intensity of such methyl groups, due to the contribution of three protons per signal (compared to one for ^{15}N-based screening) and advantageous relaxation properties, reduces protein demand significantly, particularly that for larger proteins up to >70 kDa [34].

The NMR based screening approach provides numerous advantages for drug discovery as it is the most sensitive method in detecting ligand-target interactions among currently available techniques. Further, it does not require any knowledge of a protein function, and, thus, no target-specific set-up is necessary. Therefore, protein targets that have been identified purely on a genomics-based research can immediately be screened by NMR. It can also concomitantly enable to determine the binding constant (K_D) [35]. NMR spectra separate individual components, allowing the direct screening and deconvolution of mixtures from natural products or combinatorial chemistry. In addition to the mere binary binding information, NMR can provide crucial structural information for both target and ligand with atomic resolution for the subsequent optimization of weak initial hits into strongly binding drug candidates.

2.1. Receptor-Based Screening

A receptor-based method is able to observe and compare NMR parameters of the receptor molecule resonance in the presence or absence of compound mixture. Thus far, a receptor-based method has focused mostly on proteins or other macromolecules. This method incorporates site-specific characterization afforded by assigned protein NMR spectra along with a priori knowledge of the protein's 3-D structure (either from X-ray crystallography or NMR) [36,37]. By identifying perturbations of assigned protein resonance, not only ligands are identified, but also their binding sites are localized. The characterization of site-specific binding can suggest a strategy for the fragment-based lead generation, in which lower affinity molecular fragments binding to distinct subsites can be linked or elaborated to yield higher affinity compounds [38,39]. Localization of binding sites also enables to immediately distinguish specific binding from nonspecific binding. The receptor-based methods do not rely on fast exchange to retrieve the information on a bound state, and its result permits the characterization of both higher and lower affinity hits.

The use of NMR for receptor-based screening requires detailed data about the physico-chemical properties of the protein target, and the screening process usually starts from a simple method and takes gradually more difficult and complex challenges. Thus, a milligram quantity of soluble, non-aggregated proteins must be purified, for which a suitable expression host is required. This is because isotope enrichment (e.g. ^{13}C, ^{15}N, ^2H) is critical for the resonance assignment of typically large (>30 kDa) therapeutic targets. It should be noted that *E. coli*, the most commonly used host for protein expression, is often not an option for preparing a large amount of mammalian proteins that are toxic to the host when overexpressed. After sufficient quantities of labeled protein are prepared, it must be ensured that the sample is stable enough during subsequent sequential resonance assignments. Although new data acquisition approaches promise to accelerate resonance assignment, it still can be a relatively lengthy process (a few weeks) for large *monomeric* proteins (>30 kDa) routinely encountered in drug discovery research. Regrettably, the time required for NMR assignment of such targets certainly favors other approaches, such as X-ray crystallography, which can provide high-resolution structural information in a relatively short time.

2.2. Ligand-Based Screening

The ligand-based screening method compares NMR parameters of the compound mixture in the presence and absence of receptor molecules. However, this approach renders the molecular weight of the receptor irrelevant. The ligand-based approach is more sensitive when the screening is carried out with large receptor molecules. However, it is not necessary to produce milligram quantity of isotope-labeled receptors. Depending on the detailed method, less than a milligram of unlabeled protein is usually sufficient for screening (receptor concentration is often ≤ 1 μM, and no receptor assignment is necessary). Therefore, this method allows a medicinal chemist to evaluate new targets rapidly. This is important not only for adapting to constantly shifting priorities in drug discovery and also less time is required for screening

A potential caveat of the ligand-based approach is the inability to localize the binding site of the small molecule hits on the receptor. Also, the ligand-based approach heavily relies on the exchange-mediated transfer of bound state information to the free state (Fig. **2**) [40]. In summary, the ligand-based method has a merit in identifying weakly bound ligands (rapid exchange) and using large ligand molar excesses. However, this method has the risk of identifying ligands that are nonspecifically bound to receptors.

2.3. Experiment for Target-Based Screening

The binding of a ligand to a protein often results in changes of its conformation and molecular dynamics, in addition to the alterations of the local environment on the binding surface (Fig. **2**, Global effects). NMR spectroscopy has long been a choice of techniques for the characterization of both high- and low-affinity macromolecular complexes [41]. A common approach is to label the target receptor protein with stable isotopes such as ^{15}N and ^{13}C. Binding of a ligand or a macromolecule alters the chemical environment around the binding site, resulting in the chemical shift of magnetic nuclei at this site. This change can be observed using $^{15}N/^1H$ and/or $^{13}C/^1H$ correlation spectra [42,43]. The valuable information on intermolecular interactions can then be obtained from chemical-shift mapping, even in the absence of NMR assignments in a system that consist of a macromolecular receptor with low-molecular-weight ligand. This approach can be fully exploited once sequence-specific resonance assignments for the receptor protein are available. The polypeptide backbone assignments can be obtained form protein structures with sizes of up to ~150 kDa, with the use of suitable selective-isotope-labeling schemes [34]. The backbone assignments can be supplemented with assignments for selected amino-acid side chains. Combining with the sequence-specific NMR assignments, chemical-shift mapping can then provide unambiguous identification of the location of binding sites for small or large molecules on a receptor protein.

2.3.1. Chemical-Shift Perturbations

The chemical shift is the most important NMR parameter. It defines the location of a NMR line along the radiofrequency axis in the spectrum. Chemical shift is measured by comparing to that of a reference compound (tetramethylsilane [TMS]), and expressed in ppm (parts per million). The chemical shift is very sensitive to the chemical environment of a nucleus, and the magnitude of the shift depends on many factors, including atoms involved in tertiary contacts within the folded protein and intermolecular interactions in complexes as well as solvent properties and temperature. The target-observed screening approach relies on chemical-shift changes as indicators for intermolecular binding. After assignment, these changes can be mapped upon the protein structure, revealing its binding sites [44,45] and guiding structure-based ligand optimization. This forms the basis of the 'SAR-by-NMR' approach [46,47], in which optimized ligands are constructed by linking weakly binding two (or more) fragments. The approach continues to be successfully used in drug discovery and development [48-51].

The standard two-dimensional (2-D) experiment of ^{15}N, ^{1}H correlation spectra has recently been complemented by ^{13}C, ^{1}H-heteronuclear single quantum coherence (HSQC)-based screening on proteins bearing ^{13}C-labeled methyl groups [34]. The selective labeling, using inexpensive starting materials such as carbon-13, avoids the detrimental effects of ^{13}C–^{13}C homonuclear coupling while greatly reducing the cost of sample preparation. This application is benefited from the higher proton multiplicity and slower transverse relaxation of the methyl groups. Consequently, intensity was three-times greater than the average value by other methods. On the other hand, the methyl group-based screening suffers from assignment difficulties and lower spectral dispersion. Note that the distribution and accessibility of (polar) amide and (non-polar) methyl groups are not equivalent, possibly leading to different screening sensitivities. Equating chemical-shift perturbations with spatial proximity to the ligand may be misleading, especially in the presence of large shielding anisotropies in a ligand, or because of conformational changes in the target. Therefore, an alternative has been proposed, in which the binding site and the orientation of the ligand bound can be determined more reliably [52] without the need for a complete re-determination of the complex structure. It is based on the comparison of chemical-shift changes in the proximity of the binding site induced by a series of closely related ligands. It is worth noting that the chemical shift perturbation may change the variation of pH and/or salt concentration.

2.3.2. Protein Dynamics

Protein dynamics are vital to its functions, and alterations in 3-D due to specific molecular interactions will lead to changes to its dynamics. The dynamics and their alterations can often be readily measurable in autorelaxation that is one of the most extensively explored methods for accessing the dynamics of the protein backbone. The spin properties of ^{15}N, ^{13}C or ^{2}H sites in proteins, including both longitudinal and transverse relaxation rates (R_1 and R_2), are sensitive to the dynamics of proteins, particularly, the local flexibility [53]. Upon binding, changes will occur in the relaxation rates of the nuclear spins at the interactive interface. In addition, the transferred cross-correlated relaxation [54] and transferred residual dipolar coupling are also important parameters for probing the ligand–proteins interactions [55] in the cotext of their binding geometries. In addition, a number of novel methods have recently been introduced to probe the side chain dynamics [56,57]. This is important because the side chains are always related to the protein surface properties and their motions that are important for ligand binding. The protein-targeted experiment can give information on the interaction interface and structure [58].

2.4. Experiment for Ligand-Based Screening

The ligand-based methods are established on the basis of ligand binding to a target protein, which alters the molecular tumbling rates of ligands in solution. In general, most of the NMR screening libraries consists of small molecules (< 500 Da) that always exhibit slow relaxation (T_1, T_2) and fast diffusion. When binding to a large receptor, nuclear spin relaxation time and molecular self-diffusion coefficient of the ligand will be dramatically reduced. These changes enable the ligand–protein interaction to be accessed by directly observing the ligand NMR signals. When the resonance of ligand molecules is measured, the choice of physical mechanisms manifested in measurable NMR experiments is diverse. This includes longitudinal, transverse, DQ relaxation, diffusion coefficients, and inter- and intra-molecular magnetization transfer. The latter includes transferred NOE, NOE pumping and reverse NOE pumping, saturation transfer, and WaterLOGSY experiments.

The major benefit of the ligand-based screening is that there is no need to isotopically enrich the target macromolecule, nor is there any upper limit to the size of target that can be screened. Also, unlike the target-based SAR-by-NMR method, the identity of a ligand contained in a mixture of compounds can often be obtained directly from the screening data without the need to deconvolute the mixture. The disadvantages are that ligands with high affinity will be missed and that no information regarding ligand-binding sites is directly available from the screening data. The ligand must be in medium to fast exchange between the free and bound states for the binding interaction to be detected; thus, ligands in slow exchange will appear as non-binders. Recently, a competition binding method has been developed, which allows the detection of high affinity ligands and their binding site locations, making it possible for very efficient HTS.

2.4.1. Measuring Relaxation

A relaxation-based method, which depends on increment of apparent molecular weight of ligand binding to protein, is more sensitive than diffusion experiment [25]. In particular, the paramagnetically induced relaxation approach, such as SLAPSTIC (the spin-labeled attached to protein side chains as a tool to identify interacting compounds), offers a sensitive method for ligan-target protein interaction mapping [27]. In this technique, the magnetic moment of electron is thousands times larger than that of the proton, and, thus, the amount of the biological targets required will be reduced by 1 to 2 orders of magnitude. In addition, because relaxation is a spin property, a spin or nucleus that closed to the target protein or binding site will experience a larger relaxation-rate change (similar to chemical shift). Therefore, the relaxation rate measurement can also provide information of the ligand orientation at the binding site. One-dimensional (1-D) relaxation-based screening usually requires subtraction of a reference spectrum of the target-free ligand mixture, and the measurement must be taken to avoid artifacts due to binding-induced line shape distortions, chemical-shift changes, and residual target signals [25]. The T_2-based ligand screening may also be performed by comparing the width of spectral lines [59] that increases upon binding. Complications may arise from superposition of line-broadening due to chemical exchange or slow conformational motions (on the microsecond or millisecond timescale), which are often relevant to induced-fit during intermolecular interactions and damped upon binding. Thus, a size-dependent decrease in T_2 time may be compensated for by a flexibility-dependent reduction in line-broadening. In contrast, the recently proposed concept of transferred T_1 time [60] relies on the fact that relaxation rates measured on a fast-exchanging ligand in the free state also contain contributions from its local dynamics in the bound state, as immobilization of flexible regions of the ligand generally increases relaxation rates.

2.4.2. Diffusion Experiment

Measuring self-diffusion coefficient is a convenient way to study the ligand binding without prior separation of the components [61-63]. Because diffusivity is a principle property of the whole molecule, all NMR signals of a ligand are decayed identically in the experiment with pulsed-field-gradients (PFG). In solution containing a target protein and ligands, the bound ligands will have small diffusion coefficients and their NMR signal intensities will be less attenuated in the PFG-NMR diffusion experiment. The pseudo 2-D diffusion edited experiments (diffusion-ordered spectroscopy, DOSY), where molecules are separated according to their individual diffusion coefficients, are particularly useful for deconvoluting mixtures. Alternatively, 1-D diffusion filters may be combined with standard 2-D correlation experiments (e.g., the diffusion-modulated gradient COSY (correlation spectroscopy)) [64] and diffusion-encoded total correlation spectroscopy (TOCSY; DECODES) [29] to elucidate spin systems and facilitate mixture analysis. Diffusion filtering or editing might also be enhanced by a combination with other screening techniques such as Nuclear Overhauser effect (NOE) pumping [65,66]. However, the decay of transferred NOE (trNOE) can also be used as an effective way to gain insight into ligand–protein interactions during diffusion time [65-67]. In a promising new application for combinatorial chemistry, diffusion filtering has recently been used in solid-state magic-angle spinning (MAS) NMR to identify and distinguish resin-bound molecules from impurities [68].

2.4.3. NOE/trNOE Experiment

Nuclear Overhauser effect (NOE) adopts the concept of both inter- and intra-molecular magnetization transfers [65-67]. Because magnitude of NOE enhancement between two nuclei spins are exponentially related to the internuclear distance (r^{-6}), NOE related experiments have been widely used for determining the three dimensional structure of protein and ligand–protein complex, orientations of the binding domain in protein, and ligand inside or toward the binding site, as well as for deriving dynamic information for the ligand–protein interactions. Furthermore, NOE experiments work well for high affinity interactions, where a stable ligand–protein structure is formed. However, the low affinity binding or weak interactions of ligand can be monitored by trNOE spectroscopy [69,70]. The trNOE provides information about orientation of the ligand at the protein binding site. The trNOE is well established as a unique source of structural information on the bound ligand, where spin-diffusion should be suppressed to obtain optimal structural precision [71,72]. The observation of trNOE has proved to be equally useful in the fast screening of ligand mixture [59], forming the basis of 'bioaffinity NMR' [73]. The technique relies on the size-dependence of the intramolecular NOE, which shows slow NOE build-up with a weak positive maximum for free ligands and rapid build-up with a strong negative maximum for the bound state. If dissociation of the ligand occurs quickly enough (*i.e.* $K_D > 10^{-7}$ M), a sufficient percentage of observable free ligand will retain intense negative trNOE as a 'memory' of the bound state, indicating binding occurred. If the residence time is too short (*i.e.* $K_D > 10^{-3}$ M), however, trNOE will build up too weakly for detection [74]. Routinely, trNOE is recorded as a 2-D NOESY (NOE spectroscopy) spectrum with a short mixing time favoring build-up of trNOE over direct NOE and relaxation filtering suppresses residual target signals.

Screening by trNOE imposes an upper limit on the ligands (< 1 kDa), which must fall into the positive NOE regime. For borderline cases, this regime may be extended, for example, by increasing the temperature [75]. Another problem is the possible cancellation of positive direct and negative transferred NOE, which could be resolved by lengthy 3-D experiments such as 3-D TOCSY–trNOESY [74]. A major drawback of this method is the requirement of a high ligand concentration. Therefore, it is not suitable for the identification of ligands with poor solubilities. Another disadvantage is the presence in the spectrum of strong diagonal peaks that hampers the observation of cross peaks between ligand resonances with similar chemical shifts. In addition, the strong diagonal peaks may introduce t_1 noise or a baseline problem that may interfere with the observation of weak cross peaks.

2.4.4. NOE Pumping and Reverse NOE Pumping

Chen and Shapiro [65,66] successfully carried out a 1-D NOE pumping experiment and its related reverse NOE pumping experiment for primary NMR screening. In the NOE pumping experiment, magnetization transfer from a receptor to a ligand was observed, and in the reverse NOE pumping, the reverse process of magnetization transfer from the ligand to the receptor was monitored. A diffusion filter was introduced in the NOE pumping experiment before the NOE mixing time to destroy completely the ligand magnetization while preserving most of the receptor magnetization. During the mixing time, the Zeeman magnetization of the receptor relaxes and part of the magnetization is transferred via intermolecular cross relaxation to the ligand. The choice of the optimal mixing time depends on the protein longitudinal relaxation value. The reverse NOE pumping involves the acquisition of two experiments using the pulse sequences [66]. The first experiment uses a transverse relaxation filter, followed by the NOE mixing time and the detection pulse. The relaxation filter suppresses all of the protein signals while inverting the ligand proton signals. The ligand magnetization will be partially transferred via intermolecular cross relaxation to the receptor during mixing time. A second experiment is then recorded where the order of the mixing time and transverse relaxation filter is reversed. In this experiment, no net magnetization transfer from the ligand to the receptor takes place. Therefore, the magnetization of a binding molecule decays at a slower rate, compared to the first experiment. A subtraction of the two spectra, recorded with interleaved acquisition, results in a spectrum containing only the signals of the molecules interacting with the protein.

2.4.5. Saturation Transfer

In early 1960s, the saturation transfer method was developed, and it is still being applied to NMR screening with successful results [76-78]. The saturation transfer difference (STD) usually starts from saturating one of the resonances (peaks) of protein [79,80], and continues until a stead state of saturation is achieved. Magnetization transfer mechanisms (e.g., spin diffusion) and cross-relaxation occur during the saturation period. Because of spin diffusion, all signals of the protein will be saturated. Most importantly, cross-relaxation carries the saturation effect to the ligand that interacts with the target protein. By taking the difference in two spectra obtained with or without saturation, one can get a spectrum containing the resonance of a protein and its binding ligand. NMR peaks of the nonbinding ligands in the sample are not affected by the saturation. Therefore, the STD experiment is useful in identifying all potential ligands that interact with protein in the mixture [78,81]. The STD can be coupled with standard 2-D correlation experiments such as TOCSY, COSY, and HMQC, to alleviate spectral overlap [81]. The distance-dependence of the observed signal intensities identifies those nuclei closest to the target–ligand interface. This 'epitope mapping' [72,77] may be seen as a ligand-based complement to the 'SAR-by-NMR' approach. Remarkably, application can extend the STD screening to solid-state NMR [78], in which the target may be immobilized on controlled-pearl glass to facilitate target protein recovery.

2.4.6. WaterLOGSY

WaterLOGSY method (Water–Ligand Observation via Gradient SpectroscopY) is a powerful tool for primary screening of compound mixtures by NMR. It utilizes intermolecular magnetization-transfer difference from saturation of water resonance, instead of protein resonance [82]. Because water plays an essential role in determining protein surface, water signals saturated the protein and the bound compounds. Because of the fast exchange between bound water and bulk water, the large quantity of bulk water magnetization increases the sensitivity of the WaterLOGSY experiment. Titration binding experiments can be performed in order to extract an approximate value of the K_D of weak affinity ligands. However, particular care must be taken in the analysis of the titration experiments, since the signal intensity in the Water-LOGSY spectra contains an off-setting effect deriving from the hydration of the free ligand. This contribution can be calculated in a complementary experiment recorded for the ligand in the absence of the protein. The subtraction of this experimentally derived contribution from the measured Water-LOGSY signals in the presence of the protein permits the calculation of the binding constant [83]. A drawback of this method is that a very low-affinity ligand cannot be detected when the ratio of total ligand concentration by total protein concentration is high, due to the competing effect arising from hydration of the free ligand. This effect is more pronounced for protons adjacent to ligand exchangeable protons.

2.4.7. Competitive Binding Studies

Competitive binding studies are based on biochemical concept for determining the ligand binding specificity and affinity. Lately, the competition binding method has been incorporated into several strategies for ligand-based NMR screening. In this approach, displacement of a reporter ligand by test compounds (ligands) is observed. The precondition of rapid exchange between free and bound forms only applies to the reporter ligand, meaning that there is no upper affinity limit. The lower affinity limit can be fine tuned by adjusting the concentration of the reporter ligand, relative to its affinity. The competition-based approach is ideal for experiments that do not require a separate reference sample (e.g. WaterLOGSY and STD) [79,84]. Unfortunately, the competitive binding method is unable to use directly for screening compounds from mixture without deconvolution. The need to identify a suitable reporter ligand and the large amounts of reporter protein is required for a full-scale screen. Further, only a ligand with a binding site that overlaps with reporter ligand can be detected by a competition assay.

2.4.8. Fluorine-Based Method

Similarly to the ^{1}H nucleus, the ^{19}F nucleus has a number of unique properties that render a highly effective relaxation probe for NMR screening [85]. The absence of endogenous ^{19}F in biological molecules permits clean observation of ligand spectra, thus, obviating the need for relaxation filters and/or difference spectroscopy to eliminate receptor or large solvent signals. ^{19}F occurs at 100% natural abundance, and has a gyromagnetic ratio 0.94; therefore, the sensitivity of ^{19}F NMR is competitive with that of ^{1}H. The ^{19}F-based technique is applicable for transverse relaxation method, as the chemical shift range of ^{19}F is much larger than that of ^{1}H ($\approx$ 900 ppm) [86]. This large chemical shift range implies high sensitivity of the ^{19}F chemical shift to changes in microenvironment. ^{19}F relaxation is also useful for secondary screening aimed at estimating exchange rates and equilibrium dissociation constants [87,88]. More recently, cross-correlation between the ^{19}F aromatic chemical shift anisotropy (CSA) and ^{1}H-^{19}F DD relaxation mechanisms has been exploited to improve the accuracy of K_D estimates [85]. A negative aspect of ^{19}F relaxation as a primary screening method is the need to assemble a sufficiently large library of fluorinated compounds as well as the intrinsic lack of diversity. This creates difficulties when attempting to choose a diverse set of compounds for a screening library. However, useful screening information can still be obtained by examining the ^{19}F relaxation of a small set of compounds. Dalvit and co-workers demonstrated this approach by a fluorine chemical shift anisotropy and exchange for screening (FAXS) strategy [89,90]. The study suggested that the relaxation properties of a small set of the ^{19}F "spy" compound report on the binding of a large set of high affinity binders via competitive displacement.

2.4.9. Competition-Based Fluorine Screening

This technique is a combination of competition screening with ^{19}F-screening and the principle based on the fact that a binding of test ligands is observable via displacement of a fluorinated reporter ligand [89]. The use of ^{19}F avoids an overlap between reporter and test ligand spectra, which are potential sources of problems in other competition-based methods. The amounts of both protein and test ligand required in the competition-based ^{19}F-screening are usually much lower than those of other NMR screening methods. The demands on the solubility of the ligand are concurrently low and the size of mixtures is limited only by the need to identify hits by mixture deconvolution. In earlier studies, a typical screening experiment was conducted in 5–10 minutes on mixtures of ~20 compounds, and the introduction of dedicated hardware for ^{19}F detection increased this throughput further [89]. The ^{19}F screening of a small library of fluorinated compounds could be used to identify suitable reporter ligands. The main test library need not contain fluorinated compounds, and, thus, the problems of limited diversity that are associated with fluorinated libraries are avoided.

2.4.10. Spin Label Method

The spin labels attached to protein side chains as a tool to identify interacting compounds (SLAPSTIC) can be used to identify and characterize intermolecular interactions. Jahnke and colleagues have developed and successfully applied the spin labeling method for NMR-based screening of compound libraries for several different applications [26,27,38,91]. The experiment is based on relaxation enhancement of ligand resonances caused by its proximity (typically up to 15–20 Å apart) to a spin-labeled group (e.g., paramagnetic organic nitroxide radicals such as TEMPO). The spin labels can be covalently attached either to the protein or the first ligand. The method is an extension of the traditional $T1\rho$ relaxation approach, where one forces amplification of the bound state relaxation properties via covalently attached spin labels to protein side chains such as lysine, tyrosine, cysteine, histidine and methionine [26,27]. A limiting condition is that at least one residue of this type should be as close as possible to the binding site, but must not interfere with ligand binding. It is a very sensitive technique for NMR screening: it allows reduction of the protein amount by 1–2 orders of magnitude, compared to the $T1\rho$ relaxation approach on a non-modified protein target. Second-site NMR screening can be efficiently performed with the use of spin labeling [26,38]. This method utilizes a spin-labeled compound as a first-site ligand. Screening this complex with a library of compounds allows identifying compounds that bind simultaneously to the first spin-labeled ligand of the neighboring binding site (i.e., second site). The second-site ligand can then be identified from quenching of their NMR signals by the spin-labeled first ligand. A spin label can also be attached to a known ligand. This altered first ligand then serves to screen for new ligand binding in a second proximal site. Again, it must be confirmed that the addition of the spin label does not compromise either receptor binding site.

2.4.11. Isotope Filtering/Editing Techniques

An isotope filtering/editing NMR technique can be used for isotopic (^{13}C, ^{15}N, etc) labeling to selectively detect the spectrum of the labeled protein or ligand, and the spectral signals of the other components are filtered out [92]. By a combining filtering/editing approach with conventional 2-D or 3-D experiments, the structure of bound proteins or ligands can be directly obtained. To maintain high sensitivity, the target component has to be completely labeled with NMR active isotopes, followed by high performance of isotope filtering to suppress the unlabeled component. This type of experiment can also be carried out with direct recognition of the labeling nuclear spin, ^{13}C or ^{15}N NMR. Because of low gyromagnetic ratio (γ) of ^{13}C and ^{15}N, NMR based on this technique is less sensitive than proton ^{1}H NMR.

2.4.12. Double-Quantum Relaxation Approach

The double-quantum (DQ) relaxation method requires the creation of DQ coherence via antiphase magnetization, and then conversion it to single quantum (SQ) coherence for detection. Therefore, this approach is appropriately defined as transverse SQ and DQ relaxation filtering. The spectra can be recorded in a 1-D or 2-D version. The sensitivity improvement of 1-D versus 2-D of the DQ experiment is only $\sqrt{2}$. In addition, the problem of overlapping spectra and the absence of information about the spin networks in the 1-D spectrum limits its utilization to only simple mixtures comprised of a few compounds. The scalar connectivity observed in the 2-D DQ spectrum allows the direct identification of molecules interacting with a target protein without deconvolution of the complex mixture. Significant multiple solvent suppression is achieved with coherence selection gradients tilted at the magic angle [93]. This becomes particularly relevant in the observation of diluted samples used in NMR screening, where the dynamic range of the receiver needs to be maximized in an optimal way. An advantage of using samples dissolved in H_2O is that it allows one to clearly observe the NH doublet resonance. The NH proton signals are in a spectral region with little or no overlap with other signals and, therefore, can be used for assignment purposes in complicated mixtures. Furthermore, the chemical shift difference of a ligand NH proton free in solution or when hydrogen bonded to a protein is typically large. A drawback of this experiment is the requirement of two doublets in the spectrum of the molecule to be screened.

2.5. Utilization of NMR-Based Screening in Fragment Based Drug Discovery

In NMR-based screening applications, independent NMR experimental can be shaped by the strategy used to direct information flows through the lead development and discovery process [94-97]. The approach articulates the selection of compound libraries to be screened, the choice of screening experiments and conditions, and the means by which information from the screen is used to find more desirable molecules. NMR-based screening is usually integrated into a wide variety of other techniques, both in designing libraries (e.g., virtual screening, physicochemical property-based selection, retrosynthetic fragmentation of known drugs) and follow-up screening of potential hits (e.g., crystallography, molecular modeling, combinatorial chemistry, and an enzymological approach). Many different permutations of these methods have been used. For example, HTS has been used to validate hits identified by NMR-based screening, and NMR has been used to validate hits identified by HTS. The integrated use of molecular modeling, NMR screening, *in silico* virtual screening, X-ray crystallography, and an enzymological approach is proven to be very effective.

NMR and protein X-ray crystallography have been used extensively in FBDD, because these techniques are not only highly sensitive in detecting low affinity fragment (ligand) binding, but also give information about the fragment protein interactions [98,99]. FBDD is primarily a new lead discovery approach, in which low molecular-weight compounds (120–250 Da) are screened relative to a HTS assay. Since fragment-based hits are typically weak inhibitors (10 μM – mM), high concentration is needed for this approach. Compared with HTS hits, fragments are usually simple and less functionalized compounds with correspondingly low affinity [5,100]. However, fragment hits typically possess high 'ligand efficiency' (binding affinity per heavy atom) with drug-like properties; therefore, they are highly suitable for optimization into clinical candidates. Once the interactions between ligand and protein are structurally validated and understood, the subsequent chemical optimization may result in a high success rate of developing effective drugs.

Recently, physico-chemical properties of fragment (ligand) libraries have been received particular attention. For example, in order for fragments to be screened at high concentrations, they need to have high aqueous solubility and stability. The fragment hits usually be obeyed a 'rule of 3', which is defined as the molecular weight is < 300, hydrogen bond donors (HBD) ≤3, hydrogen bond acceptors (HBA) ≤3, and CLogP ≤3 [101]. It clearly indicates that a 'rule of 3' could be useful for constructing fragment libraries for the discovery of effective leads and distinguishes fragments from the larger, 'rule of 5' properties for drug-like compounds [102]. The following section of review describes the NMR based screening applications on FBDD approach in lead discovery and its classified into four categories: fragment linking, fragment optimization, fragment evolution; and fragment self-assembly.

2.5.1. Fragment Linking

In the fragment linking approach, hits identified by primary screening are mapped to specific binding sites, and their relative orientations are determined for the fragments that bind in close proximity to one another. SAR-by-NMR guides the design of linkers between fragments using NMR derived structural information. For example, contacts between fragments are determined using an NOE method [47], and the location of binding sites and the orientation of protein bound to ligands are established with chemical shift perturbation data [52,103]. By linking two weakly binding ligands via positions that are not crucial for interaction with the protein (as determined by STD NMR), an optimized ligand can be obtained much more quickly [79]. The increase in potency by optimally linking two fragments is assumed to be additive, as the free energy of the linked molecule in binding to the target protein is approximately equal to the sum of the free energies in binding to each of the two fragments (*i.e.*, mM × mM = μM) [104].

2.5.1.1. SAR-by-NMR

The SAR-by-NMR is a fragment linking method, which comprises a screening for binders at a first site and a second screening for binders at a proximal site. The experiment is carried out in the presence of saturating amounts of a first-site binder. However, this method has thus far been demonstrated only for protein targets, although it can be applicable to nucleic acid targets. The schematic protocol is outlined in Fig. (**3**) [47]. In the first step, a library of small organic molecules is screened to identify molecules that bind to the target protein (Fig. **3**, Step 1). To elucidate ligand-induced changes in chemical shift, binding is determined by comparing 2-D ^{1}H–^{15}N HSQC spectra of the ^{15}N-enriched target protein in the absence or presence of ligand. Binding constants for two identified ligands can be determined by monitoring chemical shift changes as a function of ligand concentration [46]. Once a lead compound is identified, it should be optimized by determining binding constants or activity measurements (Fig. **3**, Step 2). Subsequently, a low affinity of second ligand is identified, based on amide ^{1}H/^{15}N chemical shift changes for a different set of residues in either the initial screen or in a second screen carried out in the presence of the optimized ligand from the first screen (Fig. **3**, Step 3). In Fig. (**3**), Step 4, the identification of related analogues was carried out as for the first ligand, followed by optimization of the second ligand. Following identification of the two lead fragments, the ternary complex orientation and their location were determined experimentally by either an isotope-edited or trNOE approach. Based on this information, the two lead fragments (ligand) are synthetically linked together in a manner that two ligand moieties of the linked molecule maintains the spatial orientation to each other and to the target protein to produce the final high-affinity ligand (Fig. **3**, Step 5) [47].

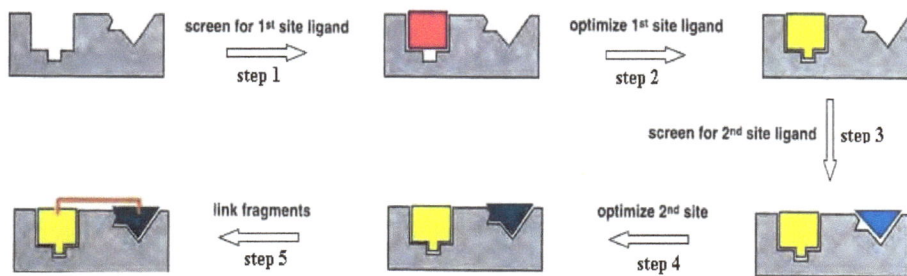

Fig. (3). General schematic outline of SAR-by-NMR method (*Adopted from [47]*).

The advantages of this method is quite clear: lack of background signals from test compounds because of ^{15}N spectral editing; applicability to any class of compound (providing that the aqueous solubility is greater than about 1 mM); concomitant identification of ligand binding site location; rapid SAR development; and simple binding assays without a rigorous functional assay. The requirements for ^{15}N-labeled target protein, backbone ^{1}H–^{15}N resonance assignments (which are typically limited to proteins with < 30 kDa), and knowledge of the target protein's structure are the main disadvantages of this technique. In addition, the 2-D ^{1}H–^{15}N HSQC experiment used by this method typically requires a longer acquisition time than various other 1-D NMR screening methods.

2.5.1.1.1. Stromelysin Inhibitor

Stromelysin belongs to the family of the matrix metalloprotease (MMP) enzymes, and involved in matrix degradation and tissue remodeling. The deregulation of stromelysin (*i.e.*, overexpression or lack of it) is associated with arthritis and tumor metastases. Although peptide-based inhibitors of stromelysin were synthesized on the basis of substrate specificity, many of them exhibited poor bioavailability [105]. Numerous natural product-based stromelysin inhibitors have been identified, including pycnidione and tetracycline derivatives. However, they are only moderately effective in inhibiting the stromelysin activity [106]. Therefore, efforts have been made to identify small molecules that can effectively inhibit the stromelysin activity [107]. However, traditional HTS of 115,000 compounds failed to identify a promising lead.

Fesik and co-workers tried to discover potent nonpeptide inhibitors by NMR-based screening approach of SAR-by-NMR (Fig. **4**) [46]. The conditions set-out in this approach were: (i) ligand molecules should be inhibiting autolytic degradation; (ii) lead compounds should be soluble enough to saturate the protein; and (iii) lead compounds should be small enough so that they are not to occlude nearby binding sites of the target protein. Based on these criteria, acetohydroxamic acid (Fig. **4**, compound **1**) was selected as the first ligand. Ligand binding was detected by acquiring sensitivity-enhanced ^{15}N-HSQC spectra on 400 μL of 0.3 mM stromelysin in the presence or absence of compound(s). Acetohydroxamic acid bound to the protein weakly with the K_D value of 17 mM. The substrate specificity of stromelysin was also used to direct the screening process. The search was focused on hydrophobic compounds that might be predisposed to bind to the largely hydrophobic S'$_1$ stromelysin subsite (Fig. **4**). Several biphenyls and biphenyl analogs with high mM to low mM affinities were identified, and limited SAR results were derived from a series of biphenyl analogs. To design linked compounds, the ternary structure of stromelysin with acetohydroxamic acid and one of the biaryl screening hits (K_D =0.02 mM) (Fig. **4**, compound **2**) was determined using 3-D NMR methods.

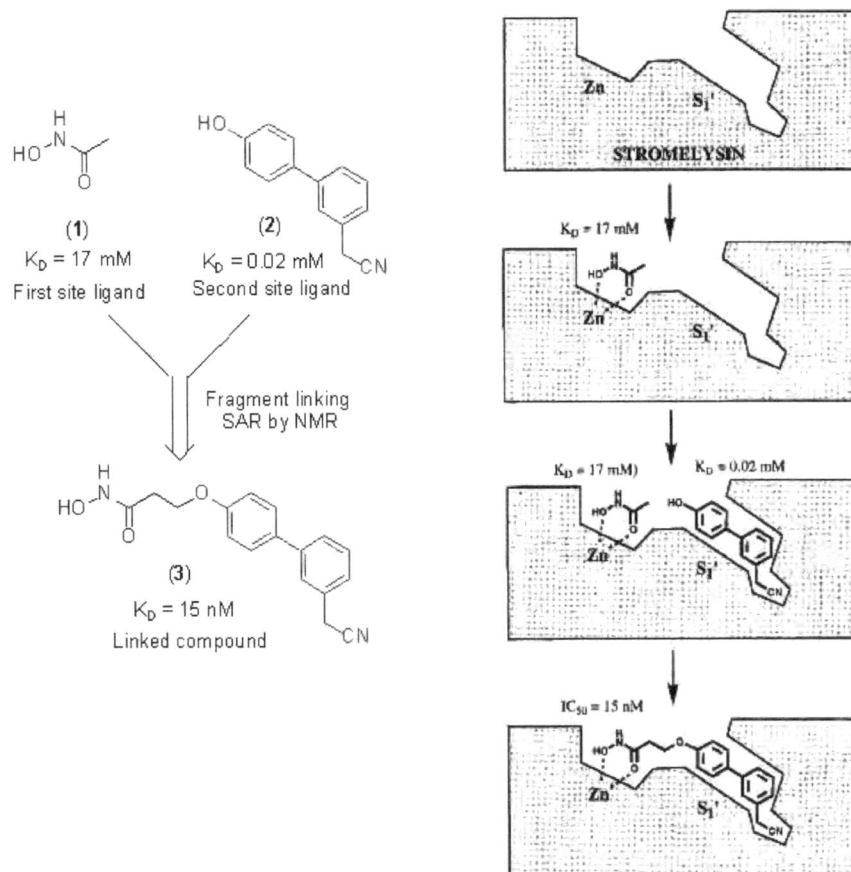

Fig. (4). Chemical structure of stromelysin inhibitors identified from a SAR-by-NMR fragment linking approach and an outline of the SAR-by-NMR approach for discovery of stromelysin inhibitors. (*Reprinted with permission from [46], copyright 1997 American Chemical Society*).

NOE data indicated that the biaryl compounds have desired binding affinity on S'$_1$ subsite, and both the compounds have the defined relative orientations with each other (Fig. **4**). Based on this structural data and the available biphenyl SAR, a series of linked compounds exemplified by Fig. (**3**), **4** were synthesized. The linker length was varied from one to four carbon atoms, with two carbon atoms found to be optimal. This linked ligand compound has an IC$_{50}$ value of 15 nM. The linked compounds bind to the protein in the expected manner, as judged by the solution structure determined for one of the binary complexes. Furthermore, they have evaluated the importance of linker length by determining the enthalpic and entropic contributions to the binding energy as a function of the number of linking carbon atoms. The results of the isothermal calorimetric (ITC) experiments showed that the presence of acetohydroxamic acid increases in the binding energy of biaryl ligands by ~ 1.3 kcal.mol^{-1}. This gain in energy is enthalpic in nature and can be attributable, at least in part, to a direct dispersion interaction between the two ligands. For the linked compounds, enthalpic contributions to the binding energy depend critically on the linker length, whereas the entropic contributions show virtually no dependence [108]. The significant gains in enthalpy observed for a compound which linked the hydroxamate to the biaryl with a two methylene (-CH$_2$CH$_2$-) bridge was not observed for compounds with longer linkers due to a difference in the position of the biaryl moiety in the binding pocket. Data from this study showed that the 2-carbon linker resulted in a significant gain in enthalpy compared to other linker lengths (3 and 4 carbon atoms), underscoring the importance of the linker in designing tethered compounds [108].

The biaryl hydroxamate inhibitor showed potent activity *in vitro* but lack of activity in the animal model, probably due to the poor pharmacokinetic properties of the alkylhydroxamate moiety [107]. To overcome this problem, the same group attempted to improve the physicochemical or pharmacokinetic profile of lead compounds by NMR-based screening of the fragments approach (Fig. **5**) [109]. To identify alternative fragments to acetohydroxamate, they screened fourteen putative zinc-binding fragments for their ability to bind to stromelysin in an NMR-based screen, in which chemical shift changes were monitored in the $^1H/^{15}N$ HSQC spectrum of stromelysin. The study suggested that 1-naphthylhydroxamate was suitable fragment for replacing acetohydroxamate. The large naphthyl group could significantly change the pharmacokinetic properties of the linked inhibitors (e.g., potentially hindering hydrolysis of the hydroxamate group or potentiating biliary excretion), thereby increasing the bioavailability of the series. In addition, 1-naphthylhydroxamate (Fig. **5**, compound **4**) ($K_D = 50$ μM) has higher intrinsic affinity, compared to acetohydroxamate ($K_D = 17$ mM). The result of the ^{13}C-edited/^{12}C-filtered NOESY experiment suggested that the naphthylhydroxamate binds to a hydrophobic pocket near the S1 subsite, leaving the S_1' subsite open to the biarylhydroxamate group. On the basis of the observed inter-fragment distance of about 5 Å, compounds containing two to five atom methylene/ether linkers were synthesized and examined. Compound **6** (Fig. **5**), the most active compound in this series, exhibited an IC_{50} of 340 nM, 6-fold weaker than the lead biarylhydroxamate **3** ($IC_{50} = 57$ nM). However, the naphthyl-biaryl-hydroxamate **6** exhibited a superior pharmacokinetic profile, compared to the biaryl hydroxamate **3**. The oral bioavailability for compound **3** was virtually nonexistent. However, the linked naphthylhydroxamate **7** exhibited a C_{max} of 28 μM and a half-life of nearly 2 h. These data clearly indicate that the naphthyl-biaryl-hydroxamate series of compounds represent lead MMP inhibitors with markedly improved bioavailability and, eventually, compound **7** (Fig. **5**) emerged as the most active one with the IC_{50} value of 62 nM [109].

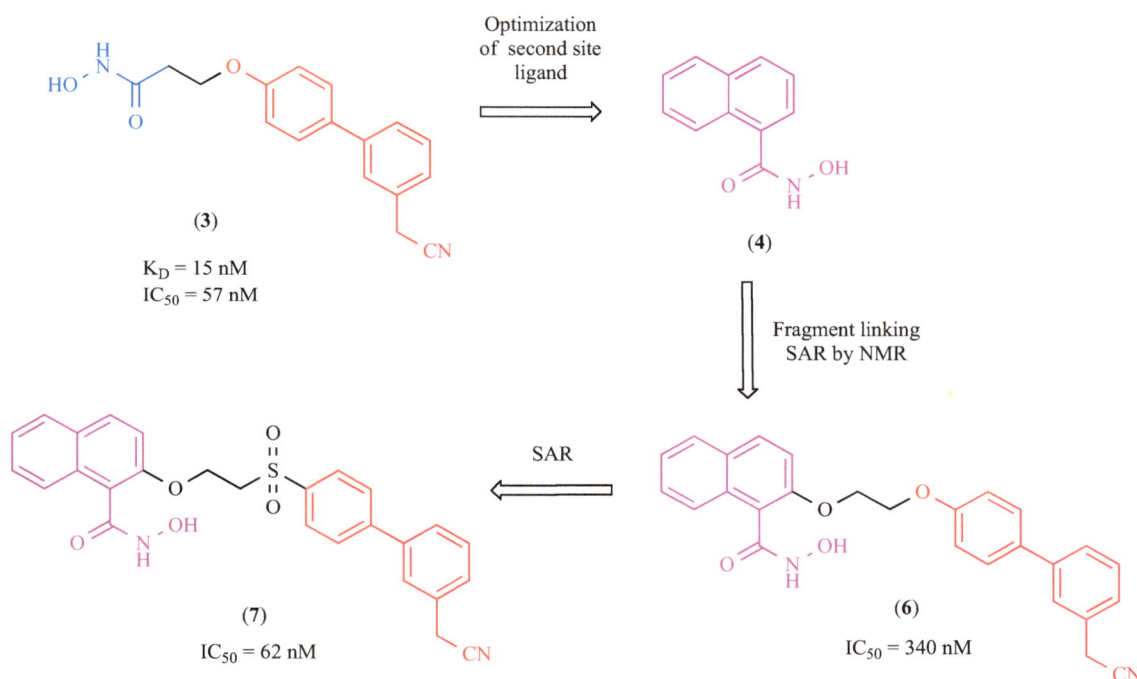

Fig. (5). Chemical structure of stromelysin inhibitors identified from a SAR-by-NMR fragment linking approach.

2.5.1.1.2. Matrix Metalloproteinase 13 (MMP-13) Inhibitors

A conceptually similar process on another MMP was studied by Chen *et al.* [110], who first identified a non-mechanism-based inhibitor of MMP-13 (also known as collagenase-3) (Fig. **6**). As mentioned earlier, the MMP family is comprised of several enzymes, where MMP-13 was recently classified as a family member on the basis of differential expression in normal breast tissues and breast carcinoma [111-113]. The MMPs are generally categorized based on their substrate specificity, where the collagenase subfamily of MMP-1, MMP-8, and MMP-13 selectively cleaves native interstitial collagens (types I, II, and III) [111-113]. Chen *et al.* [110] screened 58,079 compounds, and identified 223 hits as MMP-13 inhibitors. Based on the structural similarity and oral bioavailability profile, CL-82198 (Fig. **6**) was identified as an initial lead for further analysis. This compound showed selectivity towards MMP-13, but did not inhibit the related metalloproteases, MMP-1, MMP-9, or TACE. NMR studies revealed that the compound binds largely to the S_1' pocket of MMP-13, and does not interact with the catalytic zinc. The $^1H-^{15}N$ HSQC experiment for the MMP-13:CL-82198 complex was studied in the presence of WAY-151693 (non-specific inhibitor of MMP-13) [114]. The presence of WAY-151693 displaced all of CL-82198 as evident by the distinct differences in the ^{1}H-^{15}N HSQC spectra, suggesting that both compounds bind to the S$_1$' pocket. Further, an NMR study indicated that S$_1$' pocket of MMP-13 is deeper than that of other MMPs, providing an explanation for the observed specificity. Based on this information, Chen *et al.* [110] hybridized two hits molecules, CL-82198 and WAY-152177 (an analog of WAY-151693), into a single molecule WAY-170523 (Fig. **6**). This hybrid molecule showed high affinity for MMP-13, while maintaining selectivity against MMP-1, MMP-9, and TACE [110].

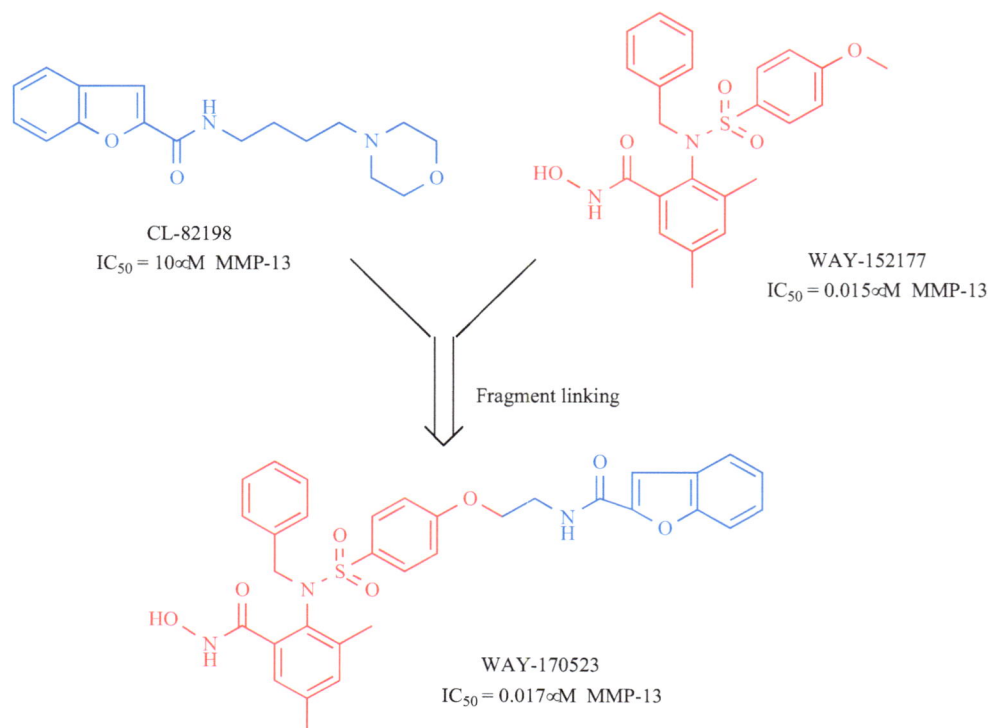

Fig. (6). Chemical structure of MMP-13 inhibitors identified by a fragment linking approach.

2.5.1.1.3. Human Papillomavirus E2 Protein Inhibitors

After failed to find an effective inhibitor of human papillomavirus (HPV) E2-DNA binding by HTS of more than 100,000 compounds, Hajduk *et al.* [115] opted for an NMR-based approach to identify ligands for the DNA-binding domain of the E2 protein (Fig. **7**). HPVs are a group of small DNA tumor viruses that infect a wide variety of mammalian cells. E2 protein, which is responsible for anogenital warts and cervical carcinomas, contains a DNA-binding domain (DBD) [116]. The E2 protein regulates the transcription of several viral genes through its DBD domain and, in conjunction with E1 protein, is required for viral replication [117,118].

The important aspect of NMR based drug discovery is the identification of lead compounds that possess high-affinity for proteins or other therapeutic targets, which can be used as starting points in the design. It is not always possible to find a suitable lead with high potency by conventional HTS of large libraries of compounds. For initial ligand compound identification, Hajduk *et al.* [115] screened a library of small organic molecules (~2000 compounds) by recording 2-D ^{15}N/^{1}H HSQC spectra of uniformly ^{15}N-labeled E2 DBD in the presence and absence of potential ligands, which led to the identification of biphenylcarboxylic acids (Fig. **7**, compound **8**) and biphenyl ether carboxylic acids (Fig. **7**, compound **9**). These compounds bound weakly to two distinct sites on the E2 protein, which resulted in significant chemical shift changes in the backbone amides of residues 305-310, 312, 313, 358-360, 364, and 371 of the protein. The initial lead ligands were further examined by a filter-binding assay to evaluate their ability to inhibit the binding of E2 to DNA, and found that compounds **8** and **9** inhibited E2-DNA interaction at the 1 mM concentration. Further structural studies were performed on the complex of E2 and compound **8** using 2-D isotope-edited

NMR to confirm the ligand binding site and provide structural basis of the inhibition of DNA binding site. The result of the NMR experiment on NOEs suggested that compound **8** bound to the hydrophobic groove between the DNA recognition helix and the adjacent loop, and the biaryl moiety made hydrophobic contacts with residues Leu306, Leu309, Leu313, Val358, Ile360, and Pro361 of the E2 protein. This site is consistent with the observed amide chemical shift perturbations and, thus, indicates that the inhibition exhibited by this class of compounds is due an occlusion of the DNA-binding site on the E2 protein.

To make further optimization, Hajduk *et al.* [115] synthesized and examined two sets of compounds for their affinity for the E2 DBD to develop a limited SAR. The authors found that 3,5-dichloro biphenylcarboxylic acid (Fig. **7**, compound **10**) ($K_D = 0.06$ mM, $IC_{50} = 0.15$ mM) and biphenyl ether analog of penta-2,4-dienoic acid (Fig. **7**, compound **11**) ($K_D = 0.35$ mM, $IC_{50} = 75$ μM) were most active among this series of compounds. Since the two initial leads bind to the same site, the SAR was merged into a single compound (Fig. **7**, compound **12**), which showed the IC_{50} value of 10 μM by a filter-binding assay. This level of activity was not found in a traditional high-throughput assay of more than 100,000 compounds. For the E2 protein target, the ability to identify weak affinity ligands for a focused chemistry effort was a distinct advantage afforded by NMR [115].

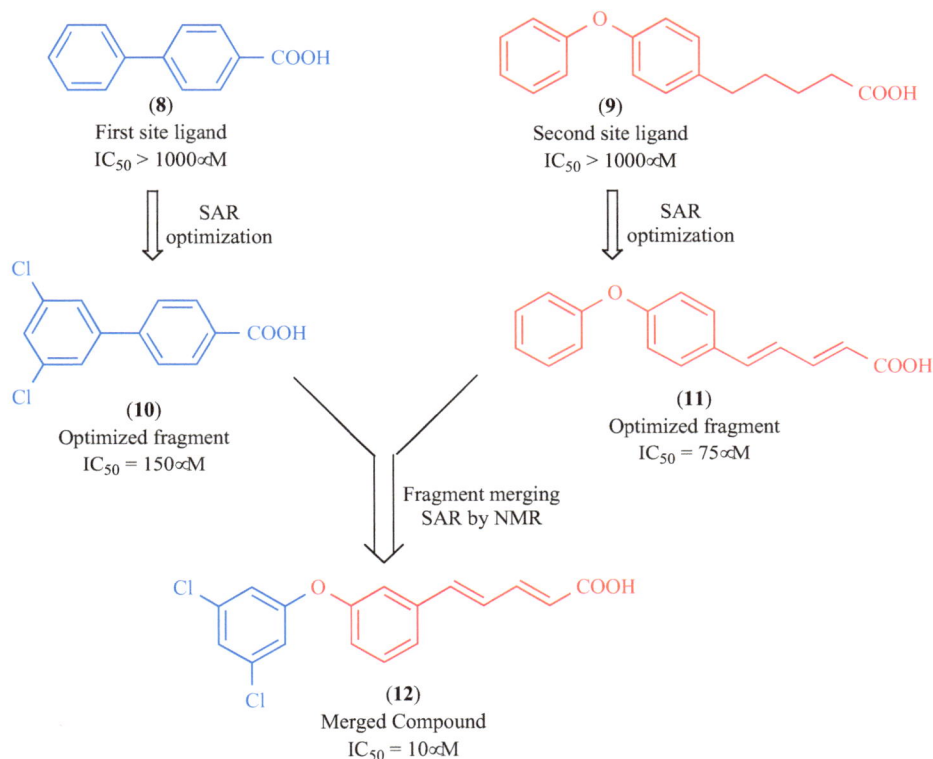

Fig. (7). Chemical structure of HPV E2 protein inhibitors identified by a fragment merging approach.

2.5.1.2. Fragment Linking by a Spin-Label Guided Approach

Spin labels have a dramatic influence on the relaxation rate of neighboring protons, and this can be exploited for the second-site screening [26,27,38]. In this technique, the ligand resonances that do not undergo line broadening or chemical shift changes upon binding to a spin-labeled protein are probably not in contact with protein, thus indicating they are possibly sites for attaching linkers. Screening for a second-site binder can be carried out in the presence of the spin-labeled first ligand, and the distance between the bound ligands can be calculated from the differential relaxation rate enhancements. This information can be helpful in deciding the attachment points for a linker. In order to find Bcl-XL inhibitors, Jahnke *et al.* carried out an NMR-based spin-label guided fragment approach (Fig. **8**) [91,119]. In parallel, an ELISA-based high-throughput screen on Bcl-XL/Bax protein interaction was also carried out, from which a biaryl compound was identified as a potential lead (Fig. **8**, compound **13**) ($IC_{50} = 180$ μM). The SAR study by medicinal chemistry could not be significantly improved the activity. To identify more potent compounds by second-site screening and subsequent linkage, compound **13** was spin labeled (TEMPO labled) at its basic amine, to yield compound **14**. This spin-labeled compound was successfully used to identify second-site ligands [91]. Paramagnetic relaxation enhancements on the protein site were measured to locate the binding site of compound **14** on Bcl-XL. For this purpose, HSQC spectra of Bcl-XL were recorded with increasing amounts of compound **14**, and the signal intensity in the HSQC spectra was analyzed. Data obtained from this study suggested that compound **13** bound nearby to the site targeted by Bak binding site. Furthermore, ITC studies carried out with the homologous protein Bcl-2 proved that two molecules of the parent compound **13** were bound to one molecule of Bcl-2. The binding ratio between the small molecule and protein target is 2:1. Based on this information, Jahnke *et al.* [119] linked the first-site (Fig. **8**, compound **13**) and second-site ligands together (Fig. **8**, compound **14**) to generate a more potent compound. The SAR showed that ionic and polar groups decreased in binding affinity, whereas hydrophobic groups could be tolerated. Therefore, the authors decided to generate a linker consisting of alkyl side chain (2- to 4-carbon atom chain length). The most potent compound was **15** with 3-carbon atoms chain length, which showed IC_{50} 10 μM as determined by a Bcl-XL/Bax ELISA assay. NMR titration of Bcl-XL with linked compound **15** led to quenching and disappearance of Bcl-XL signals close to the binding site. Unfortunately, the titration could not be performed to high excess of compound **15**, since this compound is poorly water-soluble. Therefore, the affinity of compound **15** to Bcl-XL could not be determined by NMR.

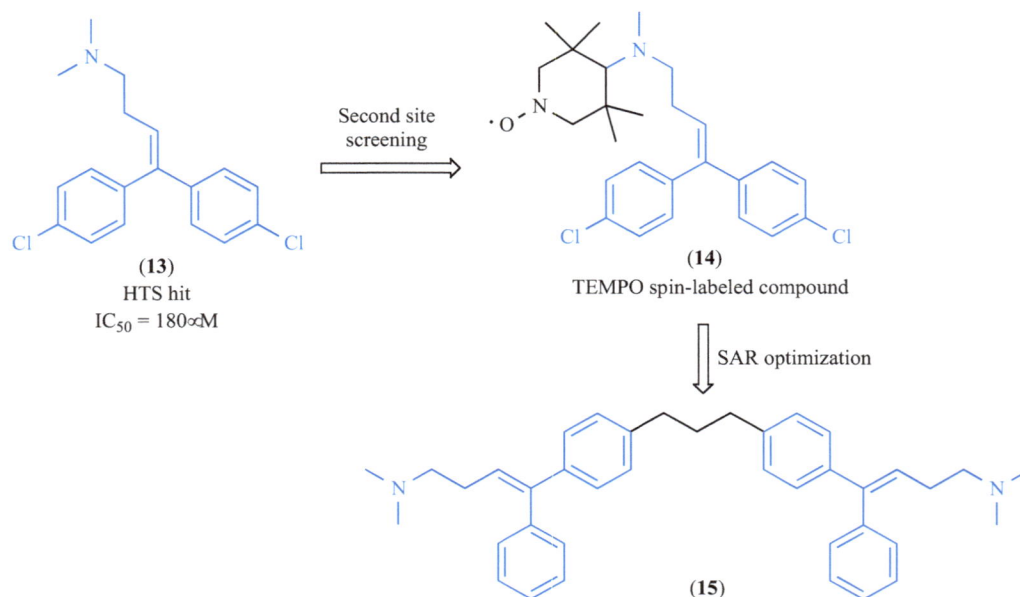

Fig. (8). Chemical structure of Bcl-XL ligands identified by a spin label-guided fragment linking approach.

Using the SLAPSTIC-based method and a reverse screening approach, Jahnke *et al.* [119] identified the 3-methylpyridine initial ligand that can inhibit the tubulin polymerization activity (Fig. **9**). To identify a second-site ligand that binds simultaneously and in the vicinity of 3-methylpyridine, a TEMPO spin label was covalently attached to 3-methylpyridine. It should be noted that the spin-labeled compound is twice as large as the original compound (Fig. **9**, compound **16**). A small library of these compounds was screened for those binding to a proximal binding site of tubulin, from which compounds **17**, **18**, and five others were identified as binding to the second site (Fig. **9**). All ligands identified bound to tubulin, even in the absence of 3-methylpyridine. However, relaxation rates were increased by adding spin label, suggesting that the second-site ligands bound simultaneously to the spin-labeled 3-methylpyridine (Fig. **9**, compound **16**).

Subsequently, a transferred NOE experiment was carried out for these seven compounds together with 3-methylpyridine and tubulin. This experiment was aimed to observe possible intermolecular NOEs between 3-methylpyridine and at least one of the second-site ligands. However, intermolecular NOEs involving resonances of two of the seven second-site ligands in the mixture were observed. This suggests that tubulin actually has at least three different and proximate sub-sites: 3-methylpyridine binds to subsite 1, and the fragments identified in the second-site screening bind to subsites 2 and 3. Based on this observation, Jahnke *et al.* [119] synthesized three linked molecules (Fig. **9**, compounds **19-21**), and their affinities for tubulin were determined by NMR T1ρ-experiments. For comparison, the results of T1ρ-experiments measuring binding affinity of epothilone A, a natural ligand that binds to tubulin at the paclitaxel binding site, were included. Two of the three linked compounds, **20** and **21**, bound to tubulin significantly better than epothilone A. Binding to the target protein at the low micromolar range was observed by severe line broadening. This example shows how ligands for a second and even a third binding site can be successfully identified by the spin-label approach. Furthermore, the observation of interligand NOEs provides valuable information for the designing of an optimal linker, even in the absence of a reliable molecular model of the ternary complex.

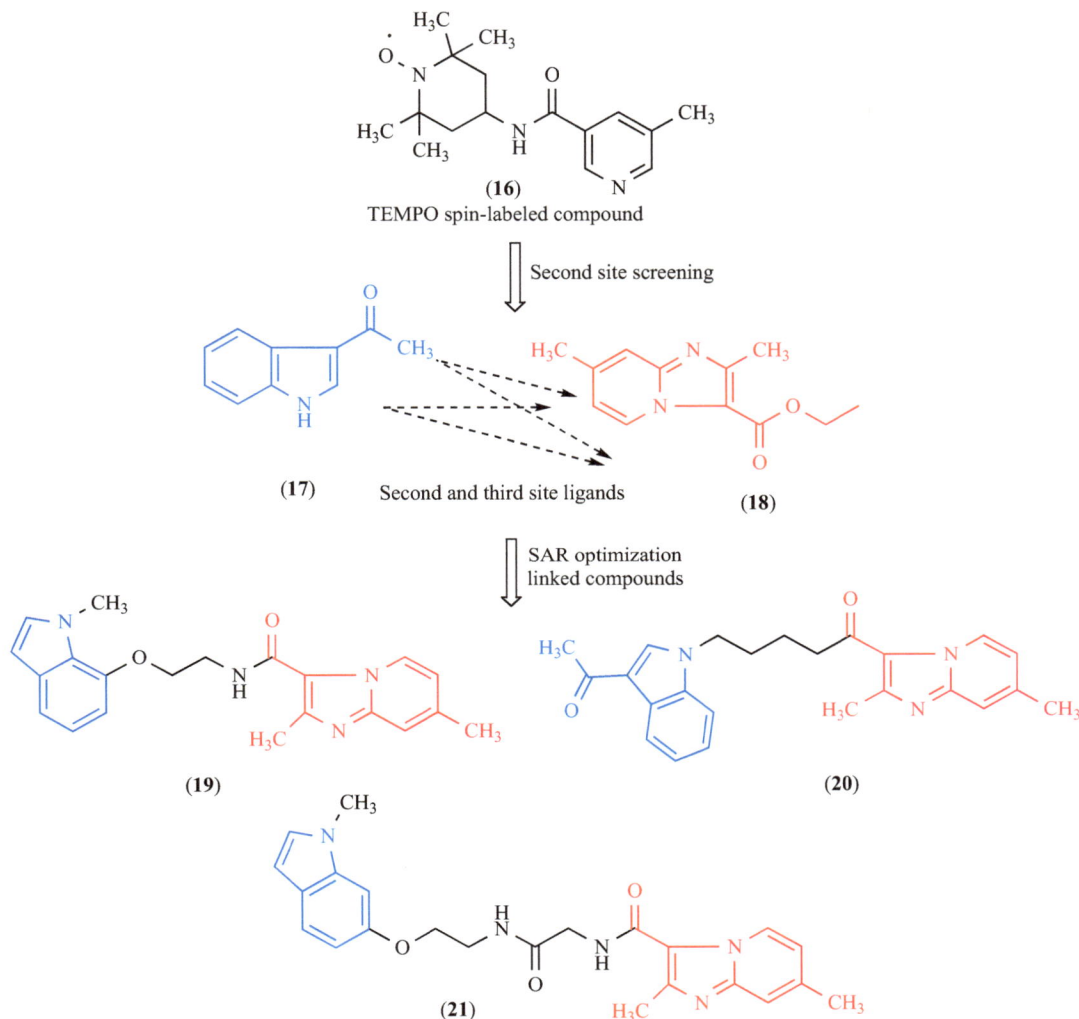

Fig. (9). Chemical structure of tubulin polymerization inhibitor identified by a spin label-guided fragment linking approach.

2.5.2. Fragment Evolution

The fragment evolution approach requires that a primary screening library comprised of small and simple molecules that can be further elaborated in lead generation. Fragments are selected on the basis of containing functional groups or ring systems that are capable of making additional interactions with a target, thereby, increasing potency without disrupting the binding of the core scaffold (Fig. **10**). The elaborated compounds are screened either by an enzymatic assay or a biophysical technique that can measure binding affinity when an enzymatic assay is not available. Iterative screening and follow-up compound generations are carried out until a low micromolar inhibitor is discovered. SHAPES [59] and needle screening strategies [117,120] are good examples of this approach. The SHAPES library contains commercially available drug-like scaffolds that are used to bias the selection of follow-up compounds.

Needle screening uses small molecular probes (< 300 Da), as they should be able to penetrate into the subpockets of active sites. The needle library is designed for each target using molecular modeling to determine minimal structural elements predicted to make essential contacts with the protein. SAR and structural information of the bound leads are often used to direct lead optimization.

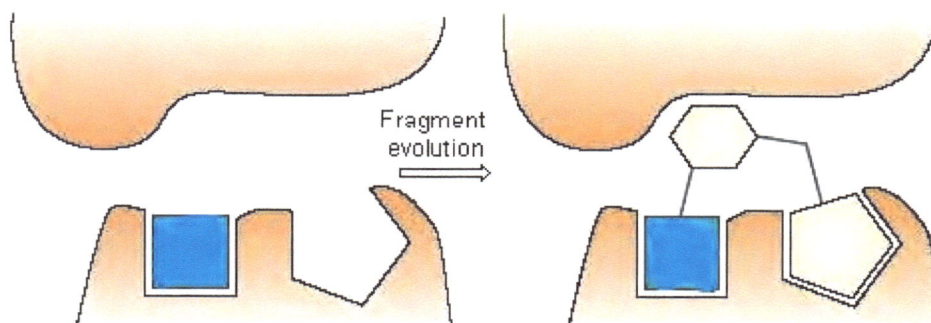

Fig. (10). A schematic sketch for a fragment evolution approach. (*Reprinted with permission from [40], copyright 2004 Nature Publishing group a division of Macmillan Publishers Limited*).

2.5.2.1. c-myc Inhibitors

The deregulation of the c-*myc* protooncogene is one of the most common genetic abnormalities associated with cancer [121]. The c-*myc* pathway remains central in tumor progression, and is an important therapeutic target for the treatment of solid tumors [122]. A small molecule-mediated disruption of the interaction between DNA and the helix-loop helix-leucine zipper domain of c-*myc* is found to be quite daunting. This is, in part, because a pocket suitable for small-molecule binding is not apparent on the c-*myc* DNA binding domain. A potentially effective alternative target is the trans-activating domain of c-*myc*. Also, as the interacting proteins with c-*myc* become well-understood [123], inhibition of the downstream effectors of c-*myc* may yield potent therapeutics. Yet, another approach may be to modulate the upstream pathways that control c-*myc* expression or function. It turned out that the far-upstream element (FUSE) and far-binding protein (FBP) are essential for c-*myc* expression. A low level of FBP resulted in a decrease in c-*myc* expression and cell growth. FBP binds to single-stranded DNA (ssDNA) by capturing exposed DNA bases in a hydrophobic pocket. This suggests that a small molecule could be designed to occupy this pocket and inhibit FBP-mediated c-*myc* expression.

Based on this hypothesis, Huth *et al.* [124] screened small molecules that can inhibit DNA binding of FBP, by targeting its ssDNA binding pocket (Fig. **11**). The authors identified six potential ligands that bind to FBP3/4 from screening 929,000 compounds using three different approaches: (i) HTS-NMR, (ii) virtual ligand screening, and (iii) affinity selection/mass spectrometry (ASMS). All of the six compounds contained acidic moieties that may mimic the phosphate backbone of ssDNA, and interacted favorably with the positively charged residues on FBP3/4. Since two of the six compounds possessed a good aqueous solubility profile, the binding affinity of FBP3/4 was determined by NMR. Compound **22** ($K_D = 0.35$ mM) contained two carboxylic acid groups, while compound **23** ($K_D = 1.7$ mM) contained a tetrazole group that might serve as an isostere for a carboxylate. NOE studies performed on a complex of compound **22** and the KH3 domain of FBP. NOEs from the ligand to Ile18, Ile25, Ile36, and Phe38 of the protein confirmed that the ligand bound to the ssDNA binding site. Furthermore, the data suggested that the ligand binding surface was composed of helices R1 and R2 as well as a $\beta 2$ strand. The carboxylate group of ligand mimics the phosphate group of the ssDNA while the other carboxylate group mimics the O4 atom of one T nucleotide, and interferes with substrate binding by preventing the interactions of the central DNA bases in the core FBP recognition sequence. However, further optimization of this compound by SAR yielded several very active analogs. For example, compound **24** contains benzyl groups linked in a *para* orientation from the central phenyl ring, whereas compound **22** is in the *meta* orientation. Compound 25, which contains single carboxylic acid functionality, binds to FBP3/4 with a K_D of 0.6 mM, suggesting that single carboxylic acid functionality is required for binding (Fig. **11**) [124].

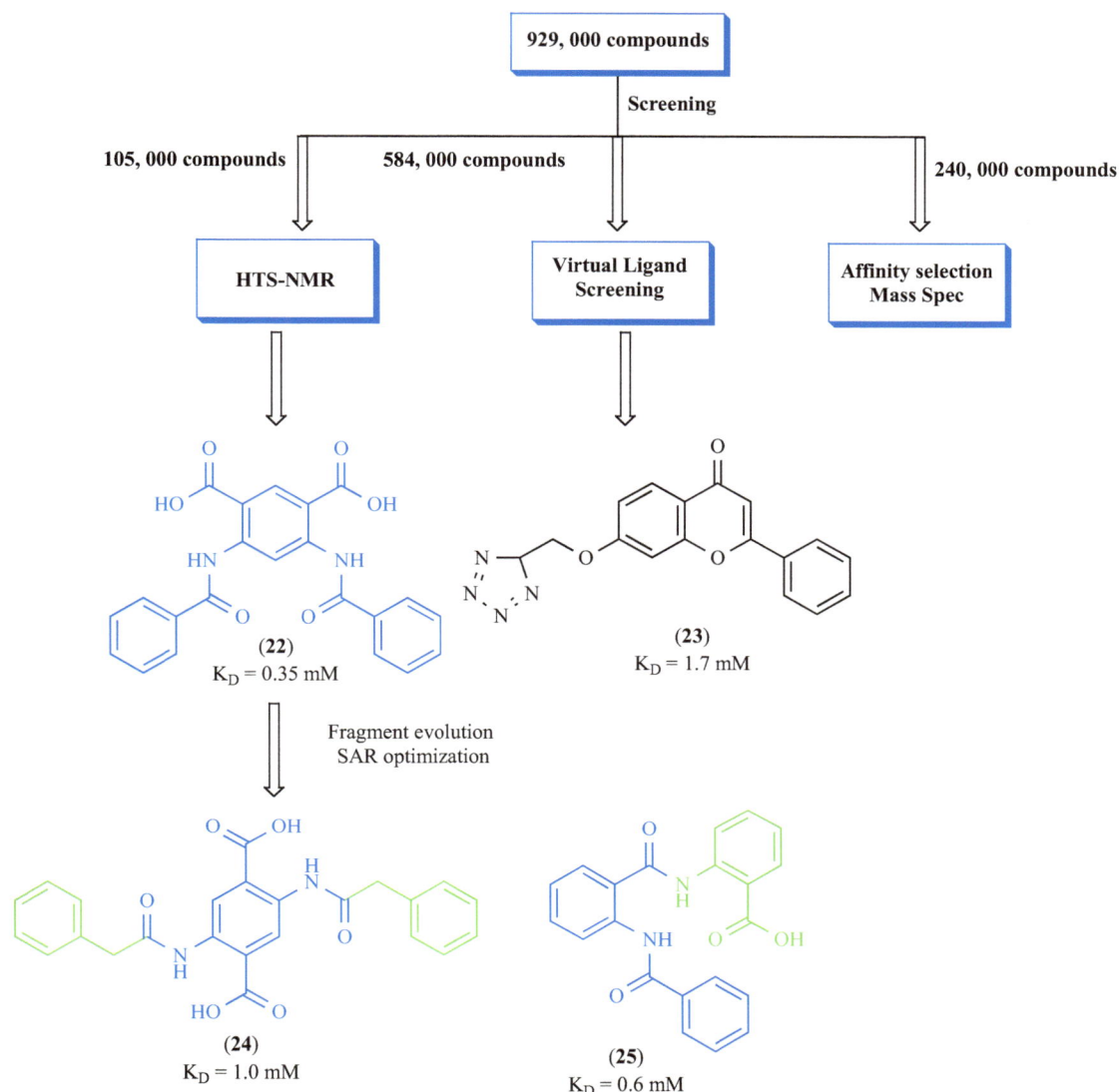

Fig. (11). Chemical structure of FPB3/4 inhibitors identified by a fragment evolution approach.

2.5.2.2. Survivin Inhibitors

Survivin, a member of IAP family (inhibitors of apoptosis proteins), plays an important role in cell division, and shows a pronounced accumulation at mitosis. The elevated level of survivin is often associated with a decrease in the cell-death signaling pathway, tumor development, poor prognosis, and shorter patient survival rates [125-127]. Thus, survivin is an attractive cancer therapeutic target [127].

Structure of survivin has been determined by NMR and X-ray crystallography: it contains a zinc-binding fold similar to that found in other members of the IAP family and a long amphipathic α-helix C-terminal to its sole BIR domain [128,129]. The interface of the dimer is composed of a several hydrophobic residues from its N-terminus as well as residues connecting its BIR domain to the C-terminal helix. However, neither the biochemical mechanism of survivin is well-understood, nor a robust functional assay has been established. Therefore, it is currently not possible to identify lead compounds that can inhibit survivin by a HTS method. To overcome this problem, Wendt *et al.* [130] used HTS-NMR and ASMS to screen 257,000 compounds. As a result, they identified two ligands (Fig. **12**). The ^{1}H/^{13}C-HSQC spectral data showed two distinct chemical shift perturbation for the lead compounds **26** and **27**. Compound **27** (K_D = 75 μM) shifted residue Leu64, suggesting that this compound binds at or near the Hid/Smac binding site on the protein. Compound **26** (K_D = 5.7 μM) shares roughly a similar L-shaped aromatic structural motif;

however, a new set of residues including Leu98, Leu6, and Leu14, was also shifted. This may suggest that they have different binding sites, and the affected residues by compound **26** may be located near the interface of the two survivin monomers. Competition binding experiments suggested that the two compounds could bind simultaneously to the protein. For further optimization, the authors synthesized several analog of compound **26** by a SAR-based approach, and found that compounds **28** ($K_D = 0.086$ µM) and **29** ($K_D = 0.037$ µM) were the most active. The HSQC spectrum of compound **28** showed perturbation of Leu64, which is part of the BIR binding site. Furthermore, NOEs were observed between the amide N-methyl and residues Gln92, Val89, Lys90, Phe93, and Leu96, as well as between the piperidine acetyl methyl and residues Ala41, Ile74 and Leu87, positioning this group in the extended cleft adjacent to the C-ring.

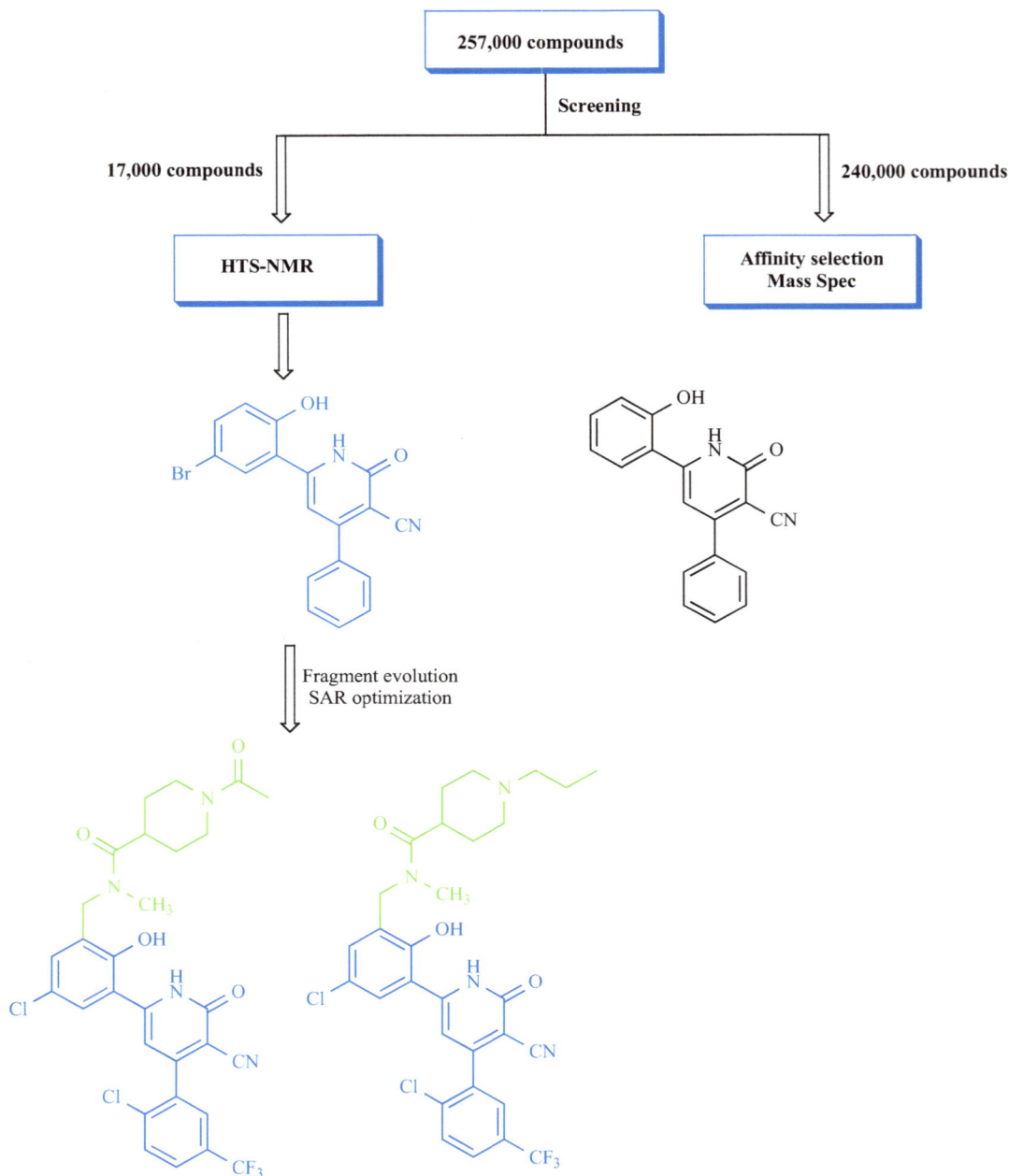

Fig. (12). Chemical structure of survivin inhibitors discovered by a fragment evolution approach.

2.5.2.3. B-cell Lymphoma Protein Inhibitor

The B-cell lymphoma (Bcl) family of proteins, which comprises both antiapoptotic members (e.g., Bcl-2, Bcl-xL, and Bcl-w) and proapoptotic members (e.g., Bak Bax, and Bad), are a group of α-helical proteins with extensive sequence and structural similarity. The antiapoptotic proteins Bcl-2 and/or Bcl-XL are overexpressed in many different types of tumors, and contribute to tumor initiation, progression, and resistance to cancer therapies [131-133]. Therefore, Bcl-2 and Bcl-XL are attractive targets of anticancer agents. Using an NMR-based fragment linking and evolution approach, Fesik and colleagues [134-137] developed ABT-737, a small-molecule that inhibits Bcl-2, Bcl-XL, and Bcl-w antiapoptotic proteins (Fig. **13**). In the study, the authors screened commercially and proprietarily available chemical libraries to identify small molecules that bind to the hydrophobic BH3-binding groove of Bcl-XL [135]. 4-Fluoro-biphenyl-4-carboxylic acid (Fig. **13**, compound **30**) (K_D = 300 µM) and 5,6,7,8-tetrahydro-naphthalen-1-ol (Fig. **13**, compound **31**) (K_D = 4.3 mM) bound to Bcl-XL [135]. Their data revealed that the carboxyl group of compound **30** was crucial for effective binding, as compounds lacking the carboxyl were less active. The structure of the Bcl-XL/**30**/**31** complex revealed two hydrophobic pockets separated by the Phe97 side chain. The binding sites for compounds **30** and **31** correspond to the region of the protein for binding to Bad and Bak. (For a matter of convenience, the binding sites for **30** and **31** will be referred to sites 1 and 2, respectively.) The SAR by the NMR approach is based on the linkage of proximal fragments to achieve high-affinity binding. Substitution of an acylsulphonamide for the biphenyl carboxyl group maintained the correct positioning of the acidic proton while providing an optimal trajectory to the site 2 that avoids steric interference by Phe97. Site directed parallel synthesis led to the discovery that a 3-nitro-4-(2-phenylthioethyl) aminophenyl group (Fig. **13**, compound **32**) spans the binding sites and efficiently occupies the site 2 through hydrophobic collapse and subsequent π-π stacking.

To further optimize the binding to the site 2, Petros *et al.* [135] synthesized a library of nitrophenyl acylsulfonamides (125 compounds), among which compound **33** was identified with improved affinity for Bcl-XL (K_D = 0.036 µM). The NMR-based structural analysis of compound **33** bound to Bcl-XL showed that the position of the thiophenyl moiety allows for the enhancement of the putative π-stacking with Phe97 and Tyr194. Thus, starting from two weakly binding ligands (Fig. **13**, compound **30**, K_D = 300 µM and compound **31**, K_D = 4.3 mM), SAR-by-NMR was successfully applied to develop a small molecule that can inhibit Bcl-xL at a nanomolar range. However, compound **33** showed reduced affinity for Bcl-XL in the presence of 1% human serum (K_D = 2.5 µM). Interestingly, human serum albumin (HSA) was later identified as a competitive binder for compound **33**. The NMR-based structural analysis of the HSA/**33** complex showed that the biphenyl moiety interacted with HAS in the deep hydropobic pocket that was inaccessible to solvent. However, the analogous site 1 of Bcl-XL was observed to be solvent-accessible. Further, while the HSA binding site did not appear capable of accommodating any further steric bulk at the biphenyl, the hydrophobic site 1 of Bcl-XL appeared to have space for an extra substituent on the biphenyl. The binding mode of the 2-(phenylthio)-aminoethyl moiety was also examined. The HSA binding site was hydrophobic and solvent-inaccessible at this position, whereas the site 2 of Bcl-XL was solvent-accessible and, thus, would allow for functionalization at the ethyl carbons.

Based on these observations, a library of functionalized biphenyl compounds was created to gain greater efficiency on altering the steric environment and polarity of the biphenyl. Consistent with the observation of the deep hydrophobic pocket of HSA, charged R groups were most effective for reducing HSA affinity. Compound **34** was found to bind to Bcl-XL with K_i = 0.01 µM, and with K_i = 0.6 µM in the presence of Bcl-XL and 1% HSA. The morpholine moiety was then incorporated into the derivatization of the α-methylene of the 2-(phenylthio)-aminoethyl group. While this introduced a chiral ligand, the *R* enantiomers were found to have higher Bcl-XL affinity, consistent with the position of the diastereotopic methylene protons in the Bcl-XL/**33** complex. Compound **35** showed lower affinity for HSA, with K_i = 0.36 µM in the presence of Bcl-XL and 10% HSA. Surprisingly, compound **35** inhibited Bcl-XL with K_i = 8 nM, an improvement of two orders of magnitude over compound **34**.

Verified the therapeutic potential of targeting both Bcl-XL and Bcl-2, Fesik and colleagues [136] designed a compound for binding to Bcl-2 without a decrease in Bcl-XL affinity. Bcl-2 and Bcl-X differ by only three residues in their binding sites and the key difference exploited by Fesik *et al.* [135] was the presence of Met108 in Bcl-2, which had greater conformational flexibility in the site 1 groove than the corresponding Leu108 in Bcl-XL. The conformational flexibility of Met108 allowed the small molecule to gain access to the deep hydrophobic groove of Bcl-2 that was not observed in the structure of Bcl-XL. NMR-based structural studies of a ligand containing site 1 phenethyl-substituted benzo-thiazole docked

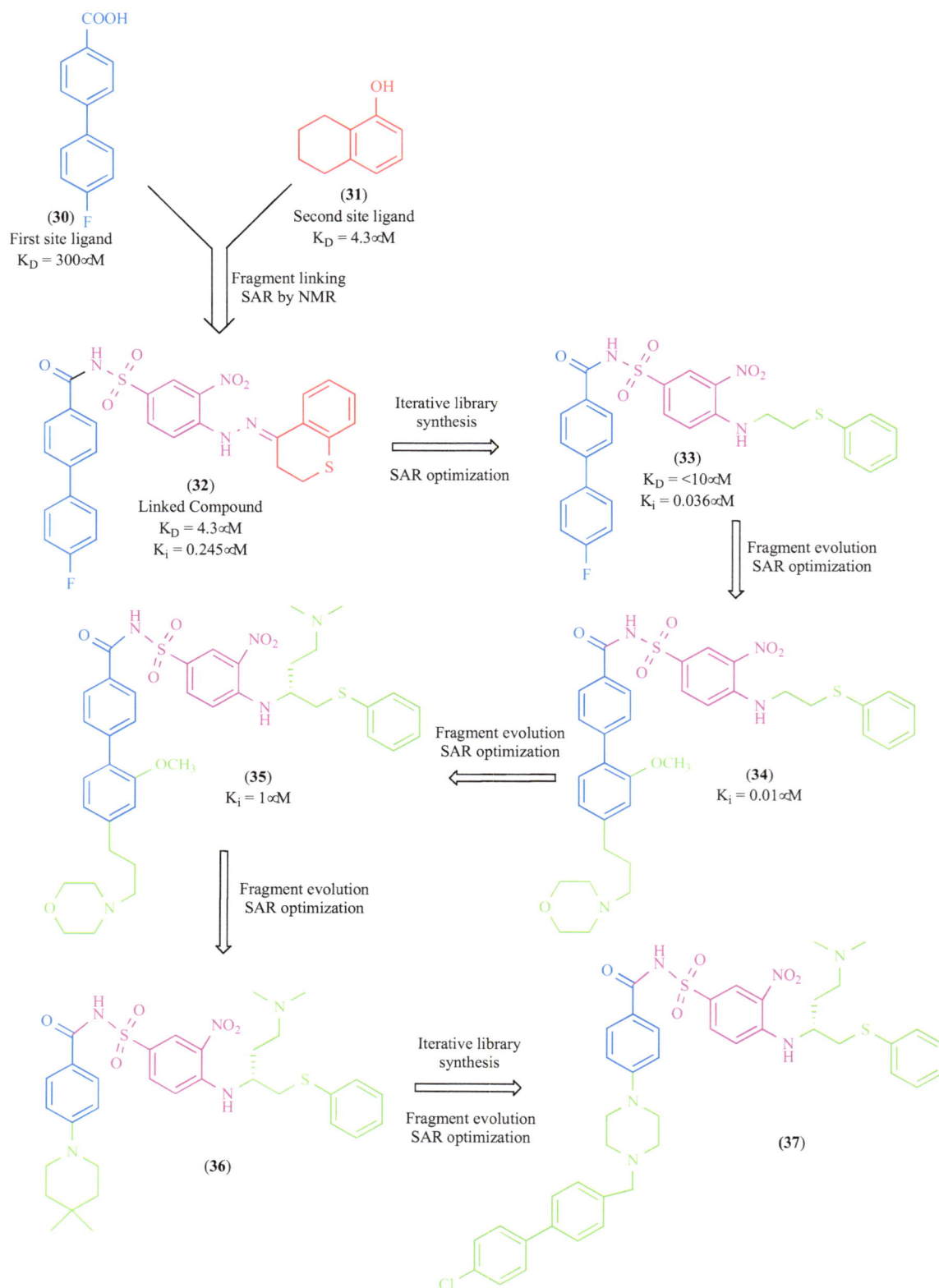

Fig. (13). Chemical structure of B-cell lymphoma protein inhibitor identified by a fragment linking and evolution approach.

with Bcl-2 and Bcl-XL revealed that the phenethyl moiety occupied the hydrophobic groove of Bcl-2 while maintaining a surface orientation on the face of the Bcl-XL binding site. Consequently, Fesik *et al.* [134] surmised that the site 1 fragment could be modified to enhance Bcl-2 binding without reduction of affinity for Bcl-XL. Fesik *et al.* [136] had previously recognized compound **36** contained a site 1 N-phenyl piperidine in the place of the biphenyl analog compound **35**, which inhibited Bcl-2 with K_i = 67 nM. However, screening of several analogous N-phenyl compounds revealed that a piperazine ring allowed for better Bcl-2 affinity, while maintaining Bcl-XL affinity. Among this line of compounds, compound **37** was the most active with K_i < 1 nm for Bcl-2 and K_i < 0.5 nM for Bcl-XL (Fig. **13**) [134-137].

2.5.2.4. Heat Shock Protein 90 Inhibitors

Numerous oncoproteins including Cdk4, Akt, BCR-ABL, mutated p53, and v-src are known to be patron proteins of the heat shock protein 90 (HSP90); thus, inhibitors of this class would have broad effects across many tumor types [138,139]. Huth *et al.* [140] synthesized novel HSP90 inhibitors by an NMR-based parallel fragment drug design (Fig. **14**). For identification of an appropriate initial ligand, the authors screened a library of 11,520 compounds with an average molecular weight of 225 Da. The screening was carried out by targeting the N-terminal domain of human HSP90 by the NMR based fragment evolution and linking approach (Fig. **14**). The chemical shift changes of Leu, Val, and Ile methyl groups in the presence of compounds were detected by the $^1H/^{13}C$ HSQC approach [34]. The study resulted in identification of two initial ligands, the first of which was aminotriazine (Fig. **14** (a), **38**) with a K_D = < 10 μM, K_i = 0.32 μM, and binding efficiency index (BEI) 21. The other ligand aminopyrimidine (Fig. **14** (a), **39**) has the value of 20 μM, 18 μM, and 27, K_D, K_i and BEI, respectively. In that study, K_i was determined by a Fluorescence Resonance Energy Transfer (FRET) assay. The authors took a FBDD approach to improve the binding affinities of these two lead compounds. The naphthyl-substituted aminotriazine (Fig. **14** (a), **38**) was found to induce a conformational change, which resulted in opening up a larger binding site, although the key hydrogen bonds from the triazine ring to the water molecules and D79 were unaffected. The exocyclic amino groups of compound **39** and the adenine ring of ADP both formed hydrogen bonds with the side chain of D79 as well as with the water molecule. The data obtained from this study suggested that compound **38** bound to an open confirmation where a larger binding pocket is formed. To further improve the binding affinity of compound **39**, several aromatic substitutions were carried out at the four positions of the aminopyrimidine. Among this series of compounds, **40** was emerged as the most active compound with K_i = 0.32 μM and BEI = 27. To identify a novel substituent for the adjacent site, a library of 3,360 compounds with an average molecular weight of 150 Da was screened by 2-D NMR. In this screening, the authors specifically looked for the ability of compounds binding to HSP90 in the presence of saturated compound **39**. The most potent hit identified in this screen was compound **41**, which contained a furanone moiety and bound to HSP90 with a K_D value of 150 μM in the presence of compound **39**. The binding of **41** was found to be co-operative, as the observed K_D was >5,000 μM in the absence of compound **39**.

In the structure-based fragment-linking strategy of furanone (Fig. **14** (b), **41**) to aminopyrimidine (Fig. **14** (b), **39**), the ternary structure of the complex formed by these compounds with HSP90 was solved by ^{13}C-edited and ^{12}C-filtered NOESY spectrum. It was found that binding of HSP90 by compound **41** did not change the conformation of the binding site for compound **39**. Interactions between the furanone moiety of compound **41** and the Leu103, Leu107, Phe138, and Val150 of HSP90 residues were observed by NOEs. Further, the data from these studies also confirmed that compounds **39** and **41** could bind to both open and closed conformations of the HSP90 N-terminal domain. Based on this geometric orientation, a methylsulfonamide linker was suggested to link the aminopyrimidine to the *m*-position of the furanone phenyl ring to generate compound **42**, which showed 10 times greater potency and favorable interactions for the closed confirmation between HSP90 and the furanone at the linked state. Medicinal chemistry elaboration was performed to find an active molecule for the open confirmation, in which an acetylene linker from the aminopyrimidine to the *o*-position of the phenyl ring of 3-(benzylideneamino)oxazolidin-2-one (Fig. **14** (b), **43**, an analog of compound **41**) satisfied this geometry with HSP90. This modification resulted in a five-fold increase in potency. X-ray crystallographic studies confirmed that this inhibitor binds to HSP90 in the open conformation [141].

(a)

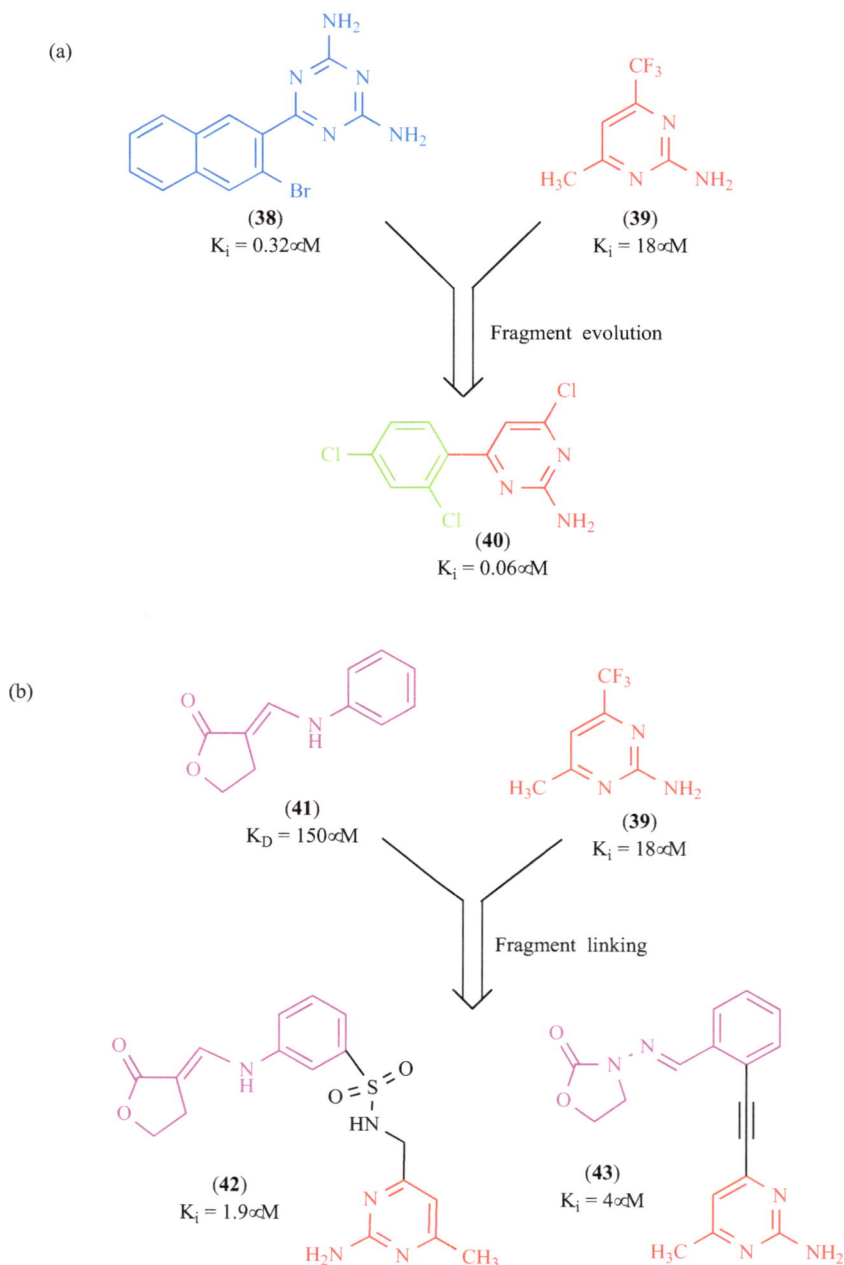

(38)
$K_i = 0.32 \propto M$

(39)
$K_i = 18 \propto M$

Fragment evolution

(40)
$K_i = 0.06 \propto M$

(b)

(41)
$K_D = 150 \propto M$

(39)
$K_i = 18 \propto M$

Fragment linking

(42)
$K_i = 1.9 \propto M$

(43)
$K_i = 4 \propto M$

Fig. (14). Chemical structures of HSP90 inhibitors discovered by (**a**) fragment evolution and (**b**) fragment linking approaches.

2.5.3. Fragment Optimization

Fragment optimization strategies usually focus on screening hits with low micromolar concentrations, or on improving affinities for core scaffold. In the latter case, simpler fragments may be added and productively optimized. Although the fragments added may have low intrinsic affinities for target proteins, they typically possess binding specificity sufficient to serve as viable anchors for subsequent derivatization by SAR (Fig. **15**) [40]. Further, the potency of the core fragment can also be directly optimized by a conventional medicinal chemistry approach. The following example shows how a fragment can be successfully optimized to a more effective lead compound for cancer therapy.

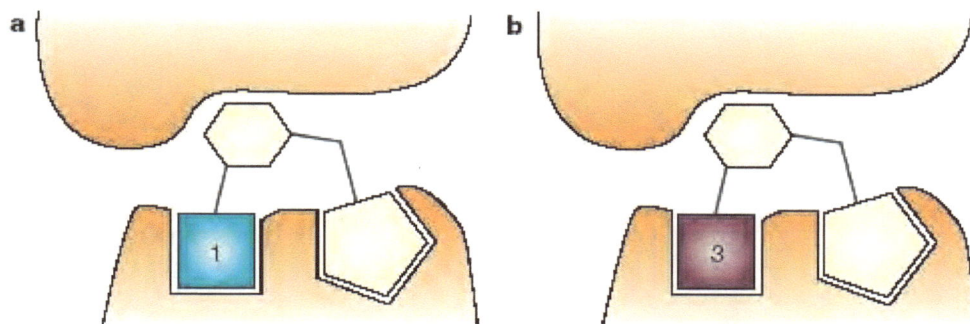

Fig. (15). A schematic outline of fragment optimization approaches: (**a**) The existing lead molecule discovered by a FBDD approach; and (**b**) further optimization by a medicinal chemistry SAR approach. (*Reprinted with permission from [40], copyright 2004 Nature Publishing group, a division of Macmillan Publishers Ltd.*).

2.5.3.1 Urokinase Inhibitors

Hajduk *et al.* [50] described an NMR-based discovery of 2-aminobenzimidazoles as novel inhibitors of urokinase (Fig. **16**). The serine protease urokinase, which is also called urokinase-type Plasminogen Activator (uPA), plays a central role in the cascade mechanism leading to basement membrane degradation and tumor metastasis [142]. Urokinase is synthesized as a zymogen form (*i.e.*, prourokinase or single chain urokinase), and is activated by proteolytic cleavage between Leu158 and Ile159. Elevated levels of urokinase and several other components of the plasminogen activation system are found to be correlated with many different types of maligancay, including tumors in breast, lung, bladder, stomach, cervix, kidney, and brain [143]. Therefore, there is immense medical interests in the development of potent urokinase inhibitors that can serve as cancer therapeutic agents. All of the known small molecule-based urokinase inhibitors contain either an amidine or guanidine group with pKa values greater than 9.0. The higher pKa of these inhibitors limits the oral bioavailability in the biological fluid, underscoring the urgent need of developing alternative small molecule-based urokinase inhibitors with favorable pharmacokinetic properties. Hajduk *et al.* [50] prepared ^{15}[N]urokinase from mammalian cells using selected ^{15}N-labeled amino acids as precursors. Ligand binding was detected at 30 °C by acquiring sensitivity-enhanced ^{15}N HSQC spectra on 400 µL of protein sample in the presence or absence of compounds to be screened, which were added as solution in perdeuterated DMSO. Since urokinase is a protease, autolytic degradation of the protein had to be inhibited during the screening process. Therefore, the autolytic protease inhibitor phenylguanidine (Fig. **16**, compound **44**) was added in the reaction mixture. Under these conditions, more than 3,000 compounds were examined for binding selectively to ^{15}N-labeled urokinase, by monitoring changes in the amide chemical shifts of the protein upon addition of the test compounds. NMR-based analysis of changes in the chemical shift suggested that 2-aminobenzimidazole (Fig. **16**, compound **45**) displaced phenylguanidine (Fig. **16**, compound **44**). Further NMR binding experiments confirmed that 2-aminoben-zimidazole (K_D = 50 µM; IC_{50} = 200 µM) and phenylguanidine bind competitively to the same site on urokinase. It should be noted that the commercially available analogues (e.g., benzimidazole, benzoxazole, and benzotriazole) did not bind to urokinase up to 5 mM, clearly indicating the critical importance of the aminoimidazole moiety for urokinase activity. Lead optimization by medicinal chemistry SAR of this molecule (Fig. **16**, compound **45**) yielded a new template compound (Fig. **16**, compound **46**), which showed a suitable pKa value, and also lowered the IC_{50} value to 10 mM, The pKa value of ~ 7.5 for compound **46** indicates that it may be uncharged at physiological pH and, thus, provides a better chance of making more efficient anticancer therapeutics.

The X-ray crystallography study of 2-amino-5-hydroxybenzimidazole (Fig. **16**, compound **46**) complexed with urokinase revealed that the electrostatic and hydrophobic interactions stabilized complex formation. Further, this may also increase the potency of other inhibitors by accessing the nearby subsites. Attaching a phenyl group to fill this pocket improved its potency by 70-fold [140]. The resulting compound is orally bioavailable and represents a novel scaffold for constructing urokinase inhibitors with improved pharmacokinetic properties.

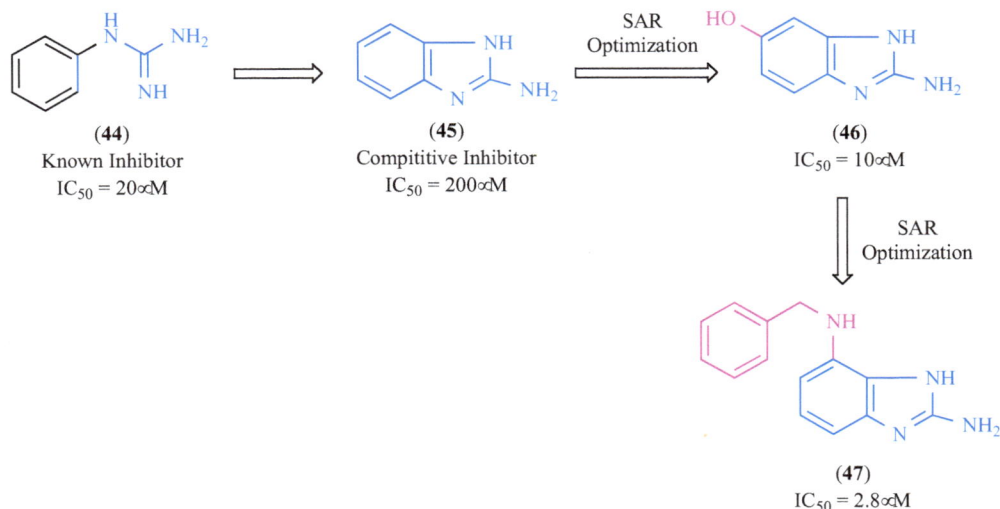

Fig. (16). Chemical structures of urokinase inhibitors discovered by a fragment optimization approach.

2.5.3.2. X-linked Inhibitor of Apoptosis Protein

Huang *et al.* [144] recently reported that the fragment optimization-based design and synthesis of small molecule for X-linked inhibitor of apoptosis protein (XAIP) (Fig. **17**). The XIAP Bir3 domain (baculovirus IAP repeat 3) binds directly to the N-terminal of caspase-9, thus, inhibiting programmed cell death [145-147]. This interaction can be displaced by SMAC (second mitochondrial activator of caspases), for which the N-terminal tetrapeptide region (NH2-AVPI, Ala-Val-Pro-Ile) is responsible [147,148]. In a fragment optimization approach, Huang *et al.* [144] replaced individual amino acid residues in a peptide with a non-amino acid chemical molecule by iterative manner. A virtual compound library was obtained by coupling the selected amino acid with low molecular weight, drug-like, synthetically accessible scaffolds. Subsequently, the library elements were docked against the target to select those compounds with the best fit for the binding site. Following chemical synthesis of top scoring compounds, these chemicals were experimentally examined by NMR spectroscopy. The use of NMR is pivotal to this approach, given that only high-micromolar binders are expected at this stage. Hit compounds are subsequently used for a second round of *in silico* derivatization, followed by synthetic medicinal chemistry of top scoring compounds. The approach can be repeated until desired potency is achieved. Huang *et al.* [144] created and screened an initial virtual library containing 1,493 derivatives of L-alanine, a critical amino acid in SMAC peptides. Based on docking results, the authors selected, synthesized, and examined by NMR the 15 most active compounds for their ability to bind to the Bir3 domain of XIAP [144]. The authors chose compound **48** with a weak inhibitor activity (K_D = 200 μM) for further studies. Their studies showed that compound **48** possessed a similar binding property with the SMAC peptide, particularly in mimicking the interactions by the first three amino acids in AVPI. Therefore, a modification of compound **48** at position 2 of the 4-phenoxybenzene scaffold was proposed in a second iteration. The selection of an additional scaffold mimicking the interactions would be provided by the isoleucine residue of AVPI, into the P2 subpocket. Based on this hypothesis, a second virtual library of compound **48** derivatives (about 900 compounds) was designed, and ranked by *in silico* docking method. Subsequently, the top scored compound **49** (Goldscore was 63.0) was synthesized and examined by NMR, which led to the synthesis of compound **50**. Compound **50** bound to the surface of Bir3 by occupying each of the two subpockets that are occupied by the SMAC peptide AVPI. This was subsequently confirmed by a second docking study and the NMR chemical shift mapping. Compound **50** showed tightest binding affinity (K_D = 1.2 μM) to Bir3 domain of XIAP, which was further confirmed by an isothermal titration calorimetry (ITC) study. Moreover, compound **50** showed cellular activity as an apoptotic inducer (IC$_{50}$ = 16.4 μM) in the MDA-MB231 breast cancer cell line (Fig. **17**) [144].

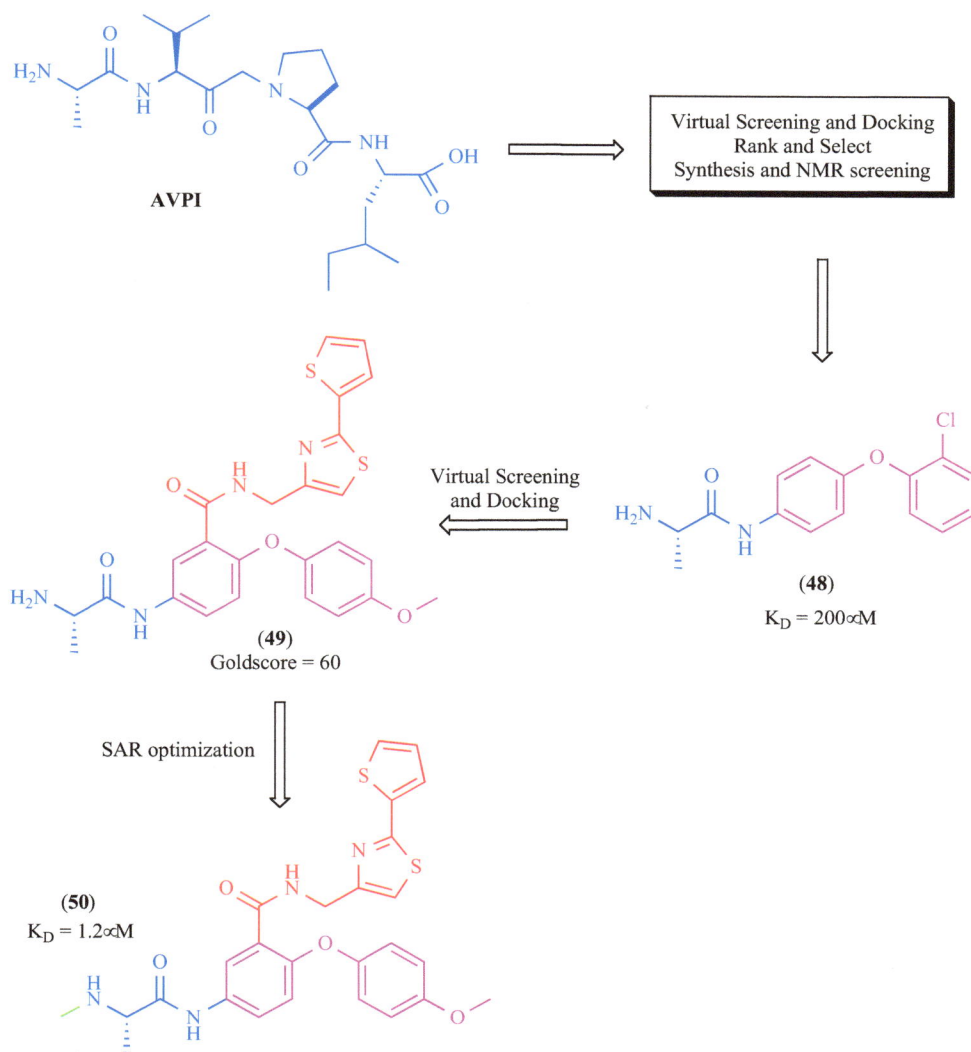

Fig. (17). Chemical structure of XAIP inhibitors discovered by a fragment optimization approach.

2.5.4. In Situ Fragment Assembly

Reactive fragments are capable of self assembly in the presence a protein, and generate a *de novo* lead [149]. In this strategy two separate fragments are linked together to form a larger and more active inhibitor in the presence of a target protein. NMR or crystallography can be used to identify fragments that bind to nearby sites in a mutually compatible manner. Competition studies can also provide useful indirect evidence to select combinations of fragments that bind close proximity, which may be good candidates for linking each other. Recently, several groups explored the use of target proteins to select and combine pairs of fragments *in situ* [150,151]. In effect, the protein assembles its own inhibitor by selecting fragments that can cross-link to each other when brought into close proximity. Although this method has a promise, it has not yet been successfully used in developing anticancer therapeutics.

2.6. Miscellaneous

2.6.1. INPHARMA

In SBDD, the relative orientation of two competitive ligands is important in designing a high-affinity drug candidate from weakly bound fragments. NOEs can be transferred from one ligand to the competitive ligand if the two are undergoing rapid exchange. The magnetisation of the first ligand is transferred to the protons of the target protein, where it spreads over the interaction surface by spin-diffusion. After the placement of the first ligand, the second ligand is selected, and a correlation peak between both ligand resonances changes are examined. Subsequently, this information can be used for designing a linker between the two ligands. This method is described as protein-mediated interligand NOEs for pharmacophore mapping (INPHARMA) [152], which may be applicable to any combinations of ligands that are weakly bound to a target protein.

Carlomangno and colleagues [153] demonstrated a system containing mixture of tubulin proteins and two ligands (Fig. **18**). Tubulin is the target of several drugs that interfere either positively or negatively with the tubulin polymerization process. Since the disruption of microtubule dynamics results in the impediment of the cell-division process and eventually cell death, this is one of the major therapeutic strategies against human solid tumors. Paclitaxel, a small organic molecule that promotes tubulin polymerization, is widely used for the treatment of several tumors including ovarian cancer. Epothilone and discoder-molide, which compete with paclitaxel for the same binding pocket, are found to be more potent than paclitaxel [154]. In an attempt to further improve the potency of these compounds, Carlomangno *et al.* [155] applied the INPHARMA method using a mixture of epothilone A and baccatin III in the presence of tubulin. The water soluble baccatin III, a precursor of paclitaxel that lacks a C13 side chain, binds to tubulin in a transiently formed complex, and delivers tr-NOEs. The complex Epo A–tubulin is only transiently formed during the mixing time of a NOESYexperiment, and tr-NOEs can be observed for a 100:1 mixture of Epo A and tubulin. The authors observed NOESY spectra at different mixing times (20, 40, 70, 100, and 200 ms) for a mixture of epothilone A (0.6 mm), baccatin III (0.6 mm), and tubulin (12 mm) in D_2O. Intermolecular NOE cross-peaks between the aromatic protons of baccatin III and several methyl groups of Epo A were clearly observable in the section of the NOESY spectrum recorded at the mixing time of 70 ms. In a control experiment, it was recorded a NOESY spectrum of the mixture of Epo A and baccatin III in the absence of tubulin with the mixing time of 400 ms; however, this spectrum showed no intermolecular NOEs. Therefore, the authors concluded that the intermolecular NOEs between Epo A and baccatin III, observed in the presence of tubulin, were mediated by protons of tubulin [155].

Epothilone A   Baccatin III

Fig. (18). Chemical structures of tubulin polymerization inhibitors.

2.6.2. In Silico Approach

The *in silico* screening approach has become an important tool for identification of small molecules that can block active sites of target proteins. However, a large percentage of the potential ligands identified by *in silico* screening eventually turns out to be false positives. Therefore, additional, complementary screening methods are needed to an *in silico* screening protocol. NMR spectroscopy has the potential of validating the results of *in silico* screening. Indeed, Grandy *et al.* [156] successfully identified a small molecule inhibitor that effectively inhibits Wnt signaling pathway and suppressed the growth of prostate cancer PC-3 cells (Fig. **19**). The Wnt signaling pathways are regulated by a family of secreted Wnt glycoproteins. In this pathway, Wnt molecules interact with the seven-transmembrane Frizzled (Fz) proteins [157] by binding to the N-terminal cysteine rich-domain [158]. The signal is then transduced into the cell through an internal sequence of Fz, the C-terminus to the seventh transmembrane domain, which binds directly to the postsynaptic density-95/discs large/zonula occludens-1 (PDZ) domain of the cytoplasmic protein Dishevelled (Dvl) [159]. Dvl then transduces the Wnt signal to downstream substrates.

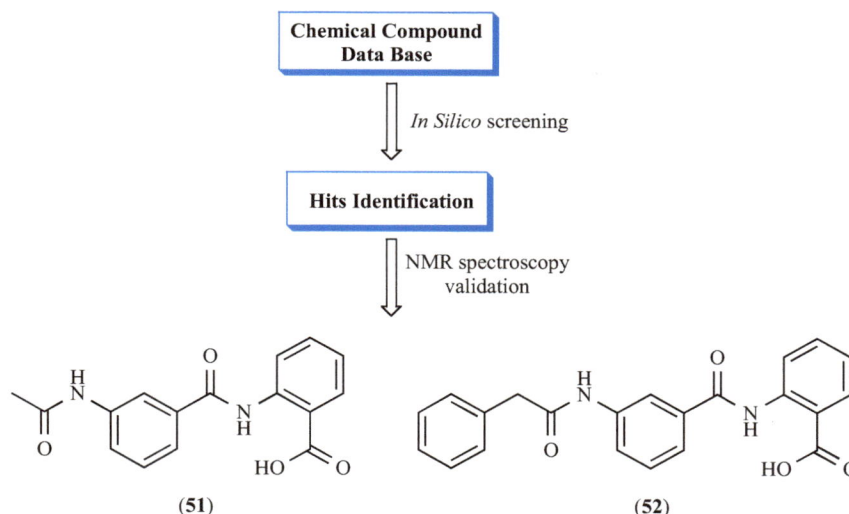

Fig. (19). Chemical structures of Dvl PDZ domain inhibitors.

The up-regulation of Dvl proteins has been reported in many cancers, including those of breast, colon, prostate, mesothelium, and lung (non-small cell) [157-159]. For initial hit identification, Grandy *et al.* [156] screened three sets of small molecule libraries provided by the data base of NCI (National Cancer Institute), Chemical Diversity Inc. (ChemDiv, San Diego, CA), and Sigma-Aldrich, using the UNITY module in the Sybyl software. The initial hits were docked to the Fz protein receptor site. The confirmation of the complexes was then scored by using Flex X module in Sybyl. To validate the docking results, the authors performed ^{1}H-^{15}N correlated HSQC spectra by titrating various concentrations of the small molecules (Fig. **19**, compounds **51**, **52**) into samples of ^{15}N-labeled Dvl PDZ domain. The spectra of chemical shift perturbations suggested that several small molecules bound to the conventional C-terminal peptide binding groove of the PDZ domain, as judged by the chemical shift perturbations due to ligand binding [160]. Compound **52** bound to the groove between the βB sheet and the αB helix. To verify accuracy of the binding, the Grandy *et al.* [156] performed fluorescence anisotropy to measure the binding affinities of the inhibitor identified (Fig. **19**, compound **52**) to the Dvl PDZ domain. The K_D value was 10.6 μM. The binding affinity between compound **52** and the Dvl PDZ domain was comparable with the binding affinity of the Dpr peptide to the Dvl PDZ domain. This result suggested that compound **52** and Dpr peptide competed for the same site on the surface of the Dvl PDZ domain. The benzene ring on one end of the compound mimics the Val side chain of Dpr, and the benzene ring on the other end of the compound lies in the same region as the Met side chain. The C-terminal carboxylate ion of Dpr and that of the compound lie in the same region, where they can interact similarly with the PDZ domain. Finally, it was demonstrated that one of the best inhibitors identified in the combination of *in silico* screening and NMR could effectively inhibit the Wnt signals *in vivo* [156].

2.6.3. NMR as a Tool of Fragment Screening

The activity of the PI3K-Akt (Protein kinase B) pathway is attributable for a wide range of cell proliferation and survival processes in many human tumors [161-163]. PI3K phosphorylates phosphatidylinositol (PI), phosphatidylinositol-4-phosphate [PI(4)P], and PI(4,5)P2 to generate PI(3)P, PI(3,4)P2, and PI(3,4,5)P3, respectively [164]. This process is negatively regulated by the tumor suppressor phosphatase-tensin homologue PTEN/MMAC [165], mutations of which are frequently found in human cancers [164,166]. The binding of PIP3 to the pleckstrin homology domains of Akt and 3-phosphoinositide-dependent kinase-1 (PDK1) co-localizes with these enzymes at the inner leaflet of the plasma membrane, inducing a conformational change in Akt [167,168]. This shift releases the auto inhibitory of the active site, allowing for PDK1 to phosphorylate at Thr308 of Akt to activate it. PDK1 also recognizes other substrate kinases in the pathway through a distinct regulatory mechanism. The N-terminal lobe of the catalytic domain of PDK1 contains a docking site, which recognizes the non-catalytic C-terminal hydrophobic motifs of substrate kinases. The binding of substrate in this so-called PIF (PDK1 Interacting Fragment) pocket allows interaction with PDK1 and, thus, enhances phosphorylation of downstream kinases [169]. This substrate recognition system allows the specific and selective control of the PI3K-Akt signaling pathway in response to extracellular events [170]. The pivotal role of PDK1 in the PI3K signaling cascade makes it an attractive target for cancer therapeutic intervention.

Stockman *et al.* [171] carried out an NMR fragment screening approach to exploit the ATP binding and PIF pocket sites on PDK1. A library of 10,237 chemical compounds, which includes diverse fragments, was screened on PDK1 kinase domain constructed and designed the way that the ATP site and the allosteric PIF pocket would all be available for potential interactions with ligands. More than 300 hits were identified from this screening. In an STD experiment, signals were observed only for those compounds that magnetization is transferred from the target protein. Thus, only ligand signals will be observed in a mixture of compounds, as no signal will be generated by non-binding compounds.

By STD NMR experiments, Stockman *et al.* [171] eventually identified compounds **53** and **54** that bind to PDK1 (Fig. **20**). Competitive NMR experiment suggested that compounds **53** and **54** bind specifically at the ATP site and PIF pocket, respectively. Compound **55**, an analog of compound **54**, was identified in a substructure search. Compound **55** accelerated the PDK1 reaction by two-fold at 300 μM. Further, the binding property of the compound on PDK1 was characterized by ^{1}H-^{15}N TROSY spectrum. The result suggested that compound **53** binds to the ATP site, while compounds **54** and **55** bind to the PIF pocket. Subsequently, molecular modeling studies confirmed that compounds **54** and **55** bind PIF pocket, and showed that the negatively charged carboxyl group was sandwiched between two aromatic-ring hydrophobic moieties. Also, the carboxyl group mimics the phosphate group of the phosphorylated substrate, and the two aromatic groups most likely mimic the two phenylalanine side chains of the FRDFD motif of the PIF pocket [171].

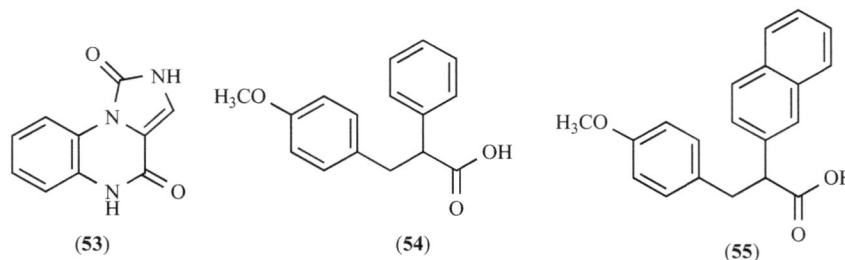

Fig. (20). Chemical structures of allosteric activators of PDK1.

3. NMR AND METABONOMICS

During the last three decades, NMR spectroscopy has played a central role in understanding of the metabolism and metabolic processes. This is referred to as magnetic resonance spectroscopy (MRS) in the biomedical field [172]. The primary definition of metabonomics is comprehensive and simultaneous systematic determination of metabolite levels in whole organisms and their changes over time, which occurs in response to a wide range of stimuli including diet, lifestyle, environment, genetic effects, and therapeutic interventions [173]. It was initially used for analyzing specific biochemical pathways; such as, energy metabolism, the Krebs cycle, and phospholipid (PL) synthesis and degradation [174]. Metabolomic NMR studies usually measure the concentrations of a large number of metabolites to determine metabolic profiles that are characteristic to the biological systems. The profiles can be used to identify biomarkers for various diseases; to analyze drug toxicity; or to determine the efficacy and selectivity of drugs *in vivo* [175].

Small pieces of intact tissue (e.g. biopsies) can also be used to study tumors by high-resolution magic-angle spinning (HR-MAS) spectroscopy. The NMR is also used for the magnetic resonance spectroscopy technique on tissue extracts, biopsies, and other *ex vivo* samples. The term MRS is used to denote studies by the same technique on living animals or patients. The *ex vivo* NMR method is complementary to an *in vivo* studies, as the former can provide additional information and help to accurately interpret data generated by *in vivo* studies.

In 1981, the first *in vivo* ^{31}Phosphorus (^{31}P) MRS study was carried out in an animal tumor model [176], followed by the first *in vivo* human tumor study within a couple of years [177]. Since then, MRS with different nuclei including ^{1}H, ^{31}P, ^{19}F, and ^{13}C, has been used widely to study cancer biology and therapeutics in cultured cells as well as in animal models [178-181]. Using ^{31}P MRS spectra, markers for tissue bioenergetics (*i.e.*, nucleotide triphosphate [NTP], inorganic phosphate [Pi]), intracellular pH (pHi), and membrane turn-over (*i.e.*, phosphorus-containing components of phospholipid membrane, phosphor-monoester [PME], and phosphodiester [PDE]) can be readily observable. PMEs comprise phosphocholine (PC) and phosphoethanolamine (PE), and PDEs glycero-phosphocholine (GPC) and glycero-phosphoethanolamine (GPE) were observed *in vitro* ^{31}P MRS spectra. ^{1}H and ^{13}C MRS can also provide information on tumor metabolism *in vivo*. Metabolites, such as lactate, lipids, and choline-containing compounds, can be observed in ^{1}H MRS [182]. ^{13}C-enriched substrates (e.g. glucose) and ^{13}C MRS may be used to study metabolic pathways (e.g. glucose metabolism) of tumors [183]. *In vivo* ^{19}F, ^{1}H and ^{13}C MRS have been used to monitor the metabolism of drugs containing those nuclei *in situ* [184,185]. Many more metabolites can be observed by HR-MAS, compared to *in vivo* MRS, since the former can resolve spectra better than the latter. This methodology has a significant merit, since pieces of intact tissue can be examined with minimal sample preparation and destruction; however, there is still a risk of sample degradation during the examination. The NMR-based metabolomics applications related to cancer biology and therapy will be reviewed in the following section.

3.1. NMR Techniques for Metabonomic Studies

Metabolic characterization of cancer tissue is to document detailed information of the metabolic composition and/or inter-sample variations. 2-D spectroscopy is a useful tool for obtaining such detailed data, for which experiments are usually carrying out in the following sequence: (i) resolving coupling constant (J), (ii) ^{1}H–^{1}H COSY, (iii) ^{1}H–^{1}H total correlated spectroscopy (TOCSY), and (iv) finally ^{1}H–^{13}C COSY. The 1-D acquisition of data has been widely used for classifying cancers. TOCSY experiments have been successfully used for the observation of differences between tissues from normal and cancerous kidney. Zekter *et al.* [186] compared different TOCSY pulse-schemes, and found that rotor-synchronized adiabatic pulses gave optimal signals. This optimized pulse-scheme was later successfully applied to prostate tissue for quantification of choline- and ethanolamine-containing metabolites in prostate tissues [187].

There is a large amount of water in human tissues, and the water resonance is usually suppressed to avoid dynamic range and signal overlap. Several types of pulse programs are available for suppressing water by utilizing different properties of the water resonance. The water suppression can be achieved by three different ways: (i) spectral editing methods based on spin relaxation or translational diffusion properties between water and solutes; (ii) presaturation methods based on actively irradiating the water signals; and (iii) tailor-excitation methods by using the PFG. The suppression of water in HR MAS experiments on cells and tissues have been explored by Chen and coworkers [188,189]. The authors found that continuous wave irradiation of the water resonance can lead to suppression of metabolite resonances due to protons exchanging with water protons, and they recom-mended a CHESS-based pulse sequence with the third selective excitation to empirically null (SEEN) the water signal.

3.2. Detection of Apoptosis

Metabolomic MRS allows distinguishing between cells that are sensitive and resistant to apoptosis, which can be achieved as a function of time after treatment, in conjunction with measurements of complementary biological parameters. Baum and colleagues demonstrated that the level of NTP decreased in a dexamethasone-sensitive CEM variant cells in response to apoptotic stimulation, but not in a dexamethasone-resistant CEM variant cells [190,191]. However, the induction of apoptosis in the Jurkat cells transfected with ERtax (which is the tax oncogene of a human leukemia retrovirus) resulted in increased NTP levels *in vitro* in response to a combination of apoptotic stimuli and estrogen, after an initial decrease for a short time [192]. Lutz *et al.* [193] studied metabolic changes in HT-29 human colorectal carcinoma cells in the context of cytokine-induced apoptosis (TNFα/interferon-γ (IFN- γ)). The authors found that the levels of NTP were progressively increasing, starting from 4 h during the 15 h of combinational treatment. However, no consistent trend was observed for PCr or Pi. Williams *et al.* [194] found that no energy charges occurred in response to apoptotic stimuli in the HL-60 human promyelocytic leukemia cell line and the CHO-K1 Chinese hamster ovary cell line, although some ATP reduction was observed. These data suggest that energy metabolism in response to apoptosis varies as a function of the cell type, the mode of apoptosis, and time after apoptotic stimulation.

3.3. NMR as a Diagnostic Tool

Damadian [195] demonstrated in early 1970's that the T_1 relaxation time in various types of neoplastic tissues was significantly longer than in the adjacent normal tissue. Based on this data, the author suggested that this phenomenon could be used to detect tumors. Subsequently, the application of MRS in oncology focused on the detection and classification of tumors [190,191,194]. Tumors generally display (i) elevated levels of phospholipid, (ii) increased glycolytic capacity including increased utilization of glucose carbons to drive synthetic processes, (iii) high glutaminolytic function, and (iv) overexpression of the glycolytic isoenzymes. The characterization of non-invasive tumor has been a major focus of *in vivo* MRS [190,191,194]. The multi-parameter dependence of the MRS signal might provide classification criteria sufficient to discriminate benign tumors from malignant tissue masses, and may even allow determining a more detailed staging in the tumor diagnosis. Using ^{31}P spectral data, four types of tumors (Morris hepatoma, GH3 prolactinoma, RIF-1, MNU-induced mammary) and three types of normal tissues (brain, liver, skeletal muscle) could be successfully classified in 62% of the samples [196]. Preul *et al.* [197] accurately classified and staged various brain tumors (glioblastoma grade II, III, IV, meningioma, metastases) in patients based on ^{1}H spectra. NMR-based examination could reproducibly show that breast cancer contained elevated level of TCC (resulting from increased PC), low GPC, and low glucose, compared to normal breast tissues or benign tumors [198-200]. Furthermore, when 91 breast tumors and 48 adjacent normal tissue specimens were examined using HR-MAS ^{1}H-NMR metabolomics, a malignant phenotype could reliably be determined from normal tissue with high sensitivity and specificity.

Similarly, prostate cancer exhibited a distinct NMR metabolic profile characterized by high TCC and PC levels, along with an increase in the glycolytic products [201,202]. Prostatic fluid from men with prostate cancer exhibited a decreased level of citrate and elevated spermine, compared to noncancer specimen. Detection of citrate and spermine in prostatic fluid by ^{1}H-NMR was correlated with Gleason score [203,204]. Metabolomic differences between healthy women and those with epithelial ovarian cancer have been investigated [205]. ^{1}H-NMR serum metabolic profiles correctly distinguished women with cancer from normal premenopausal women and those with benign ovarian disease with almost 100% certainty [205].

3.4. Monitoring the Effect of Chemotherapeutic Drugs

MRS has been used for monitoring the effect of anticancer drugs, both in patients and experimental animals. MRS is also one of the most important tools that offer significant insights into the effect of novel anticancer agents in the process of new drug development, as the metabolomic method based on *in vivo* MRS and *in vitro* ^{1}H and ^{31}P NMR spectra can help to assess tumor response to conventional cytotoxic agents as well as to novel drugs with specific molecular targets. MRS based biomarkers could potentially provide surrogate pharmacodynamic markers for use at clinical trials.

The ^{1}H-NMR on human glioma cell culture could successfully divide into drug-resistant and -sensitive groups prior to treatment with 1-(2-chloroethyl)-3-cyclohexyl-1-nitrosourea [206]. *In vivo*, ^{1}H-NMR, including HR-MAS, has been used to investigate the metabolic changes associated with the nitrosourea-mediated treatment of B16 melanoma and 3LL lung tumors grown subcutaneously in C57BL6/6J mice [207]. MRS has been used by a large number of research groups to quantitatively assess the effects of drugs on tumour growth. The somatostatin analogues octreotide and bicalutamide have been shown to significantly reduce the growth of Dunning prostate R3327 tumors in rats [208,209]. In this animal model, orchidectomy completely inhibited tumor proliferation [209,210]. Similar beneficial effects have been found for 5α-reductase inhibitors in canine benign prostate hyperplasia [210,211]. Somatostatin analogues have also been used in other tumor models such as octreotide to evaluate beneficial effects on established oestrogen-induced pituitary hyperplasia 5 or RC 160 for the treatment of liver metastases of colon tumors [212]. The application of MRS for volumetric measurements is justified for deep lying tumours, which are difficult to access by calipers.

Imatinib, an inhibitor of the oncogene BCR-ABL tyrosine kinase, decreases cell proliferation and induces apoptosis in human chronic myeloid leukemia [213]. Metabolically, imatinib interrupts the synthesis of macromolecules required for cell survival by deprivation of key substrates [214]. Investigating glucose metabolism changes in imatinib-treated BCR-ABL–positive human leukemia cells with NMR showed decreased glucose uptake by inhibition of glycolysis. However, unlike conventional therapeutics, imatinib stimulated mitochondrial metabolism leading to cell differentiation [215]. Imatinib also led to a significant decrease in PC in imatinib sensitive cells, which is correlated with a decrease in cell proliferation rate [215]. NMR-based studies also allowed analyzing metabolites in imatinib-resistance tumors, which showed the resistance to imatinib and the progress of the disease were directly correlated with a decrease in mitochondrial glucose oxidation and a nonoxidative ribose synthesis from glucose, and elevation of PC levels [214].

The depletion of high-energy phosphates (ATP and PCr) and a concomitant increase in Pi after administration of cytotoxins have been reported in several tumor models by NMR metabolic profile studies. For examples, NMR metabolic fingerprint could be used to determine the responses by many different tumors to many different anticancer agents [216-222]. Similarly, ^{31}P spectral changes have been observed in the MNU-induced mammary tumors in rats with ovaryectomy [223]. The spectral changes in these cases suggest that the breakdown of energy generating system eventually leads to tissue necrosis. The ^{31}P MRS study showed that the PE/PC ratio significantly increased after mouse mammary tumors were treated with cyclophosphamide [219]. Data from NMR-based studies also suggested that ethanolamine and choline-containing metabolites are involved in different pathways, and the PE/PC ratio could serve as surrogate marker for evaluation of cytotoxic effects by drugs.

It may be important to note the metabolic fate of 5-fluorouracil, a widely used anticancer agent, can be directly monitored using ^{19}F MRS. This is possible because the concentration of 5-fluorouracil used at clinic and the gyromagnetic ratio of the ^{19}F nucleus are relatively high. Using ^{19}F MRS, the levels of 5-fluorouracil in tissues as well as its catabolic and anabolic pathways could readily be monitored [224].

Using the ^{31}P-NMR method, Chung *et al.* [225] analyzed the tumor extracts prepared from xenografted colon cancer that had been treated with an Hsp90 inhibitor. The authors found that the extracts showed a significant increase in PC, PME/PDE ratio, valine, and PE levels, suggesting that the phospholipid metabolism was altered in response to a HSP90 inhibitor. Ronen and colleagues used *in vivo* ^{1}H MRS and *in vitro* ^{31}P and ^{13}C NMR to study the effects of SAHA, a histone deacetylase inhibitor (HDAC) on the PC3 prostate cancer line. The authors found that the levels of PC and TCC were substantially increased in SAHA-treated cell extracts, which were inversely correlated with HDAC activity [226,227].

4. SUMMARY AND FUTURE DIRECTIONS

We have reviewed here the fundamental principle and applications of NMR in the context of the development and assessment of anticancer therapeutics. We have paid a particular emphasis on the utilization of NMR in identification and further improvement of fragment leads for anticancer therapeutics, some of which are summarized in Table **2**. The versatility of the NMR technique and rich data generated by NMR spectroscopy allow scientists to continue to break a new ground for analysis and understanding of protein–ligand interactions, which provide an important new tool for the identification of new leads and, eventually, the development of effective anticancer drugs. In particular, the utilization of NMR in the

Table 2. The Utilization of NMR in the Development of Fragment-Based Anticancer Therapeutics

Target	Status	Approach
Matrix metalloproteinase (MMP)	Phase I (ABT-518)	SAR-by-NMR fragment linking
Matrix Metalloproteinase 13	Novel potent inhibitors	SAR-by-NMR fragment linking
Human papillomavirus E2 protein	Novel potent inhibitors	SAR-by-NMR fragment linking
B-cell CLL/lymphoma 2 (BCL-2), BCL-2-like 1 (BCL-XL)	Novel potent inhibitors	Spin label-guided Fragment linking
Tubulin	Novel potent inhibitors	Spin label-guided fragment linking
c-*myc*	Novel potent inhibitors	Fragment evolution
Survivin	Novel potent inhibitors	Fragment evolution
B-cell CLL/lymphoma 2 (BCL-2), BCL-2-like 1 (BCL-XL)	Preclinical development (ABT-737)	SAR-by-NMR fragment linking and evolution
HSP90	Novel potent inhibitors	Fragment linking and evolution
Urokinase	Novel potent inhibitors	Fragment optimization
X-linked inhibitor of apoptosis protein	Novel potent inhibitors	Fragment optimization

development of fragment-based drugs has become an established tool in designing optimal structure of ligand against a specific target protein, and then verifying the effectiveness of the compound. This approach will be further sophisticated and play even a greater role in developing new generations of anticancer therapeutics in the coming years.

It is clear that NMR-based metabonomics will be one of the most important applications of NMR spectroscopy in studying the interface of chemistry and biology. Most of the recent research about tumor metabolomics has come from NMR-based studies, reflecting the potential of MRS in developing tools for clinical diagnosis. This technology is applicable for all aspects of tumor management as it can be used to: (i) diagnose cancer earlier when it is still manageable; (ii) determine aggressiveness of cancer to help determine the prognosis of cancer and the directions of therapies; (iii) predict and monitor the efficacy of therapies; and (iv) monitor the toxicity of (novel) chemotherapeutic agents. Recent literatures show that metabolomic profiles are already used for the diagnosis of ovarian cancer by analyzing either sera or tumor tissues. This kind of applications will undoubtedly expand to other cancers in the near future. NMR-based studies are being undertaken for many other bodily fluids, including ascitic fluid in ovarian cancer, pancreatic secretions in pancreatic cancer, and bronchoalveolar or pleural fluid in lung cancer. If pathognomonic profiles can be identified and validated in these fluids, NMR based tumor metabolomics may save time, costs, and efforts in obtaining effective tools for early detection and treatment of many different types of cancer.

ACKNOWLEDGEMENTS

This work was supported by funds from the Natural Sciences and Engineering Research Council of Canada (NSERC), and the Northern Cancer Research Foundation (NCRF) to H.L. VRS thanks the Ontario Ministry of Research and Innovation for Postdoctoral Fellowship.

ABBREVIATIONS

Bcl	=	B-cell lymphoma
BEI	=	Binding efficiency index
CSA	=	Chemical shift anisotropy
COSY	=	Correlation spectroscopy
TOCSY; DECODES	=	Diffusion-encoded total correlation spectroscopy
DOSY	=	Diffusion-ordered spectroscopy
Dvl	=	Dishevelled
DQ	=	Double-quantum
FBP	=	Far binding protein
FUSE	=	Far upstream element
FRET	=	Fluorescence resonance energy transfer
FABS	=	Fluorine atoms for biochemical screening
FAXS	=	Fluorine chemical shift anisotropy and exchange for screening
FAXS	=	Fluorine chemical shift anisotropy and exchange for screening
FBDD	=	Fragment based dug discovery
Fz	=	Frizzled
GPC	=	Glycero-phosphocholine
GPE	=	Glycero-phosphoethanolamine
γ	=	Gyromagnetic ratio
HSP90	=	Heat shock protein 90

HSQC	=	Heteronuclear single quantum coherence
HR-MAS	=	High-resolution magic-angle spinning
HTS	=	Highthroughput screening
HPVs	=	Human papillomaviruses
HSA	=	Human serum albumin
IAP	=	Inhibitors of apoptosis family of proteins
Pi	=	Inorganic phosphate
ILOE	=	Interligand nuclear Overhauser effect
pHi	=	Intracellular pH
ITC	=	Isothermal calorimetric
MRS	=	Magnetic resonance spectroscopy
mM	=	Millimolar
NCI	=	National cancer Institute
NOESY	=	NOE spectroscopy
NMR	=	Nuclear magnetic resonance
NOE	=	Nuclear overhauser effect
NTP,	=	Nucleotide triphosphate
PC	=	Phosphocholine
PDE	=	Phosphodiester
PE	=	Phosphoethanolamine
PDK1	=	Phosphoinositide-dependent kinase-1
PL	=	Phospholipid
PME	=	Phosphomonoester
PDZ	=	Postsynaptic density-95/discs large/zonula occludens-1
PFG	=	Pulsed-field-gradients
QNP	=	Quattro nuclei probe
SMAC	=	Second mitochondrial activator of caspases
SQ	=	Single quantum
SLAPSTIC	=	Spin labels attached to protein side chain as a tool to identify interacting compounds
SAR	=	Structure-activity relationship
SBDD	=	Structure-based drug discovery
TINS	=	Target immobilized NMR screening
3D	=	Three-dimensional
TCC	=	Total choline-containing compounds
TOCSY	=	Total correlated spectroscopy
trNOE	=	Transferred NOE
TROSY	=	Transverse relaxaion-optimized spectroscopy
2D	=	Two-dimensional

| Water–Ligand Observation via Gradient SpectroscopY | = | WaterLOGSY experiment |
| XAIP | = | X-linked inhibitor of apoptosis protein |

REFERENCES

[1]	Ernst RR, Bodenhausen G, Wokaun A. Principles of nuclear magnetic resonance in one and two dimensions. Oxford: Clarendon Press 1987.
[2]	Lowe IJ. Free Induction decays of rotating solids. Phys Rev Lett 1959; (2): 285-7.
[3]	Schaefer J, Stejskal EO. Carbon-13 nuclear magnetic resonance of polymers spinning at the magic angle. J Am Chem Soc 1976; 98: 1031-2.
[4]	Ernst RR, Anderson WA. Application of Fourier transform spectroscopy to magnetic resonance. Rev Sci Instrum 1966; 37: 93-102.
[5]	Pellecchia M, Bertini I, Cowburn D, et al. Perspectives on NMR in drug discovery: a technique comes of age. Nat Rev Drug Discov 2008; 7(9): 738-45.
[6]	Fernandes PB. Technological advances in high-throughput screening. Curr Opin Chem Biol 1998; 2(5): 597-603.
[7]	Kenny BA, Bushfield M, Parry-Smith DJ, Fogarty S, Treherne JM. The application of high-throughput screening to novel lead discovery. Prog Drug Res 1998; 51: 245-69.
[8]	Kubinyi H. Combinatorial and computational approaches in structure-based drug design. Curr Opin Drug Discov Devel 1998; 1 (1): 16-27.
[9]	Kubinyi H. Structure-based design of enzyme inhibitors and receptor ligands. Curr Opin Drug Discov Devel 1998; 1 (1): 4-15.
[10]	Ratti E, Trist D. The continuing evolution of the drug discovery process in the pharmaceutical industry. Farmaco 2001; 56 (1-2): 13-9.
[11]	Pellecchia M, Sem DS, Wuthrich K. NMR in drug discovery. Nat Rev Drug Discov 2002; 1(3): 211-9.
[12]	Sem DS, Pellecchia M. NMR in the acceleration of drug discovery. Curr Opin Drug Discov Devel 2001; 4 (4): 479-92.
[13]	Stockman BJ. Flow NMR spectroscopy in drug discovery. Curr Opin Drug Discov Devel 2000; 3 (3): 269-74.
[14]	Wishart D. NMR spectroscopy and protein structure determination: applications to drug discovery and development. Curr Pharm Biotechnol 2005; 6(2): 105-20.
[15]	Hajduk PJ, Greer J. A decade of fragment-based drug design: strategic advances and lessons learned. Nat Rev Drug Discov 2007; 6 (3): 211-9.
[16]	Villar HO, Yan J, Hansen MR. Using NMR for ligand discovery and optimization. Curr Opin Chem Biol 2004; 8 (4): 387-91.
[17]	Mercier KA, Germer K, Powers R. Design and characterization of a functional library for NMR screening against novel protein targets. Comb Chem High Throughput Screen 2006; 9 (7): 515-34.
[18]	Mercier KA, Shortridge MD, Powers R. A multi-step NMR screen for the identification and evaluation of chemical leads for drug discovery. Comb Chem High Throughput Screen 2009; 12 (3): 285-95.
[19]	Baurin N, boul-Ela F, Barril X, et al. Design and characterization of libraries of molecular fragments for use in NMR screening against protein targets. J Chem Inf Comput Sci 2004; 44 (6): 2157-66.
[20]	Ferentz AE, Wagner G. NMR spectroscopy: a multifaceted approach to macromolecular structure. Q Rev Biophys 2000; 33 (1): 29-65.
[21]	Stark J, Powers R. Rapid protein-ligand costructures using chemical shift perturbations. J Am Chem Soc 2008; 130 (2): 535-45.
[22]	Stockman BJ, Dalvit C. NMR screening techniques in drug discovery and drug design. Prog Nucl Magn Reson Spectrosc 2002; 41: 187-231.
[23]	Diercks T, Coles M, Kessler H. Applications of NMR in drug discovery. Curr Opin Chem Biol 2001; 5 (3): 285-91.
[24]	Meinecke R, Meyer B. Determination of the binding specificity of an integral membrane protein by saturation transfer NMR: RGD peptide ligands binding to integrin alphaIIbbeta3. J Med Chem 2001; 44 (19): 3059-65.
[25]	Hajduk PJ, Olejniczak ET, Fesik SW. One-dimensional relaxation- and diffusion-edited nmr methods for screening compounds that bind to macromolecules. J Am Chem Soc 1997; 119: 12257-61.
[26]	Jahnke W. Spin labels as a tool to identify and characterize protein-ligand interactions by NMR spectroscopy. Chembiochem 2002; 3 (2-3): 167-73.
[27]	Jahnke W, Rudisser S, Zurini M. Spin label enhanced NMR screening. J Am Chem Soc 2001; 123 (13): 3149-50.
[28]	Barjat H, Morris GA, Smart S, Swanson AG, Williams SCR. High-resolution diffusion-ordered 2D spectroscopy (HR-DOSY) - a new tool for the analysis of complex mixtures. J Magn Reson 1995; 108: 70.
[29]	Bleicher K, Lin MF, Shapiro MJ, Wareing JR. Diffusion edited NMR: screening compound mixtures by affinity NMR to detect binding ligands to vancomycin. J Org Chem 1998; 63: 8486-90.
[30]	Lin M, Shapiro MJ, Wareing JR. Screening mixtures by affinity NMR. J Org Chem 1997; 62: 8930-1.
[31]	Russell DJ, Hadden CE, Martin GE, Gibson AA, Zens AP, Carolan JL. A comparison of inverse-detected heteronuclear NMR performance: conventional vs cryogenic microprobe performance. J Nat Prod 2000; 63 (8): 1047-9.
[32]	Logan TM, Murali N, Wang G, Jolivet C. Application of a high-resolution superconducting NMR probe in natural product structure determination. Magn Reson Chem 1999; 37: 762-5.
[33]	Pervushin K, Riek R, Wider G, Wuthrich K. Attenuated T2 relaxation by mutual cancellation of dipole-dipole coupling and chemical shift anisotropy indicates an avenue to NMR structures of very large biological macromolecules in solution. Proc Natl Acad Sci USA 1997; 94 (23): 12366-71.
[34]	Hajduk PJ, Augeri DJ, Mack JC, et al. NMR-based screening of proteins containing 13C-labeled methyl groups. J Am Chem Soc 2000; 122: 7898-904.

[35]   Fielding L. Determination of association constants (Ka) from solution NMR data. Tetrahedron 2000; 56: 6151-70.
[36]   Hajduk PJ, Meadows RP, Fesik SW. NMR-based screening in drug discovery. Q Rev Biophys 1999; 32 (3): 211-40.
[37]   Hajduk PJ, Gerfin T, Boehlen JM, Haberli M, Marek D, Fesik SW. High-throughput nuclear magnetic resonance-based screening. J Med Chem 1999; 42 (13): 2315-7.
[38]   Jahnke W, Floersheim P, Ostermeier C, *et al.* NMR reporter screening for the detection of high-affinity ligands. Angew Chem Int Ed Engl 2002; 41 (18): 3420-3.
[39]   Siriwardena AH, Tian F, Noble S, Prestegard JH. A straightforward NMR-spectroscopy-based method for rapid library screening. Angew Chem Int Ed Engl 2002; 41 (18): 3454-7.
[40]   Rees DC, Congreve M, Murray CW, Carr R. Fragment-based lead discovery. Nat Rev Drug Discov 2004; 3 (8): 660-72.
[41]   Pellecchia M, Meininger D, Dong Q, Chang E, Jack R, Sem DS. NMR-based structural characterization of large protein-ligand interactions. J Biomol NMR 2002; 22 (2): 165-73.
[42]   Hajduk PJ, Meadows RP, Fesik SW. Discovering high-affinity ligands for proteins. Science 1997; 278 (5337): 497-9.
[43]   Weigelt J, van DM, Uppenberg J, Schultz J, Wikstrom M. Site-selective screening by NMR spectroscopy with labeled amino acid pairs. J Am Chem Soc 2002; 124 (11): 2446-7.
[44]   Sakamoto T, Tanaka T, Ito Y, *et al.* An NMR analysis of ubiquitin recognition by yeast ubiquitin hydrolase: evidence for novel substrate recognition by a cysteine protease. Biochemistry 1999; 38 (36): 11634-42.
[45]   Sun C, Cai M, Gunasekera AH, *et al.* NMR structure and mutagenesis of the inhibitor-of-apoptosis protein XIAP. Nature 1999; 401 (6755): 818-22.
[46]   Hajduk PJ, Sheppard G, Nettesheim D, *et al.* Discovery of potent nonpeptide inhibitors of stromelysis using SAR by NMR. J Am Chem Soc 1997; 119: 5818-27.
[47]   Shuker SB, Hajduk PJ, Meadows RP, Fesik SW. Discovering high-affinity ligands for proteins: SAR by NMR. Science 1996; 274 (5292): 1531-4.
[48]   Hajduk PJ, Zhou MM, Fesik SW. NMR-based discovery of phosphotyrosine mimetics that bind to the Lck SH2 domain. Bioorg Med Chem Lett 1999; 9 (16): 2403-6.
[49]   Hajduk PJ, Dinges J, Schkeryantz JM, *et al.* Novel inhibitors of Erm methyltransferases from NMR and parallel synthesis. J Med Chem 1999; 42 (19): 3852-9.
[50]   Hajduk PJ, Boyd S, Nettesheim D, *et al.* Identification of novel inhibitors of urokinase via NMR-based screening. J Med Chem 2000; 43 (21): 3862-6.
[51]   Hajduk PJ, Gomtsyan A, Didomenico S, *et al.* Design of adenosine kinase inhibitors from the NMR-based screening of fragments. J Med Chem 2000; 43 (25): 4781-6.
[52]   Medek A, Hajduk PJ, Mack J, Fesik SW. The use of differential chemical shifts for determining the binding site location and orientation of protein-bound ligands. J Am Chem Soc 2000, 122, 1241-1242.
[53]   Jarymowycz VA, Stone MJ. Fast time scale dynamics of protein backbones: NMR relaxation methods, applications, and functional consequences. Chem Rev 2006; 106 (5): 1624-71.
[54]   Carlomagno T, Felli C, Czech M, Fischer R, Sprinzl M, Griesinger C. Transferred cross-correlated relaxation: application to the determination of Sugar Puker in an aminoacylated tRNAMimetic weakly bound to EF-Tu. J Am Chem Soc 1999; 121: 1945-8.
[55]   Bolon PJ, Al-Hashimi HM, Prestegard JH. Residual dipolar coupling derived orientational constraints on ligand geometry in a 53 kDa protein-ligand complex. J Mol Biol 1999; 293 (1), 107-15.
[56]   Igumenova TI, Frederick KK, Wand AJ. Characterization of the fast dynamics of protein amino acid side chains using NMR relaxation in solution. Chem Rev 2006; 106 (5): 1672-99.
[57]   Zhang X, Sui X, Yang D. Probing methyl dynamics from 13C autocorrelated and cross-correlated relaxation. J Am Chem Soc 2006; 128 (15): 5073-81.
[58]   van Nuland NA, Kroon GJ, Dijkstra K, Wolters GK, Scheek RM, Robillard GT. The NMR determination of the IIA(mtl) binding site on HPr of the Escherichia coli phosphoenol pyruvate-dependent phosphotransferase system. FEBS Lett 1993; 315 (1): 11-5.
[59]   Fejzo J, Lepre CA, Peng JW, Bemis GW, Ajay MA, Moore JM. The SHAPES strategy: an NMR-based approach for lead generation in drug discovery. Chem Biol 1999; 6 (10): 755-69.
[60]   LaPlante SR, Aubry N, Deziel R, Ni F, Xu P. Transferred ^{13}C T_1 relaxation at natural isotopic abundance: a practical method for determining site-specific changes in ligand flexibility upon binding to a macromolecule. J Am Chem Soc 2000; 122: 12530-5.
[61]   Lin M, Shapiro MJ, Wareing JR. Diffusion-edited NMR-affinity NMR for direct obervation of molecular interactions. J Am Chem Soc 1997; 119: 5249-50.
[62]   Johnson CS. Diffusion ordered nuclear magnetic resonance spectroscopy: principles and applications. Progr NMR Spectrosc 1999; 34: 203-56.
[63]   Chen A, Shapiro MJ. Affinity NMR. Anal Chem 1999; 71 (19): 669A-75A.
[64]   Gmeiner WH, Hudalla CJ, Soto AM, Marky L. Binding of ethidium to DNA measured using a 2D diffusion-modulated gradient COSY NMR experiment. FEBS Lett 2000; 465 (2-3): 148-52.
[65]   Chen A, Shapiro MJ. NOE pumping: a novel NMR technique for identification of compounds with binding affinity to macromolecules. J Am Chem Soc 1998; 120: 10258-9.
[66]   Chen A, Shapiro MJ. NOE pumping. 2. A high-throughput method to determine compounds with binding affinity to macromolecules by NMR. J Am Chem Soc 2000; 122: 414-5.
[67]   Ni F. Recent developments in transferred NOE methods. Progr NMR Spectrosc 1994; 26: 517-606.
[68]   Shapiro MJ, Chin J, Chen A, *et al.* Covalent or trapped? PFG diffusion MAS NMR for combinatorial chemistry. Tetrahedron Lett 1999; 40: 6141-3.
[69]   Feng N. Recent developments in transferred NOE methods. Prog NMR Spectrosc 1994; 26: 517-606.
[70]   Massefski W Jr, Alfred R. Elemination of multiple-step spin diffusion effects in two-dimenssional NOE spectroscopy of nucleic acids. J Magn Reson 1988; 78: 150-5.
[71]   Vincent SJ, Zwahlen C, Post CB, Burgner JW, Bodenhausen G. The conformation of NAD+ bound to lactate dehydrogenase determined by nuclear magnetic resonance with suppression of spin diffusion. Proc Natl Acad Sci USA 1997; 94 (9): 4383-8.
[72]   Maaheimo H, Kosma P, Brade L, Brade H, Peters T. Mapping the binding of synthetic disaccharides representing epitopes of chlamydial lipopolysaccharide to antibodies with NMR. Biochemistry 2000; 39 (42): 12778-88.

[73]   Meyer B, Weimar T, Peters T. Screening mixtures for biological activity by NMR. Eur J Biochem 1997; 246 (3): 705-9.

[74]   Mayer M, Meyer B. Mapping the active site of angiotensin-converting enzyme by transferred NOE spectroscopy. J Med Chem 2000; 43 (11): 2093-9.

[75]   Henrichsen D, Ernst B, Magnani JL, Wang WT, Meyer B, Peters T. Bioaffinity NMR spectroscopy: identification of an E-selectin antagonist in a substance mixture by transfer NOE. Angew Chem Int Ed Engl 1999; 38: 98-102.

[76]   Forsén S. Long range spin coupling involving methoxy groups in aromatic compounds. J Phys Chem 1963; 67: 1740.

[77]   Mayer M, Meyer B. Characterization of ligand binding by saturation transfer difference NMR spectroscopy. Angew Chem Int Ed Engl 1999; 38: 1784-8.

[78]   Klein J, Meinecke R, Mayer M, Meyer B. Detecting binding affinity to immobilized receptor proteins in compound libraries by HR-MAS STD NMR. J Am Chem Soc 1999; 121: 5336.

[79]   Mayer M, Meyer B. Group epitope mapping by saturation transfer difference NMR to identify segments of a ligand in direct contact with a protein receptor. J Am Chem Soc 2001; 123 (25): 6108-17.

[80]   Megy S, Bertho G, Gharbi-Benarous J, et al. STD and TRNOESY NMR studies on the conformation of the oncogenic protein beta-catenin containing the phosphorylated motif DpSGXXpS bound to the beta-TrCP protein. J Biol Chem 2005; 280 (32): 29107-16.

[81]   Vogtherr M, Peters T. Application of NMR based binding assays to identify key hydroxy groups for intermolecular recognition. J Am Chem Soc 2000; 122: 6093-9.

[82]   Dalvit C, Pevarello P, Tato M, Veronesi M, Vulpetti A, Sundstrom M. Identification of compounds with binding affinity to proteins via magnetization transfer from bulk water. J Biomol NMR 2000; 18 (1): 65-8.

[83]   Dalvit C, Fogliatto G, Stewart A, Veronesi M, Stockman B. WaterLOGSY as a method for primary NMR screening: practical aspects and range of applicability. J Biomol NMR 2001; 21 (4): 349-59.

[84]   Dalvit C, Fasolini M, Flocco M, Knapp S, Pevarello P, Veronesi M. NMR-Based screening with competition water-ligand observed via gradient spectroscopy experiments: detection of high-affinity ligands. J Med Chem 2002; 45 (12): 2610-4.

[85]   Peng JW. Cross-correlated 19F relaxation measurements for the study of fluorinated ligand-receptor interactions. J Magn Reson 2001; 153: 32-47.

[86]   Gerig JT. Fluorine NMR of proteins. J Magn Reson 1994; 26: 293-370.

[87]   London RE, Gabel SA. Fluorine-19 NMR Studies of Fluorobenzeneboronic Acids. 1. Interaction Kinetics with Biologically Significant Ligands. J Am Chem Soc 1994; 116: 2562-9.

[88]   London RE, Gabel SA. Fluorine-19 NMR studies of fluorobenzenboronic acids. 2. Kinetic characterization of the interaction with subtilisin Carlsberg and model ligands. J Am Chem Soc 1994; 116: 2570-5.

[89]   Dalvit C, Flocco M, Veronesi M, Stockman BJ. Fluorine-NMR competition binding experiments for high-throughput screening of large compound mixtures. Comb Chem High Throughput Screen 2002; 5 (8): 605-11.

[90]   Dalvit C, Fagerness PE, Hadden DT, Sarver RW, Stockman BJ. Fluorine-NMR experiments for high-throughput screening: theoretical aspects, practical considerations, and range of applicability. J Am Chem Soc 2003; 125: 7696-703.

[91]   Jahnke W, Perez LB, Paris CG, Strauss A, Fendrich G, Nalin CM. Second-site NMR screening with a spin-labeled first ligand. J Am Chem Soc 2000; 122: 7394-5.

[92]   Breeze AL. Isotope-filtered NMR methods for the study of biomolecular structure and interactions. Prog NMR Spectrosc 2000; 36: 323-72.

[93]   Dalvit C, Bohlen JM. Proton phase-sensitive pulsed field gradient double-quantum spectroscopy. Ann Rep NMR Spectrosc 1999; 37: 203-71.

[94]   Moore JM. NMR techniques for characterization of ligand binding: utility for lead generation and optimization in drug discovery. Biopolymers 1999; 51 (3): 221-43.

[95]   Roberts GC. Applications of NMR in drug discovery. Drug Discov Today 2000; 5 (6): 230-40.

[96]   Craik DJ, Scanlon MJ. Pharmaceutical Applications of NMR. In: Webb GA, Ed. Annual Reprots on NMR Spectroscopy. New York: Academic Press 2000; pp. 115-174.

[97]   Moore JM. NMR screening in drug discovery. Curr Opin Biotechnol 1999; 10 (1): 54-58.

[98]   Jhoti H. Fragment-based drug discovery using rational design. Ernst Schering Found Symp Proc 2007; (3): 169-85.

[99]   Jhoti H, Cleasby A, Verdonk M, Williams G. Fragment-based screening using X-ray crystallography and NMR spectroscopy. Curr Opin Chem Biol 2007; 11 (5): 485-93.

[100] Pellecchia M. Fragment-based drug discovery takes a virtual turn. Nat Chem Biol 2009; 5 (5): 274-5.

[101] Congreve M, Carr R, Murray C, Jhoti H. A 'rule of three' for fragment-based lead discovery? Drug Discov Today 2003; 8 (19): 876-7.

[102] Lipinski CA. Drug-like properties and the causes of poor solubility and poor permeability. J Pharmacol Toxicol Methods 2000; 44 (1): 235-49.

[103] McCoy MA, Wyss DF. Alignment of weakly interacting molecules to protein surfaces using simulations of chemical shift perturbations. J Biomol NMR 2000; 18 (3): 189-98.

[104] Jencks WP. On the attribution and additivity of binding energies. Proc Natl Acad Sci USA 1981; 78 (7): 4046-50.

[105] Niedzwiecki L, Teahan J, Harrison RK, Stein RL. Substrate specificity of the human matrix metalloproteinase stromelysin and the development of continuous fluorometric assays. Biochemistry 1992; 31 (50): 12618-23.

[106] Bols M, Binderup L, Hansen J, Rasmussen P. Inhibition of collagenase by aranciamycin and aranciamycin derivatives. J Med Chem 1992; 35 (15): 2768-71.

[107] Wilhelm SM, Shao ZH, Housley TJ, et al. Matrix metalloproteinase-3 (stromelysin-1). Identification as the cartilage acid metalloprotease and effect of pH on catalytic properties and calcium affinity. J Biol Chem 1993; 268 (29): 21906-13.

[108] Olejniczak ET, Hajduk PJ, Marcotte PA, et al. Stromelysis Inhibitors Designed from Weakly Bound Fragments, Effects of Linking and Cooperativity. J Am Chem Soc 2002; 45: 5628-39.

[109] Hajduk PJ, Shuker SB, Nettesheim DG, et al. NMR-based modification of matrix metalloproteinase inhibitors with improved bioavailability. J Med Chem 2002; 45 (26): 5628-39.

[110] Chen JM, Nelson FC, Levin JI, et al. Structure-based design of a novel, potent, and selective inhibitor for MMP-13 utilizing NMR spectroscopy and computer-aided molecular design. J Am Chem Soc 2000; 122: 9648-54.

[111] Browner MF, Smith WW, Castelhano AL. Matrilysin-inhibitor complexes: common themes among metalloproteases. Biochemistry 1995; 34 (20): 6602-10.

[112] Morphy JR, Millican TA, Porter JR. Matrix metalloproteinase inhibitors: Current status. Curr Med Chem 1995; 2: 743-62.

[113]   Zask A, Levin JI, Killar LM, Skotnicki JS. Inhibition of matrix metalloproteinases: structure based design. Curr Pharm Des 1996; 2: 624-61.

[114]   Moy FJ, Chanda PK, Chen JM, *et al.* NMR solution structure of the catalytic fragment of human fibroblast collagenase complexed with a sulfonamide derivative of a hydroxamic acid compound. Biochemistry 1999; 38 (22): 7085-96.

[115]   Hajduk PJ, Dinges J, Miknis GF, *et al.* NMR-based discovery of lead inhibitors that block DNA binding of the human papillomavirus E2 protein. J Med Chem 1997; 40 (20): 3144-50.

[116]   Beckter TM, Stone KM. Genital human papillomavirus infection: a growing concern obstet gynecol. Clin North Am 1987; 14: 389-96.

[117]   Mohr IJ, Clark R, Sun S, Androphy EJ, MacPherson P, Botchan MR. Targeting the E1 replication protein to the papillomavirus origin of replication by complex formation with the E2 transactivator. Science 1990; 250 (4988): 1694-9.

[118]   Prakash SS, Grossman SR, Pepinsky RB, Laimins LA, Androphy EJ. Amino acids necessary for DNA contact and dimerization imply novel motifs in the papillomavirus E2 trans-activator. Genes Dev 1992; 6 (1): 105-16.

[119]   Jahnke W, Florsheimer A, Blommers MJ, *et al.* Second-site NMR screening and linker design. Curr Top Med Chem 2003; 3 (1): 69-80.

[120]   Boehm HJ, Boehringer M, Bur D, *et al.* Novel inhibitors of DNA gyrase: 3D structure based biased needle screening, hit validation by biophysical methods, and 3D guided optimization. A promising alternative to random screening. J Med Chem 2000; 43 (14): 2664-74.

[121]   Grandori C, Cowley SM, James LP, Eisenman RN. The Myc/Max/Mad network and the transcriptional control of cell behavior. Annu Rev Cell Dev Biol 2000; 16: 653-99.

[122]   Jain M, Arvanitis C, Chu K, *et al.* Sustained loss of a neoplastic phenotype by brief inactivation of MYC. Science 2002; 297 (5578): 102-4.

[123]   Park J, Wood MA, Cole MD. BAF53 forms distinct nuclear complexes and functions as a critical c-Myc-interacting nuclear cofactor for oncogenic transformation. Mol Cell Biol 2002; 22 (5): 1307-16.

[124]   Huth JR, Yu L, Collins I, *et al.* NMR-driven discovery of benzoylanthranilic acid inhibitors of far upstream element binding protein binding to the human oncogene c-myc promoter. J Med Chem 2004; 47 (20): 4851-7.

[125]   Ambrosini G, Adida C, Altieri DC. A novel anti-apoptosis gene, survivin, expressed in cancer and lymphoma. Nat Med 1997; 3 (8): 917-21.

[126]   Altieri DC. Survivin, versatile modulation of cell division and apoptosis in cancer. Oncogene 2003; 22 (53): 8581-9.

[127]   Altieri DC. Validating survivin as a cancer therapeutic target. Nat Rev Cancer 2003; 3 (1): 46-54.

[128]   Muchmore SW, Chen J, Jakob C, *et al.* Crystal structure and mutagenic analysis of the inhibitor-of-apoptosis protein survivin. Mol Cell 2000; 6 (1): 173-82.

[129]   Sun C, Nettesheim D, Liu Z, Olejniczak ET. Solution structure of human survivin and its binding interface with Smac/Diablo. Biochemistry 2005; 44 (1): 11-7.

[130]   Wendt MD, Sun C, Kunzer A, *et al.* Discovery of a novel small molecule binding site of human survivin. Bioorg Med Chem Lett 2007; 17 (11): 3122-9.

[131]   Cory S, Adams JM. The Bcl2 family: regulators of the cellular life-or-death switch. Nat Rev Cancer 2002; 2 (9): 647-56.

[132]   Borner C. The Bcl-2 protein family: sensors and checkpoints for life-or-death decisions. Mol Immunol 2003; 39 (11): 615-47.

[133]   van Delft MF, Huang DC. How the Bcl-2 family of proteins interact to regulate apoptosis. Cell Res 2006; 16 (2): 203-13.

[134]   Oltersdorf T, Elmore SW, Shoemaker AR, *et al.* An inhibitor of Bcl-2 family proteins induces regression of solid tumours. Nature 2005; 435 (7042): 677-81.

[135]   Petros AM, Dinges J, Augeri DJ, *et al.* Discovery of a potent inhibitor of the antiapoptotic protein Bcl-xL from NMR and parallel synthesis. J Med Chem 2006; 49 (2): 656-63.

[136]   Wendt MD, Shen W, Kunzer A, *et al.* Discovery and structure-activity relationship of antagonists of B-cell lymphoma 2 family proteins with chemopotentiation activity *in vitro* and *in vivo.* J Med Chem 2006; 49 (3): 1165-81.

[137]   Bruncko M, Oost TK, Belli BA, *et al.* Studies leading to potent, dual inhibitors of Bcl-2 and Bcl-xL. J Med Chem 2007; 50 (4): 641-62.

[138]   Cullinan SB, Whitesell L. Heat shock protein 90: a unique chemotherapeutic target. Semin Oncol 2006; 33 (4): 457-65.

[139]   Solit, DB, Rosen N. Hsp90: a novel target for cancer therapy. Curr Top Med Chem 2006; 6 (11): 1205-14.

[140]   Huth JR, Sun C. Utility of NMR in lead optimization: fragment-based approaches. Comb Chem High Throughput Screen 2002; 5 (8): 631-43.

[141]   Huth JR, Park C, Petros AM, *et al.* Discovery and design of novel HSP90 inhibitors using multiple fragment-based design strategies. Chem Biol Drug Des 2007; 70 (1): 1-12.

[142]   Duffy MJ. Urokinase-type plasminogen activator and malignancy. Fibrinolysis 1993; 7: 295-302.

[143]   Andreasen PA, Kjoller L, Christensen L, Duffy MJ. The urokinase-type plasminogen activator system in cancer metastasis: a review. Int J Cancer 1997; 72 (1): 1-22.

[144]   Huang JW, Zhang Z, Wu B, *et al.* Fragment-based design of small molecule X-linked inhibitor of apoptosis protein inhibitors. J Med Chem 2008; 51 (22): 7111-8.

[145]   Fesik SW, Shi Y. Structural biology. Controlling the caspases. Science 2001; 294 (5546): 1477-8.

[146]   Wang Z, Cuddy M, Samuel T, *et al.* Cellular, biochemical, and genetic analysis of mechanism of small molecule IAP inhibitors. J Biol Chem 2004; 279 (46): 48168-76.

[147]   Shiozaki EN, Chai J, Rigotti DJ, *et al.* Mechanism of XIAP-mediated inhibition of caspase-9. Mol Cell 2003; 11 (2): 519-27.

[148]   Srinivasula SM, Hegde R, Saleh A, *et al.* A conserved XIAP-interaction motif in caspase-9 and Smac/DIABLO regulates caspase activity and apoptosis. Nature 2001; 410 (6824): 112-6.

[149]   Ramstrom O, Lehn JM. Drug discovery by dynamic combinatorial libraries. Nat Rev Drug Discov 2002; 1 (1): 26-36.

[150]   Hochgurtel M, Biesinger R, Kroth H, *et al.* Ketones as building blocks for dynamic combinatorial libraries: highly active neuraminidase inhibitors generated via selection pressure of the biological target. J Med Chem 2003; 46 (3): 356-8.

[151]   Hochgurtel M, Kroth H, Piecha D, *et al.* Target-induced formation of neuraminidase inhibitors from *in vitro* virtual combinatorial libraries. Proc Natl Acad Sci USA 2002; 99 (6): 3382-7.

[152]   Orts J, Griesinger C, Carlomagno T. The INPHARMA technique for pharmacophore mapping: a theoretical guide to the method. J Magn Reson 2009; 200 (1): 64-73.

[153]   Sanchez-Pedregal VM, Reese M, Meiler J, Blommers MJ, Griesinger C, Carlomagno T. The INPHARMA method: protein-mediated interligand NOEs for pharmacophore mapping. Angew Chem Int Ed Engl 2005; 44 (27): 4172-5.

[154]   He L, Orr GA, Horwitz SB. Novel molecules that interact with microtubules and have functional activity similar to Taxol. Drug Discov Today 2001; 6 (22): 1153-64.

[155]   He L, Jagtap PG, Kingston DG, Shen HJ, Orr GA, Horwitz SB. A common pharmacophore for Taxol and the epothilones based on the biological activity of a taxane molecule lacking a C-13 side chain. Biochemistry 2000; 39 (14): 3972-8.

[156]   Grandy D, Shan J, Zhang X, *et al.* Discovery and characterization of a small molecule inhibitor of the PDZ domain of dishevelled. J Biol Chem 2009; 284 (24): 16256-63.

[157]   Bhanot P, Brink M, Samos CH, *et al.* A new member of the frizzled family from Drosophila functions as a Wingless receptor. Nature 1996; 382 (6588): 225-30.

[158]   Dann CE, Hsieh JC, Rattner A, Sharma D, Nathans J, Leahy DJ. Insights into Wnt binding and signalling from the structures of two Frizzled cysteine-rich domains. Nature 2001; 412 (6842): 86-90.

[159]   Wong HC, Bourdelas A, Krauss A, Lee HJ, Shao Y, Wu D, Mlodzik M, Shi DL, Zheng J. Direct binding of the PDZ domain of Dishevelled to a conserved internal sequence in the C-terminal region of Frizzled. Mol Cell 2003; 12 (5): 1251-60.

[160]   Wuthrich K. Protein recognition by NMR. Nat Struct Biol 2000; 7 (3): 188-9.

[161]   Klippel A Reinhard C, Kavanaugh WM, Apell G, Escobedo MA, Williams LT. Membrane localization of phosphatidylinositol 3-kinase is sufficient to activate multiple signal-transducing kinase pathways. Mol Cell Biol 1996; 16 (8): 4117-27.

[162]   Andjelkovic M, Jakubowicz T, Cron P, Ming XF, Han JW, Hemmings BA. Activation and phospho-rylation of a pleckstrin homology domain containing protein kinase (RAC-PK/PKB) promoted by serum and protein phosphatase inhibitors. Proc Natl Acad Sci USA 1996; 93 (12): 5699-704.

[163]   Auger KR, Serunian LA, Soltoff SP, Libby P, Cantley LC. PDGF-dependent tyrosine phosphorylation stimulates production of novel polyphosphoinositides in intact cells. Cell 1989; 57 (1): 167-75.

[164]   Datta K, Bellacosa A, Chan TO, Tsichlis PN. Akt is a direct target of the phosphatidylinositol 3-kinase. Activation by growth factors, v-src and v-Ha-ras, in Sf9 and mammalian cells. J Biol Chem 1996; 271 (48): 30835-9.

[165]   Maehama T, Dixon JE. The tumor suppressor, PTEN/MMAC1, dephosphorylates the lipid second messenger, phosphatidylinositol 3,4,5-trisphosphate. J Biol Chem 1998; 273 (22): 13375-8.

[166]   Steck PA, Pershouse MA, Jasser SA, *et al.* Identification of a candidate tumour suppressor gene, MMAC1, at chromosome 10q23.3 that is mutated in multiple advanced cancers. Nat Genet 1997; 15 (4): 356-62.

[167]   Cantley LC. The phosphoinositide 3-kinase pathway. Science 2002; 296 (5573): 1655-7.

[168]   Rameh LE, Cantley LC. The role of phosphoinositide 3-kinase lipid products in cell function. J Biol Chem 1999; 274 (13): 8347-50.

[169]   Biondi RM, Cheung PC, Casamayor A, Deak M, Currie RA, Alessi DR. Identification of a pocket in the PDK1 kinase domain that interacts with PIF and the C-terminal residues of PKA. EMBO J 2000; 19 (5): 979-88.

[170]   Biondi RM. Phosphoinositide-dependent protein kinase 1, a sensor of protein conformation. Trends Biochem Sci 2004; 29 (3): 136-42.

[171]   Stockman BJ, Kothe M, Kohls D, *et al.* Identification of allosteric PIF-pocket ligands for PDK1 using NMR-based fragment screening and 1H-15N TROSY experiments. Chem Biol Drug Des 2009; 73 (2): 179-88.

[172]   Cohen JS, Lyon RC, Daly PF. Monitoring intracellular metabolism by nuclear magnetic resonance. Methods Enzymol 1989; 177: 435-52.

[173]   Nicholson JK, Lindon JC, Holmes E. 'Metabonomics': understanding the metabolic responses of living systems to pathophysiological stimuli via multivariate statistical analysis of biological NMR spectroscopic data. Xenobiotica 1999; 29 (11): 1181-9.

[174]   Ross BD. The biochemistry of living tissues: examination by MRS. NMR Biomed 1992; 5 (5): 215-9.

[175]   Nicholson JK, Connelly J, Lindon JC, Holmes E. Metabonomics: a platform for studying drug toxicity and gene function. Nat Rev Drug Discov 2002; 1 (2): 153-61.

[176]   Griffiths JR, Stevens AN, Iles RA, Gordon RE, Shaw D. 31P-NMR investigation of solid tumours in the living rat. Biosci Rep 1981; 1 (4): 319-25.

[177]   Griffiths JR, Cady E, Edwards RH, McCready VR, Wilkie DR, Wiltshaw E. 31P-NMR studies of a human tumour in situ. Lancet 1983; 1 (8339): 1435-6.

[178]   Griffiths JR, Glickson JD. Monitoring pharmacokinetics of anticancer drugs: non-invasive investigation using magnetic resonance spectroscopy12. Adv Drug Deliv Rev 2000; 41 (1): 75-89.

[179]   Ronen SM, Jackson LE, Beloueche M, Leach MO. Magnetic resonance detects changes in phosphocholine associated with Ras activation and inhibition in NIH 3T3 cells1. Br J Cancer 2001; 84 (5): 691-6.

[180]   Stubbs M, Griffiths JR. Monitoring cancer by magnetic resonance. Br J Cancer 1999; 80 (Suppl 1): 86-94.

[181]   Wolf W, Waluch V, Presant CA. Non-invasive 19F-NMRS of 5-fluorouracil in pharmacokinetics and pharmacodynamic studies. NMR Biomed 1998; 11 (7): 380-7.

[182]   Shungu DC, Bhujwalla ZM, Wehrle JP, Glickson JD. 1H NMR spectroscopy of subcutaneous tumors in mice: preliminary studies of effects of growth, chemotherapy and blood flow reduction. NMR Biomed 1992; 5 (5): 296-302.

[183]   Artemov D, Bhujwalla ZM, Glickson JD. *In vivo* selective measurement of (1-13C)-glucose metabolism in tumors by heteronuclear cross polarization. Magn Reson Med 1995; 33 (2): 151-5.

[184]   He Q, Bhujwalla ZM, Maxwell RJ, Griffiths JR, Glickson JD. Proton NMR observation of the antineoplastic agent Iproplatin *in vivo* by selective multiple quantum coherence transfer (Sel-MQC). Magn Reson Med 1995; 33 (3): 414-6.

[185]   Artemov D, Bhujwalla ZM, Maxwell RJ, Griffiths JR, Judson IR, Leach MO, Glickson JD. Pharmacokinetics of the 13C labeled anticancer agent temozolomide detected *in vivo* by selective cross-polarization transfer. Magn Reson Med 1995; 34 (3): 338-42.

[186]   Zektzer AS, Swanson MG, Jarso S, Nelson SJ, Vigneron DB, Kurhanewicz J. Improved signal to noise in high-resolution magic angle spinning total correlation spectroscopy studies of prostate tissues using rotor-synchronized adiabatic pulses. Magn Reson Med 2005; 53 (1): 41-8.

[187]   Swanson MG, Keshari KR, Tabatabai ZL, Simko JP, Shinohara K, Carroll PR, Zektzer AS, Kurhanewicz J. Quantification of choline- and ethanolamine-containing metabolites in human prostate tissues using 1H HR-MAS total correlation spectroscopy. Magn Reson Med 2008; 60 (1): 33-40.

[188]   Chen JH, Sambol EB, Kennealey PT, O'Connor RB, DeCarolis PL, Cory DG, Singer S. Water suppression without signal loss in HR-MAS 1H NMR of cells and tissues. J Magn Reson 2004; 171 (1): 143-50.

[189] Chen JH, Sambol EB, Decarolis P, O'Connor R, Geha RC, Wu YV, Singer S. High-resolution MAS NMR spectroscopy detection of the spin magnetization exchange by cross-relaxation and chemical exchange in intact cell lines and human tissue specimens. Magn Reson Med 2006; 55 (6): 1246-56.

[190] Post JF, Baum E. 31P nuclear magnetic resonance studies of growth inhibition and dexamethasone resistance in human leukemic cells. Cancer Lett 1990; 51 (2): 157-62.

[191] Adebodun F, Post JF. 31P NMR characterization of cellular metabolism during dexamethasone induced apoptosis in human leukemic cell lines. J Cell Physiol 1994; 158 (1): 180-6.

[192] Rocha M, Kruger A, Van RN, Schirrmacher V, Umansky V. Liver endothelial cells participate in T-cell-dependent host resistance to lymphoma metastasis by production of nitric oxide *in vivo*. Int J Cancer 1995; 63 (3): 405-11.

[193] Lutz NW, Tome ME, Cozzone PJ. Early changes in glucose and phospholipid metabolism following apoptosis induction by IFN-gamma/TNF-alpha in HT-29 cells. FEBS Lett 2003; 544 (1-3): 123-8.

[194] Williams SN, Anthony ML, Brindle KM. Induction of apoptosis in two mammalian cell lines results in increased levels of fructose-1,6-bisphosphate and CDP-choline as determined by 31P MRS. Magn Reson Med 1998; 40 (3): 411-20.

[195] Damadian R. Tumor detection by nuclear magnetic resonance. Science 1971; 171 (976): 1151-3.

[196] Howells SL, Maxwell RJ, Howe FA, *et al.* Pattern recognition of 31P magnetic resonance spectroscopy tumour spectra obtained *in vivo*. NMR Biomed 1993; 6 (4): 237-41.

[197] Preul MC, Caramanos Z, Collins DL, Villemure JG, Leblanc R, Olivier A Pokrupa R, Arnold DL. Accurate, noninvasive diagnosis of human brain tumors by using proton magnetic resonance spectroscopy. Nat Med 1996; 2 (3): 323-5.

[198] Bathen TF, Jensen LR, Sitter B, Fjosne HE, Halgunset J, Axelson DE, Gribbestad IS, Lundgren S. MR-determined metabolic phenotype of breast cancer in prediction of lymphatic spread, grade, and hormone status. Breast Cancer Res Treat 2007; 104 (2): 181-9.

[199] Gribbestad IS, Sitter B, Lundgren S, Krane J, Axelson D. Metabolite composition in breast tumors examined by proton nuclear magnetic resonance spectroscopy. Anticancer Res 1999; 19 (3A): 1737-46.

[200] Sitter B, Lundgren S, Bathen TF, Halgunset J, Fjosne HE, Gribbestad IS. Comparison of HR MAS MR spectroscopic profiles of breast cancer tissue with clinical parameters. NMR Biomed 2006; 19 (1): 30-40.

[201] Cheng LL, Wu C, Smith MR, Gonzalez RG. Non-destructive quantitation of spermine in human prostate tissue samples using HRMAS 1H NMR spectroscopy at 9.4 T. FEBS Lett 2001; 494 (1-2): 112-6.

[202] Swanson MG, Zektzer AS, Tabatabai ZL, *et al.* Quantitative analysis of prostate metabolites using 1H HR-MAS spectroscopy. Magn Reson Med 2006; 55 (6): 1257-64.

[203] Serkova NJ, Spratlin JL, Eckhardt SG. NMR-based metabolomics: translational application and treatment of cancer. Curr Opin Mol Ther 2007; 9 (6): 572-85.

[204] Kline EE, Treat EG, Averna TA, Davis MS, Smith AY, Sillerud LO. Citrate concentrations in human seminal fluid and expressed prostatic fluid determined *via* 1H nuclear magnetic resonance spectroscopy outperform prostate specific antigen in prostate cancer detection. J Urol 2006; 176 (5): 2274-9.

[205] Odunsi K, Wollman RM, Ambrosone CB, *et al.* Detection of epithelial ovarian cancer using 1H-NMR-based metabonomics. Int J Cancer 2005; 113 (5): 782-8.

[206] El-Deredy W, Ashmore SM, Branston NM, Darling JL, Williams SR, Thomas DG. Pretreatment prediction of the chemotherapeutic response of human glioma cell cultures using nuclear magnetic resonance spectroscopy and artificial neural networks. Cancer Res 1997; 57 (19): 4196-9.

[207] Morvan D, Demidem A. Metabolomics by proton nuclear magnetic resonance spectroscopy of the response to chloroethylnitrosourea reveals drug efficacy and tumor adaptive metabolic pathways. Cancer Res 2007; 67 (5): 2150-9.

[208] Siegel RA, Tolcsvai L, Rudin M. Partial inhibition of the growth of transplanted dunning rat prostate tumors with the long-acting somatostatin analogue sandostatin (SMS 201-995). Cancer Res 1988; 48 (16): 4651-5.

[209] Furr BJ. The development of Casodex (bicalutamide): preclinical studies. Eur Urol 1996; 29 (Suppl 2): 83-95.

[210] Cohen SM, Werrmann JG, Rasmusson GH, *et al.* Comparison of the effects of new specific azasteroid inhibitors of steroid 5 alpha-reductase on canine hyperplastic prostate: suppression of prostatic DHT correlated with prostate regression. Prostate 1995; 26 (2): 55-71.

[211] Hausler A, Allegrini PR, Biollaz M, Batzl C, Scheidegger E, Bhatnagar AS. CGP 53153: a new potent inhibitor of 5alpha-reductase. J Steroid Biochem Mol Biol 1996; 57 (3-4): 187-95.

[212] Qin Y, Van CM, Osteaux M, Schally AV, Willems G. Inhibitory effect of somatostatin analogue RC-160 on the growth of hepatic metastases of colon cancer in rats: a study with magnetic resonance imaging. Cancer Res 1992; 52 (21): 6025-30.

[213] Vigneri P, Wang JY. Induction of apoptosis in chronic myelogenous leukemia cells through nuclear entrapment of BCR-ABL tyrosine kinase. Nat Med 2001; 7 (2): 228-34.

[214] Serkova N, Boros LG. Detection of resistance to imatinib by metabolic profiling: clinical and drug development implications. Am J Pharmacogenomics 2005; 5 (5): 293-302.

[215] Gottschalk S, Anderson N, Hainz C, Eckhardt SG, Serkova NJ. Imatinib (STI571)-mediated changes in glucose metabolism in human leukemia BCR-ABL-positive cells. Clin Cancer Res 2004; 10 (19): 6661-8.

[216] Kamm VJ, Rietjens IM, Vervoort J, *et al.* Effect of modulators on 5-fluorouracil metabolite patterns in murine colon carcinoma determined by *in vitro* 19F nuclear magnetic resonance spectroscopy. Cancer Res 1994; 54 (16): 4321-6.

[217] Street JC, Alfieri AA, Traganos F, Koutcher JA. *In vivo* and *ex vivo* study of metabolic and cellular effects of 5-fluorouracil chemotherapy in a mouse mammary carcinoma 2. Magn Reson Imaging 1997; 15 (5): 587-96.

[218] Scheiber C, Kiss R, De Launoit Y, Fruhling J. MXT mammary tumor treated by low dose chemotherapy. A 31P-NMR and 1H-MRI study. Anticancer Res 1988; 8 (3): 403-7.

[219] Street JC, Mahmood U, Matei C, Koutcher JA. *In vivo* and *in vitro* studies of cyclophosphamide chemotherapy in a mouse mammary carcinoma by 31P NMR spectroscopy. NMR Biomed 1995; 8 (4): 149-58.

[220] Rodrigues LM, Maxwell RJ, McSheehy PM, Pinkerton CR, Robinson SP, Stubbs M, Griffiths JR. *In vivo* detection of ifosfamide by 31P-MRS in rat tumours: increased uptake and cytotoxicity induced by carbogen breathing in GH3 prolactinomas. Br J Cancer 1997; 75 (1): 62-8.

[221] Hoffer FA, Taylor GA, Spevak M, Ingber D, Fenton T. Metabolism of tumor regression from angiogenesis inhibition: 31P magnetic resonance spectroscopy. Magn Reson Med 1989; 11 (2): 202-8.

[222] Jiang Q, Chopp M, Hetzel FW. *In vivo* 31P NMR study of combined hyperthermia and photodynamic therapies of mammary carcinoma in the mouse. Photochem Photobiol 1991; 54 (5): 795-9.

[223]   Rodrigues LM, Midwood CJ, Coombes RC, Stevens AN, Stubbs M, Griffiths JR. 31P-nuclear magnetic resonance spectroscopy studies of the response of rat mammary tumors to endocrine therapy. Cancer Res 1988; 48 (1): 89-93.

[224]   Semmler W, Bachert-Baumann P, Guckel F, *et al.* Real-time follow-up of 5-fluorouracil metabolism in the liver of tumor patients by means of F-19 MR spectroscopy. Radiology 1990; 174 (1): 141-5.

[225]   Chung YL, Troy H, Banerji U, *et al.* Magnetic resonance spectroscopic pharmacodynamic markers of the heat shock protein 90 inhibitor 17-allylamino, 17-demethoxygeldanamycin (17AAG) in human colon cancer models. J Natl Cancer Inst 2003; 95 (21): 1624-33.

[226]   Sankaranarayanapillai M, Tong WP, Maxwell DS, *et al.* Detection of histone deacetylase inhibition by noninvasive magnetic resonance spectroscopy. Mol Cancer Ther 2006; 5 (5): 1325-34.

[227]   Sankaranarayanapillai M, Kaluarachchi K, Ronen SM. In: Chung Y-L, Griffiths JR, Eds. Using metabolomics to monitor anticancer drugs. 13C MRS detection of increased choline metabolism following HDAC inhibition 2007; p. 123.

Structure-Activity Relationship Studies in Drug Development by NMR Spectroscopy, 2011, 1, 145-155

Targeting NMR Parameters and Dynamics of Radiosensitizers in Solution: Theoretical Studies of Prototypical Some Bioreductive Drugs

Teodorico C. Ramalho[*], Elaine F. F. da Cunha and Marcus V. J. Rocha

Universidade Federal de Lavras, Campus Universitário – UFLA, Dept. de Química, 37200-000, Lavras-MG - Brazil

Abstract: Recently, ^{1}H, ^{31}P, and ^{19}F NMR magnetic resonance imaging MRI of radiosensitizers has been applied for measuring tumor and tissue oxygenation. In spite of the great importance of such bioreductive drugs, urthermore, surprisingly little detailed computational work on this subject has appeared. Thus, in this work a detailed computational study of BD is presented, calling special attention to the performance of various theoretical methods in reproducing the ^{13}C and ^{15}N, coupling-constant (H-N) data observed in solution. The most sophisticated approach involves density functional based Car–Parrinello molecular dynamics simulations (CPMD) in aqueous solution and averaging chemical shifts over snapshots from the trajectory. In the NMR calculations for these snapshots (performed at the B3LYP level), a small number of discrete water molecules are retained, and the remaining bulk solution effects are included via a polarizable continuum model (PCM). A similarly good accord with experiment is obtained from much less involved, static geometry optimization and NMR computation of pristine **1** employing a PCM approach. Solvent effects on chemical shift are not due to changes in geometric parameters upon solvation, but arise from the direct response of the electronic wave function to the presence of the solvent, which can be represented by discrete molecules and/or the dielectric bulk.

Keywords: ^{15}N NMR, radiosensitizers, molecular dynamics simulations, solvent effects, spin–spin coupling constants.

INTRODUCTION

Bioreductive drugs are used to combat cancer and, in spite of early progress, cancer [1,2] is still one of the most serious problems of humanity, causing many deaths and having a great and limiting influence on life quality and development of many countries. Nowadays, it is well known that the high rate of tumor cell proliferation increases oxygen consumption in tumor tissue [2]. Furthermore, structural and functional abnormalities and function in tumor vessels lead to decreased oxygen delivery to tumor tissue, [3] because of poor vascularization and potentially high oxygen demand. These hypoxic cells are resistant to radiation therapy and to some kinds of chemotherapy [4, 5]. Therefore, tumor hypoxia can be exploited for selective anticancer drug treatment using hypoxic cell cytoxins or hypoxic cell radiosensitizers [6]. Hypoxic cell radiosensitizers, such as electron-affinic compounds, behave as oxygen mimicking compounds, affecting tumor hypoxic cells, thus leading to radiation-induced damage to DNA and to other molecules [7]. Thus, normally inert compounds, which are activated by enzymes or by radiation under hypoxic conditions, will behave selectively toward tumors. Ideally, the compound is reduced reversibly so that, in normal cells, it can readily revert to its inactive form [8].

The most commonly used radiosensitizers are 5-nitroimidazole derivatives [9, 10], because the first nitroimidazole drugs such as metronidazole (Fig. **1**, compound **1**) exhibit many undesirable side-effects, research efforts are being concentrated on finding suitable alternatives, either based on the same nitroimidazole motif or with bioreducible groups other than nitro [11]. In our previous work, we studied the physical chemistry properties associated with the biological activity of radiosensitizers in aqueous and carbon tetrachloride solution [12-14]. Recently, ^{1}H, ^{31}P and ^{19}F NMR/MRI (magnetic resonance imaging) [15] of nitroimidazoles (Fig. **1**) has been applied for measuring tumor and tissue oxygenation [16]. In spite of the great importance of such radiosensitizers, the solvent and dynamic effects on their NMR parameters have not yet been investigated. Furthermore, surprisingly little detailed computational work on this subject has appeared.

*Corresponding author: Tel: +55 (35) 3829-1891; Fax: +55 (35) 3829-1271; E-mails: teo@dqi.ufla.br, teodorico@dacafe.com

Atta-ur-Rahman / M. Iqbal Choudhary (Eds.)

Ab initio computation of NMR parameters, such as chemical shifts or coupling constants, can be very useful to support recorded spectra and investigate structure–property relationships [17, 18]. In fact, calculations based on density functional theory (DFT) with a suitable choice of functionals are able to provide accurate values for many NMR properties [19]. Theoretical approaches to compute NMR parameters are now implemented in very efficient codes using computer facilities with rapidly increasing performance. This allows researchers to complement measurements on an almost routine basis with theoretical calculations that provide further information on configuration and conformations and insight into intramolecular interactions that may be the ultimate sources for the experimental trends. On the other hand, the usual approach, the computation of magnetic shieldings for isolated molecules in some static structure, obtained either from experiment or from quantum chemical optimizations, may be insufficient if comparison is sought with experiments conducted in polar solvents and at ambient temperature. Protic solvents such aswater not only provide polar media, but also can interact with solutes by way of hydrogen bonds. Modeling the concomitant effects on structures and properties of these solutes is a considerable challenge for computational chemistry [20]. Hydrogen bonds in liquids can be assessed using geometric and energetic criteria, [21] and can be crucial for solvent effects in spectroscopy [22]. Apart from shielding constants, indirect nuclear spin–spin coupling constants (SSCCs) are further important NMR parameters that play an essential role in the conformational analysis of biochemical compounds such as polypeptides [23], nucleic acids [24] and metalloproteins [25]. The dependence of measured of SSCC values on the stereochemical relationship between the coupled nuclei is well documented both experimentally and theoretically [24]. Nowadays, the need for inexpensive computational techniques capable of reliably predicting SSCCs stems from this, arguably the most important application area of high-resolution NMR spectroscopy. Since the modern tools of DFT are capable of describing large parts of the electron correlation and relativistic effects, DFT has been emerging as the most popular method for the calculation of SSCCs [19, 26, 27]. It is clear that DFT is especially suitable for larger molecules, such as biological systems, because it is an inexpensive computational technique capable of reliably predicting NMR parameters [26]. Our first goal in this study was to obtain the 15N NMR chemical shifts of the radiosensitizers depicted in Fig. (**1**), in order to complement the full spectroscopic characterization of these compounds. In addition, we intended to model and predict these and other experimental parameters using modern computational tools, calling attention to thermal and solvation effects on spectroscopic properties.

EXPERIMENTAL

NMR Spectroscopy

All 5-nitroimidazoles were supplied by Rhodia Laboratories and by the Pharmacy Department, University of Sao Paulo (USP) Fig. (**1**).

	Compound	R_1	R_2
	1	CH_2CH_2OH	CH_3
	2	$CH_2CH(OH)CH_3$	CH_3
	3	$CH_2CH_2N\diagup O$	H
	4	$CH_2CH_2SO_2CH_2CH_3$	CH_3

Fig. (1). Compounds used in this study.

^{13}C and ^{15}N NMR spectra were recorded under standard conditions on a Bruker DMX 600 spectrometer equipped with an inverse 5 mm probehead, using ca 1.0 M samples in DMSO-d6 at 300 K. ^{13}C chemical shifts are given relative to TMS using the signal of the deuterated lock solvent (at δ D 39.5) as internal reference. The ^{15}N resonances were detected indirectly via ^1H using the HMBC protocol [28] and are given relative to nitromethane as an external standard. Geometries and molecular dynamics procedure Geometries were fully optimized using the gradient corrected BP86 combination of density functionals [29] and 6–31G basis set, either in vacuo or in the presence of a polarizable continuum in the integral equation formalism model [30-33], using the dielectric constants of water or DMSO (denoted PCM(H$_2$O) or PCM(DMSO), respectively). The initial optimization of pristine **1** was started from the coordinates taken from the structure in the solid [34]. Molecular dynamics (MD) simulations were performed using the DFT-based Car-Parrinello scheme CPMD) [35]. Since the time-scale accessible to the CPMD method is very restricted, no extensive equilibration is possible and care must be taken to start from reasonably well pre-equilibrated configurations. To this end, we first prepared a classical system through an MD simulation using the consistent valence force field (CVFF) [36] with the Insight/Discover program (version 2.9.7). As usual, periodic boundary conditions (PBC) and a cutoff distance of 9.0 °A were applied [37]. The system consists of the substrate, embedded in an array of TIP3P37 water molecules in a cubic cell with a side of 11.5 Å. The MD simulation was performed at 300 K as an NVT (or canonical) ensemble. First, the initial configuration was minimized using the steepest descent and the conjugate gradient algorithm until an energy gradient of 0.01 kcal mol Å^{-1} (1 kcal = 4.184 kJ). The simulation consisted of a thermalization stage of 500 ps, followed by an additional period of 1 ns, which is sufficient to equilibrate the system completely. Using the last configuration from classical MD as the starting point, we subsequently started CPMD simulations [35] using the CPMD program [38]. The BP86 functional was used, together with norm-conserving Troullier–Martins pseudopotentials [39] in the Kleinman–Bylander form [40]. Periodic boundary conditions were imposed using cubic super cells with box sizes 11.5 Å. Kohn–Shan orbitals were expanded in plane waves up to a kinetic energy cut-off of 80 Ry, and a fictitious electronic mass of 600 a.u. was used. To allow for an increased simulation time step, which was chosen as 0.121 fs, hydrogen was replaced with deuterium. After an equilibration time of 0.5 ps, in which a temperature of 300 ps 50 K was maintained via velocity rescaling, statistical averages and snapshots for the NMR calculations were collected from subsequent unconstrained microcanonical runs 1ps long. Snapshots from the CPMD simulations were taken every 25 fs and were used for the chemical shift calculations. In addition, equilibrium geometries for isolated molecules were obtained by optimizing the forces on all atoms with the CPMD program using the setup detailed above (denoted CP-opt). NMR calculations Magnetic shieldings δ were computed for equilibrium geometries and for snapshots taken from the MD simulations at the GIAO (gauge-including atomic orbitals)–B3LYP level, [41, 42] employing basis II [43] for the solute (which is of polarized triple-zeta quality for non-hydrogen atoms and of double-zeta quality for H) and DZ basis [43] set for solvent (when water molecules were included explicitly). ^{15}N and ^{13}C NMR chemical shifts δ were calculated relative to nitromethane and benzene, respectively, optimized or simulated at the same level. The corresponding δ values are collected in Table **1** (see the section Chemical shifts below for a detailed description of the notation). No bulk solvent effects using PCM methods were evaluated for these reference compounds, as they are used in neat form experimentally, rather than dissolved in a polar solvent. The resulting ^{13}C chemical shifts were converted to the usual TMS scale using the experimental value for benzene of δ(^{13}C) = 128.5 ppm [26].

Table 1. ^{15}N and ^{13}C NMR Experimental Data (ppm) for Nitroimidazoles Compounds

Compd.	C-2	C-4	C-5	C-6	C-7	C-8	C-9	C-10	N-1	N-3	N-8	N-11
1	151.91	132.92	138.37	14.19	48.23	59.73	---	---	-217.0	-120.2	---	*
2	151.91	132.85	138.46	14.37	52.55	65.55	20.82	---	-216.2	-120.4	---	*
3	143.43	133.14	138.44	44.08	57.53	---	53.13	66.14	-211.8	-120.3	-341.7	-25.0
4	151.37	133.00	138.33	13.83	38.90	49.61	46.88	5.84	-219.0	-119.7	---	*

* Not observed.

The SSCC calculations were carried out at the B3LYP level employing the EPR-II basis set [44]. All NMR calculations were carried out using Gaussian 03 [45].

RESULTS AND DISCUSSION

Experimental Data

When we began our studies, NMR spectroscopic characterization of the known species **1-4** was incomplete. Specifically, ^{15}N chemical shifts, to our knowledge, were known only for one example, **2** [46]. We have now obtained the full set of δ (^{13}C) and δ (^{15}N) data (Table **2**). As expected, all the nitroimidazoles show a single peak each for N-1 and N-3, around δ 220 ppm and δ 120 ppm, respectively, in good agreement with other experimental results for similar compounds [46, 47]. The δ ^{15}N value for N-3 is almost invariant along the series, whereas a slight variation (up to ca 7 ppm) is found for N-1, which bears the different substituents, R_1. Interestingly, the chemical shift of this nitrogen atom appears to be more sensitive to the substituent R_2 at the adjacent C atom than to the directly attached R (compare, e.g., N1 entries for **3** and **4** in Table **2**). The $N(O_2)$ (nitro group) resonance was not detected, except in **3** (see below). The ^{13}C data have been reported before [46, 47], and are collected in Table **2**. For **1**, **2** and **4** the carbon C-2 signal is almost invariably around 151 ppm. Also, the carbon atoms in the imidazole ring adjacent to the nitro group do not show significant differences in the series **1-4** (133 and 138 ppm for C-4 and C-5, respectively).

Table 2. ^{15}N and ^{13}C NMR Computed the GIAO-BP86/II Level

Level of Approximation[a]	C-2	C-4	C-5	N-1	N-3
δ_e (//CP-opt)	154.88	137.85	142.01	-207.73	-87.26
δ_e (//BP86)	151.45	134.05	140.49	-210.26	-93.65
δ_e (//BP86/PCM(H_2O))	153.27	136.45	140.17	-220.03	-102.87
δ_e (//BP86/PCM(DMSO))	153.35	136.41	140.21	-219.97	-103.14
δ_e (PCM(H_2O)//BP86/PCM(H_2O))	157.82	137.66	142.25	-213.49	-116.50
δ_e (PCM(DMSO) //BP86/PCM(H_2O))	157.73	137.69	142.09	-213.84	-116.02
δ_e (PCM(H_2O)//BP86/PCM(DMSO))	157.85	137.60	142.30	-213.35	-116.79
δ_e (PCM(DMSO) //BP86/PCM(DMSO))	157.74	137.63	142.12	-213.76	-116.56
δ_e (//CP-opt(1w))	154.97	132.28	140.99	-219.64	-119.96
δ_e (//BP86(1w))	157.26	134.70	143.33	-217.63	-122.28
δ^{300K} (//CPMD)	154.27	136.74	143.43	-221.00	-104.53
δ^{300K} (//CPMD(1w))	155.94	133.47	143.01	- 210.90	-114.55
δ^{300K} (//CPMD(H_2O))[d]	154.20	139.43	141.11	-216.87	-118.48
δ^{300K} (CPMD//CPMD(H_2O)[d])	152.69	136.71	139.79	-220.70	-98.12
δ^{300K} (CPMD/PCM(H_2O)//CPMD(H_2O)[d])	158.06	139.06	141.11	-217.53	-114.45
δ^{300K} (CPMD(H_2O)[d]/PCM(H_2O) //CPMD(H_2O)[d])	157.13	138.36	142.66	-217.31	-117.39
δ Experimental	*151.91*	*132.92*	*138.37*	*- 217.00*	*-120.20*

[a]Notation "level of chemical shift computation// level of geometry optimization or MD simulation".

Chemical shifts among the nitroimidazole derivatives, metronidazole (**1**) is a proto-typical representative, which has been widely used in the treatment of anaerobic protozoan and bacterial infections [1, 48]. We therefore chose this species as the target for extensive tests of the computational methodologies available for computation of NMR properties, calling special attention to the chemical shifts of the C and N atoms in the five membered heterocycle. From static calculations for **1** in vacuo, we did not find significant differences in the chemical shifts between structures optimized with the same functional (BP86), but using different basis sets and methodologies as implemented in the CPMD and G03 programs (compare δ (//CP-opt) and (//BP86) values in Table **3**). Both sets of data agree within ca 6 and 5 ppm for δ (^{15}N) and δ (^{13}C), respectively. This variation is due to minor geometric differences between both optimized structures, as illustrated in the bond lengths in Table **4**, which can change by 1-2 pm. Ongoing from the static equilibrium values to the thermal averages in the gas phase, the concomitant changes in the δ values are small for carbon and large for nitrogen atoms, up to 2.2 and 19.9 ppm, respectively (compare δ (//CP-opt) and δ 300K(//CPMD) values in Table **3**). That the dynamic effects are larger on ^{15}N than on ^{13}C chemical shifts is in part related to the general observation that the former responds more sensitively than the latter to changes in the chemical environment, as reflected in the different chemical shift ranges of both nuclei. In addition, the presence of a lone pair adjacent to unsaturated bonds at N-3 gives rise to a small energy gap between magnetically coupled MOs, thus producing large paramagnetic contributions to the magnetic shielding constant of this nucleus [43], and it is usually these paramagnetic contributions that are sensitive to structural and electronic effects. Both the N-1 and N-3 chemical shifts from δ 300K(CPMD) approach closer to the experimental values than those from static calculation (Table **3**). The theoretical methodologies used to describe solvent effects on spectroscopic properties of molecules can be classified into several groups: continuum models, [31, 49] discrete cluster models [50] and molecular dynamics, usually with classical propagation of the nuclei [51, 52]. Each has its advantages and drawbacks, and careful validation is necessary in order to identify a particular method (or a combination thereof) that affords reliable results. In this paper, we assess these methods for the chemical shifts of the prototypical radiosensitizer **1**.

Table 3. Bound Length Values for Metronidazole (1) in Å

Level of Approximation	N1-C2	N1-C5	N1-C10	N3-C2	N3-C4	C4-C5	C2-C9	C5-N6
δ_e (//CP-opt)	1.3936	1.3882	1.4949	1.3293	1.4031	1.3879	1.4733	1.4190
δ_e (//BP86)	1.3749	1.3985	1.4726	1.3415	1.3612	1.3888	1.4927	1.4293
δ_e (//BP86/PCM(H_2O))	1.3695	1.3983	1.4768	1.3492	1.3597	1.3918	1.4910	1.4195
δ_e (//BP86/PCM(DMSO))	1.3698	1.3984	1.4768	1.3490	1.3598	1.3917	1.4909	1.4199
δ_e (//CP-opt(1w))	1.3707	1.3978	1.4731	1.3446	1.3570	1.3835	1.4887	1.4250
δ_e (//BP86(1w))	1.3720	1.3989	1.4737	1.3456	1.3615	1.3865	1.4918	1.4314
δ^{300K} (//CPMD)	1.3792	1.4045	1.4796	1.3442	1.3602	1.3898	1.4960	1.4318
δ^{300K} (//CPMD/1w)	1.3770	1.4003	1.4808	1.3463	1.3628	1.3864	1.4948	1.4340
δ^{300K} (//CPMD(H_2O))^d	1.3682	1.3993	1.4839	1.3513	1.3484	1.3932	1.4841	1.4032
δ Experimental^a	1.3512	1.3829	1.4752	1.3332	1.3581	1.3365	1.4791	1.4138

^{a} x-Ray Structure paper.

As a continuum approach, we used the well-known polarizable cotinuum model (PCM). This method has the advantages over earlier, simpler variants that the cavity is not limited to spherical or ellipsoidal cases, but has a realistic shape, and that the electrostatic problem is solved exactly without resorting to truncated multipolar exanpansions [53].

The PCM method can be applied at two stages, first in the geometry optimization and second during the evaluation of the magnetic shielding constants [33, 54, 55]. The results in Table **3** show that both types of solvent effect play an important role. For the equilibrium chemical shifts, we found changes of about 2 and 10 ppm for δ (^{13}C) and δ (^{15}N), respectively, on going from the gas-phase geometry to that optimized in a continuum with the dielectric parameters of water compare δe(//BP86) and δ[//BP86/PCM(H_2O)] entries in Table **3**. On the other hand, when the solvent effect is, in addition, introduced in the NMR calculation, the differences increase to about 6 ppm for δ (^{13}C), and up to ca 23 ppm for N-3 compare δ (//BP86) and δ[PCM(H_2O)//BP86/PCM(H_2O)] values. As expected, the solvent effect is most pronounced for N-3 with its lone pair (see the discussion of the paramagnetic contributions above). As the NMR spectra were recorded in DMSO solution, we also used the PCM method with DMSO as dielectric, for both optimization and NMR calculation steps. As expected, virtually no difference between the two solvents is obtained in the PCM calculations compare, for instance, δ[PCM(DMSO)//BP86/PCM(DMSO)] and δ[PCM(H_2O)//BP86/PCM(H_2O)] entries in Table **3**. Upon inspection of the structural parameters, we did not find significant effects of a surrounding continuum: on going from gas to solution phase we noted just a slight decrease in the N-1—C-2 and N-1—C-10 bonds lengths compare re(//BP86) and re[//BP86/PCM(H_2O)] data in Table **4**.

Table 4. ^{15}N and ^{13}C NMR (ppm) Computed with PCM(H_2O)//BP86/PCM(H_2O) in Different Scaling Factors (⊗) for Metronidazole

Scaling Factor	C-2	C-4	C-5	N-1	N-3
1.0	158.56	137.98	142.97	-213.61	-116.63
1.1	158.12	137.81	142.59	-213.54	-116.58
1.2	157.82	137.66	142.25	-213.49	-116.50
1.3	156.04	136.64	140.95	-216.72	-108.14
1.4	155.67	136.53	140.77	-217.32	-106.78
δ Experimental	*151.91*	*132.92*	*138.37*	*- 217.0*	*- 120.2*

It is well known that the cavity definition in continuum solvation methods can have a large impact on the computed properties [20, 49]. In the PCM approach that we applied, the cavity is defined in terms of spheres centered on atoms and with radii, $\Re_\zeta$, proportional to van der Waals radii (see Equation 1) [31].

$$\Re_\zeta = \otimes \Re_\zeta^{vdW} \tag{1}$$

In order to determine how the computed NMR constants depend on the cavity size, we performed additional PCM calculations using other cavity scaling factors. The results are reported in Table **5**. With lower f, i.e. with a smaller cavity, the solvent effect is amplified. As expected, the ^{15}N chemical shift values are more strongly influenced by cavity size than ^{13}C chemical shifts, and the PCM chemical shifts can be sensitive to the size of the cavity, in particular forN-3. It is worth noting that the chemical shift for this atom, obtained with the standard value for $f = 1.2$, agrees fairly well with experiment. However, a significant increase in δ^{15}N), by more than 8 ppm, is obtained on changing f slightly from 1.2 to 1.3. Apparently, the PCM data are associated with a notable uncertainty when the lone pair at this N atom is located close to the boundary region.

Table 5. **Calculated Nitrogen-14 NQC Constants χ (in MHz) and Asymetry Parameter η for Nitro Group of the Compounds 1 and 3 in Gas Phase**

	χ	η
1	1.1801	1.9969
3	1.1992	1.9731

This lone pair, however, can form specific interactions with the solvent; in particular, it can act as a hydrogen-bond acceptor. This effect cannot be realistically modeled in a PCM approach, and discrete solvent molecules have to be included. As a first model, we considered **1** with one explicit water molecule bound to N-3, denoted (1w) in Tables **3** and **4**. It is reasonable to assume that the traces of water present in the solvent of the NMR experiment, DMSO (where no special efforts were undertaken to eliminate these traces), will be readily available to form such hydrogen-bonded complexes. The presence of this single water molecule not only affects the chemical shift of N-3, where it is attached, but also that of C-2 and N-1, compare, for instance, δ(//BP86) and δ[//BP86(1w)] entries in Table **3**. Qualitatively, the solvent effects on ^{15}N and ^{13}C are very similar when assessed by a continuum model or by a single water molecule compare δ[//BP86(1w)] and δ[PCM(H$_2$O)//BP86/PCM(H$_2$O)] data in Table **3**. Of the chemical shifts obtained with these two models, those using the explicit solvent turned out to fit slightly better to the experimental data.

When, however, dynamic effects are included for this cluster model, all nuclei are noticeably deshielded, up to ca 2 and 9 ppm for ^{13}C and ^{15}N, respectively, contrasting δ[//CP-opt(1w)] and δ300K [//CPMD(1w)] values, which reduces the accord between theory and experiment somewhat. Clearly, a single solute–solvent cluster in a particular arrangement is not necessarily a good description of the actual situation of the dissolved molecule. We modeled the latter by performing CPMD simulations of 1 embedded in a periodic water box, followed by quantum-mechanical evaluations of the chemical shift averaged over snapshots along the trajectory. In the NMR calculations, smaller (nonperiodical) sections of the full (periodic) system were taken. Similar approaches have been adopted before, for instance, to study the gas-to-liquid shift in water [56] or to model transition metal chemical shifts in aqueous solution [57].

The number of solvent molecules to be included in the NMR computations was determined as follows. First, the number of water molecules involved in hydrogen bonds with **1** was determined based on purely geometric criteria, assuming an H-bond if the distance between an N or O atom from **1** and an O atom from the water is <3.5 °A and the OH bond is directed towards the acceptor atom such that the X—H—O angle is >140° [60]. The resulting average number was 4.8, with minimum and maximum values of 3 and 7, respectively. On average, roughly one water molecule is attached to N-3 and each of the O atoms from the nitro group, and the alcoholic function in the side-chainR1 donates and accepts one H-bond (Fig. **2**) (for a representative snapshot). The magnetic shielding constants were calculated for a typical snapshot, keeping just the substrate and its H-bonded water molecules. In subsequent calculations, more and more water molecules from the actual snapshot were included, adding them successively based on the shortest distance between the O (water) atom and the center of the imidazole ring. It turned out that the δ values of interest did not change significantly when more than seven water molecules in total were present. This number was therefore chosen for a total of 40 snapshots, over which the shieldings were averaged.

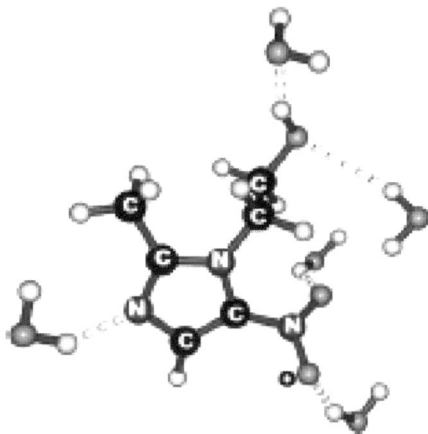

Fig. (2). Typical snapshot from a CPMD simulation of **1** in water, showing hydrogen-bonded solvent molecules.

On going from the gas phase to water, the correspondingly averaged $\delta 300K$ (^{13}C) values are little affected (compare//CPMD and//CPMD(H$_2$O) values in Table **3**). For the $\delta 300K$ (^{15}N) of N-3, we observe a significant shielding of about 14 ppm, similar to the solvent effect estimated with static PCM calculations (see above). When the water molecules were deleted in NMR snapshots from the solution, affording unsolvated metronidazole (**1**) in the geometries of solvated ones, the resulting $\delta 300K$ averages (CPMD//CPMD(H$_2$O) values in Table **3**) are similar to those of gaseous **1** (//CPMD entries), suggesting that it is not the geometric change on salvation that is responsible for the gas-to-liquid shifts which would be an indirect effect, as found, e.g., in BH$_3$NH$_3$ [58] or [Fe(CN)$_6$]$_4$ [57], but rather the direct interaction with the solvent molecules, due to the response of the electronic wave function. The same conclusion could be inferred from the close resemblance of the δ values for the geometries optimized in vacuo and in a dielectric (see above). In a final step, we included bulk medium effects for the snapshots from the CPMD simulation in solution by way of the PCM approach discussed above. The results with and without explicit water molecules [CMPD/PCM(H$_2$O) *vs.* CPMD(H$_2$O)/PCM(H$_2$O) entries at the bottom of Table **3**] are fairly similar, although a better accord with experiment is obtained when both discrete water molecules and the dielectric are included. This agreement suggests that the CPMD simulations in water offer a reasonable description of the structure and dynamics of the solution. As far as the chemical shifts are concerned, it is noteworthy that the much simpler, PCM-based approach gives δ values close to the most sophisticated CPMD-based averages (compare the δ values for PCM(H$_2$O)//BP86/PCM(H$_2$O) and CPMD(H$_2$O)/PCM(H$_2$O)//CPMD(H$_2$O) in Table **3**) and also to experiment. For routine NMR computations for nitroimidazole derivatives, static PCM results for PCM optimized minima therefore appear to be an inexpensive, viable alternative to the more involved MD simulations.

It should be kept in mind, however, that the sensitivity to computational details (see the discussion of the cavity size above) may limit the general applicability of the PCM approach. It has also been shown, for instance, that PCM methods perform less well for ^{17}O chemical shifts [59].

nJ(N,H) Coupling Constants

One particular compound, nimorazole (**3**), is unique because it is the only compound within the series **1-4** for which the ^{15}N resonance of the nitro group could be detected experimentally (Table **2**). In the NMR experiments, an indirect detection technique was applied, which is based on the transfer of magnetization from a more sensitive nucleus, a proton in our case, to the heteronucleus in question. A prerequisite for this technique is that both nuclei need to be involved in detectable (if not necessarily resolvable) mutual indirect spin–spin coupling, which actually marks the limitation of this technique: ^{15}N nuclei that are only weakly coupled to protons elsewhere in the molecule cannot be detected.24 For N-1 and N-3 in all compounds **1-4**, there exist 2J(N,H) coupling pathways with SSCCs large enough, apparently, for indirect detection of 15N. These couplings are observed as cross peaks in the 2D (^{15}N,^{1}H) correlation spectra. For the nitro resonances, the shortest coupling pathway is via 3J(N,H) to the imidazole proton at C-4. This pathway is longer and, in addition, involves coupled nuclei in syn positions, which usually results in diminished SSCCs and thereby provides a rationale for the usual 'nonobservation' of the nitro resonance in 5-nitroimidazoles.

Compound **3** differs from the other members in that it lacks the methyl group at C-2, which bears a proton instead. Detailed analysis of the $^1H/^{15}N$ and $^1H/^{13}C$ 2D NMR spectra shows unambiguously that it is indeed the coupling to this proton (which is not present in the other derivatives) that makes the nitro resonance detectable in this case. This path is rather long, involving four-bond coupling, 4J(N,H), but the coupled nuclei are in trans positions ('W' arrangement), which could facilitate the interaction between the nuclei [59,60]. Hoping for further insights, we computed the salient SSCCs at a suitable DFT level, calling special attention to the 3J(N,H) and 4J(N,H) values for both compounds **3** and **1**. The results are summarized in Table **6**. In the light of the experiences with chemical shifts discussed in the preceding section, we employed BP86/PCM(H$_2$O) geometries, i.e. optimized in a dielectric continuum.

Table 6.   **N-H Coupling Constant J (Hz) for N-11 in 1 and 3 with both Constringed (1.3738 Å) and Optimised C4-C5 Bond Length at Level B3LYP/EPR-II**

		^{3}J(N,H6)	^{4}J(N,H2)	^{3}J(N,H4)
1				
	N-1	0.5084	1.2608	-0.7057
	N-3	1.6911	8.7176	-0.1507
	N-11	-0.1085	-0.1208	-0.0994
	N-11^a	-0.0995	-0.0990	-0.3308
3				
	N-1	1.3555	6.2893	-0.6516
	N-3	8.7566	9.2720	-0.0969
	N-11	-0.1402	0.4582	-0.0883

^{a} C4-C5 bond length constringed.

In both 1 and 3, N-1 and N-3 show the largest SSCCs with the proton at C-4, with notable values well above 1 Hz. These couplings can be detected as cross peaks in the 2D spectra recorded in the course of the indirect NMR measurements. The other couplings involving N-1 and N-3 in Table **6** can also be detected, except J(N-3,H(R1)), the absolute values of which are computed to be <0.2 Hz. The smallest value in Table **6** that can be detected is J(N-1,H(R2)) in 1, at 0.5 Hz. Apparently, couplings smaller than this value were not observed experimentally. For the nitro N atom in 1, all computed absolute J(N,H) couplings were indeed much smaller, around 0.1 Hz and below, which explains why this ^{15}N resonance was not detected.

Based on these results and on the above-mentioned analysis of the coupling paths, the absolute SSCC between the N(O$_2$) nucleus and the proton at C-2 in **3** should be at least 0.5 Hz. A much smaller absolute value is computed, however, ca 0.1 Hz (Table **6**). One computed SSCC is found in the detectable region

Table 7. Orbital Interaction Energies (Kcal/mol) at the B3LYP/6-311G (d,p) Level

Interaction	1	3
n_{N3}/σ^*_{C4-H}	2.15	2.02
n_{N3}/σ^*_{C4-C5}	5.75	5.67
n_{N3}/σ^*_{N1-C2}	8.77	8.10
n_{N3}/σ^*_{N1-C10}	0.63	0.51
n_{N3}/σ^*_{C5-N6}	0.65	0.61
n_{N3}/σ^*_{C2-C9}	0.73	- - -
$\pi_{C2-N3}/\pi^*_{C4-C5}$	32.92	26.11
π_{N6-O}/π^*_{C4-C5}	4.59	4.60
n_{N1}/π^*_{C4-C5}	32.28	29.08
Wiberg Index		
C4-C5	1.4375	1.4512

around 0.5 Hz, but this coupling involves the proton at C-4, rather than that at C-2. Based on this result, one would have erroneously predicted that the cross peak involving $N(O_2)$, observed in the 2D NMR spectrum of 3, would arise from coupling to the proton at C-4. Apparently, such small, long-range couplings transmitted through δ-systems are difficult to describe theoretically, and the modern hybrid DFT variants such as B3LYP, which has proven to be reliable for many types of SSCCs so far, [27] can even fail to predict correctly the relative magnitudes of 3J and 4J values. The mechanism of spin–spin coupling across δ-systems has long been studied theoretically [59]. Breakdown of the computed total SSCCs into the basic contributions reveals that, as expected, the Fermi contact (FC) term dominates the couplings in Table **6**. In view of the apparent disagreement with experiment, we refrain from a deeper analysis of these long-range couplings (Table **7**).

CONCLUSION

We have determined the ^{15}N chemical shifts of a number of nitroimidazole-based radiosensitizers, complementing the spectroscopic characterization of this important class of compounds. Thermal and solvent effects on the chemical shifts of metronidazole (**1**) were studied computationally with appropriate quantum-chemical methods. These effects can be notable, in particular for N-3, which bears a lone pair amenable to hydrogen bonding with a protic solvent. A DFT-based molecular dynamics simulation of the bulk aqueous solution offers a realistic description of the system, and good agreement is obtained between observed chemical shifts (^{13}C and ^{15}N) and computed values averaged over the MD trajectory. A similar accord is obtained with a much less involved approach based on geometry optimization and chemical shift calculation in a polarizable continuum.

In the second part, we computed SSCCs involving ^{15}N and ^1H nuclei, motivated by the importance of these couplings for the indirect detection of the ^{15}N chemical shifts. It turned out that small, long-range SSCCs can be a problem for approximate DFT methods, since the

Otherwise reliable B3LYP functional could not correctly reproduce the observation that in one particular molecule a 4J(N,H) value is larger than 3J(N,H). This apparent inconsistency notwithstanding, theoretical computations of NMR parameters are emerging as useful complements to NMR spectroscopic studies of radiosensitizers. Further applications along these lines are in progress.

REFERENCES

[1] Brown JM, Koong A. Therapeutic advantage of hypoxic cells in tumors: a theoretical study. J Natl Cancer Inst 1991; 83(3): 178-85.
[2] Hatherill JR. Eat to beat cancer. a research scientist explains how you and your family can avoid up to 90% of all cancers. New York: Renaissance Books 1998.
[3] Kasai S, Nagasawa H, Yamashita M, *et al*. New antimetastatic hypoxic cell radiosensitizers: design, synthesis, and biological activities of 2-nitroimidazole-acetamide, TX-1877, and its analogues. Biorg Med Chem 2001; 9(2): 453-64.
[4] Teicher BA. Hypoxia and drug resistance. Cancer Metastasis Rev 1994; 13(2): 139-68.
[5] Tomida A, Tsuruo T. Drug resistance mediated by cellular stress response to the microenvironment of solid tumors. Anti-Cancer Drug Des 1999; 14(2): 169-77.
[6] Brown JM, Wouters BG. Apoptosis, p53, and tumor cell sensitivity to anticancer agents. Cancer Res 1999; 59(7): 1391-9.
[7] Horis H, Nagasawa H. In advances in environmental science and technology. Vol. 28: Environmental Oxidants. New York: Wiley, 1994. p 62.
[8] Viodé C, Bettache N, Cenas N, *et al*. Enzymatic reduction studies of nitroheterocycles. Biochem Pharmacol 1999; 57(5): 549-57.
[9] Hori H, Jin CZ, Kiyono M, Kasai S, Shimamura M, Inayama S. Design, synthesis, and biological activity of anti-angiogenic hypoxic cell radiosensitizer haloacetylcarbamoyl-2-nitroimidazoles. Biorg Med Chem 1997; 5(3): 591-9.
[10] Chapman JD, Coia LR, Stobe CC, Engelhardt EL, Fenning MC, Schneider RF. Prediction of tumour hypoxia and radioresistance with nuclear medicine markers. Br J Cancer 1996; 27: S204-8.
[11] Ames AR, Ryan MD, Kovaic P. Mode of action of antiprotozoan agents. Electron transfer and oxy radicals. Life Sci 1987; 41(16): 1895-902.
[12] Ramalho TC, Alencastro RB, La-Scalea MA, Figueroa-Villar JD. Theoretical evaluation of adiabatic and vertical electron affinity of some radiosensitizers in solution using FEP, *ab initio* and DFT methods. Biophys Chem 2004; 110(3): 267-79.
[13] Ramalho TC, Cunha EFF, Alencastro RB. Theoretical study of adiabatic and vertical electron affinity of radiosensitizers in solution part 2: analogues of tirapazamine. J Theor Comput Chem 2004; 3(1): 1-13.
[14] Ramalho TC, Cunha EFF, Alencastro RB. Solvent effects on ^{13}C and ^{15}N shielding tensors of nitroimidazoles in the condensed phase: a sequential molecular dynamics/quantum mechanics study. J Phys Condens Matter 2004; 16(34): 6159-70.
[15] Anderson CJ, Welch MJ. Radiometal-labeled agents (non-technetium) for diagnostic imaging. Chem Rev 1999; 99(9): 2219-34.
[16] McCoy CL, McIntyre DJO, Robinson SP, Aboagye EO, Griffiths JR. Magnetic resonance spectroscopy and imaging methods for measuring tumour and tissue oxygenation. Br J Cancer 1996; 27(Suppl): S226-31.

[17] Zilm KW, Duchamp JC. In nuclear magnetic shieldings and molecular structure. Tossel JA, Ed. NATO ASI Series C386. Dordrecht: Kluwer 1993; p. 315.

[18] Kaupp M, Bühl M, Malkin VG, Eds. Calculation of NMR and ESR parameters. Theory and Applications. Weinheim: Wiley-VCH 2004.

[19] Bühl M. Density functional computation of 55Mn NMR parameters. Theor Chem Acc 2002; 107(6): 336-42.

[20] Orozco M, Luque FJ. Theoretical methods for the description of the solvent effect in biomolecular systems. Chem Rev 2000; 100(11): 4187-226.

[21] Rahman A, Stilinger FH. Molecular dynamics study of liquid water. J Chem Phys 1971; 55(7): 3336-60.

[22] JorgensenWL, Chandrasekhar J, Madura JD, Impey RW, Klein ML. Comparison of simple potential functions for simulating liquid water. J Chem Phys 1983; 79(2): 926-36.

[23] Munzarová ML, Sklenář V. DFT Analysis of NMR scalar interactions across the glycosidic bond in DNA. J Am Chem Soc 2003; 125(12): 3649-58.

[24] Grant DM, Harris RK, Eds. Encyclopedia of nuclear magnetic resonance. vols. 1-8. Chichester: Wiley 1996.

[25] Torrent M, Mansour D, Day EP, Morokuma K. Quantum chemical study on oxygen-17 and nitrogen-14 nuclear quadrupole coupling parameters of peptide bonds in α-helix and β-sheet proteins. J Phys Chem A 2001; 105(18): 4546-57.

[26] Helgaker T, Jszunski M, Ruud K. *Ab Initio* methods for the calculation of nmr shielding and indirect spin-spin coupling constants. Chem Rev 1999; 99(1): 293-352.

[27] Helgaker T, Pecul M. In Calculation of NMR and ESR parameters. theory and applications, Kaupp M, Bühl M, Malkin VG, Eds. Weinheim: Wiley-VCH 2004, p. 101.

[28] Ernst RR, Bodenhausen G. Principles of nuclear magnetic resonance in one and two dimensions. Oxford: Clarendon Press; 1987.

[29] Becke AD. Density-functional exchange-energy approximation with correct asymptotic behaviour. Phys Rev A 1988; 38(6): 3098-100.

[30] Cances TM, Mennucci B, Tomasi J. A new integral equation formalism for the polarizable continuum model: Theoretical background and applications to isotropic and anisotropic dielectrics. J Chem Phys 1997; 107(8): 3032-41.

[31] Barone V, Cossi M, Tomasi J. Geometry optimization of molecular structures in solution by the polarizable continuum model. J Comput Chem 1998; 19(4): 404-17.

[32] Cossi M, Barone V, Mennucci B, Tomasi J. *Ab initio* study of ionic solutions by a polarizable continuum dielectric model. Chem Phys Lett 1998; 286(3-4): 253-60.

[33] Cossi M, Scalmani G, Rega N, Barone V. New developments in the polarizable continuum model for quantum mechanical and classical calculations on molecules in solution. J Chem Phys 2002; 117(1): 43-54.

[34] Calva-Tejada N, Bernes S, Castillo-Blum SE, Nöth H, Vicente R, Barba-Behrens N. Supramolecular structures of metronidazole and its copper(II), cobalt(II) and zinc(II) coordination compounds. J Inorg Biochem 2002; 91(1): 339-48.

[35] Car R, Parrinello M. Unified approach for molecular dynamics and density-functional theory. Phys Rev Lett 1985; 55(22): 2471-4.

[36] Dauber-Osguthorpe P, Roberts VA, Osguthorpe DG, Wolff J, Genest M, Hagler AT. Structure and energetics of ligand binding to proteins: *Escherichia coli* dihydrofolate reductase-trimethoprim, a drug-receptor system. Proteins 1988; 4(1): 31-47.

[37] Allen MP, Tildesley DJ. Computer Simulation of Liquids. Oxford: University Press 1987.

[38] Hutter J, Deutsch AA, Bernasconi M, *et al.* CPMD Version 3.1a. Max-Planck-Institut für Festkörperforshung: Stuttgart and IBM Research Laboratory: Zürich, 1995-9.

[39] Troullier, Martins JL. Efficient pseudopotentials for plane-wave calculations. Phys Rev B 1991; 43(3): 1993-2006.

[40] Kleinman L, Bylander DM. Efficacious form for model pseudopotentials. Phys Rev Lett 1982; 48(20): 1425-8.

[41] Cheeseman JR, Trucks GW, Keith TA, Frisch MJ. A comparison of models for calculating nuclear magnetic resonance shielding tensors. J Chem Phys 1996; 104(14): 5497-509.

[42] Lee C, Yang W, Parr RG. Development of the Colle-Salvetti correlation-energy formula into a functional of the electron density. Phys Rev B 1988; 37(2): 785-9.

[43] Kutzelnigg W, Fleischer U, Schindler M. In NMR basic principles and progress. vol. 23. Berlin: Springer 1990; p. 165.

[44] Barone V. In Recent advances in density functional methods. Part I, Chong DP, Ed. Singapore World Scientific 1996, p. 287.

[45] Frisch MJ, Trucks GW, Schlegel HB, *et al.* Gaussian 03. Gaussian: Pittsburgh, PA, 2003.

[46] Chen BC, von Philipsborn W, Nagarajan, K. ^{15}N-NMR spectra of azoles with two heteroatoms helv. Chim Acta 1983; 66(5): 1537-55.

[47] Mckillop A, Wright DE, Podmore ML, Chambers RK. 4- and 5-nitroimidazoles: ^{13}C NMR assignment of structure. Tetrahedron 1983; 39(22): 3797-800.

[48] Silvestri R, Artico M, Massa S, Marceddu T, De Montis F, La Colla P. 1-[2-(Diphenylmethoxy)ethyl]-2-methyl-5-nitroimidazole: a potent lead for the design of novel NNRTIs. Bioorg. Med Chem Lett 2000; 10(3): 253-6.

[49] Cossi M, Crescenzi O. Solvent effects on [17]O nuclear magnetic shielding: *N*-methylformamide in polar and apolar solutions. Theor Chem Acc 2004; 111(2-6): 162-7.

[50] Pivnenko NS, Drushlyak TG, Kutulya LA, Vashchenko VV, Doroshenko AO, Goodby JW. Conformational analysis of some 1*R*, 4S-2-arylidene-*p*-menthan-3-ones by ^{1}H NMR spectroscopy and molecular simulation. Magn Reson Chem 2002; 40(9): 566-72.

[51] Martins TLC, Ramalho TC, Figueroa-Villar JD, Flores AFC, Pereira CMP. Theoretical and experimental ^{13}C and ^{15}N NMR investigation of guanylhydrazones in solution. Magn Reson Chem 2003; 41(12): 983-8.

[52] Sebastiani D, Parrinello M. A new ab-Initio approach for NMR chemical shifts in periodic systems. J Phys Chem A 2001; 105(10): 1951-8.

[53] Cossi M, Crescenzi O. Different models for the calculation of solvent effects on [17]O nuclear magnetic shielding. J Chem Phys 2003; 118(19): 8863-72.

[54] Cremer D, Olsson L, Reichel F, Kraka E. Calculations of NMR chemical shift - the third dimension of quantum chemistry. Isr J Chem 1993; 33: 369-85.

[55] Mennucci B, Martínez JM, Tomasi J. Solvent effects on nuclear shieldings: continuum or discrete solvation models to treat hydrogen bond and polarity effects? J Phys Chem A 2001; 105(30): 7287-96.

[56]   Malkin VG, Malkina OL, Steinebrunner G, Huber H. Solvent effect on the nmr chemical shieldings in water calculated by a combination of molecular dynamics and density functional theory. Chem Eur J 1996; 2(4): 452-7.

[57]   Bühl M, Schurhammer R, Imhof P. Peroxovanadate imidazole complexes as catalysts for olefin epoxidation: density functional study of dynamics, ^{51}V NMR chemical shifts, and mechanism. J Am Chem Soc 2004; 126(10): 3310-20.

[58]   Bühl M, Steinke T, Schleyer PvR, Boese R. Solvation effects on geometry and chemical shifts. An *Ab Initio*/IGLO reconciliation of apparent experimental inconsistencies on $H_3B \cdot NH_3$. Angew Chem Int Ed Engl 1991; 30(9): 1160-1.

[59]   Barfield M, Chakrabarty B. Long-range proton spin-spin coupling. Chem Rev 1969; 69(6): 757-78.

[60]   Barfield M. DFT/FPT studies of the structural dependencies of long-range ^{1}H,^{1}H coupling over four bonds ^4J(H,H) in propanic and allylic systems. Magn Reson Chem 2003; 41(5): 344-58.

Evaluation of Host/Guest Cyclodextrin Complex by NMR in Drug Product Development

Giorgio Zoppetti[*]

*R&D Pharma Department, IBSA Institut Biochimique SA –Via al Ponte 13-6903
Lugano 3 (CH), Switzerland*

Abstract: The application of NMR technology in drug development is generally adopted to describe the structure of a drug substance and its degradation products under standard stressed conditions. As far as drug products are concerned, standard stability conditions are recommended by international health authorities (ICH conditions), while the stability profile of a drug substance can be defined by non-standard protocols. Both proton and carbon NMR spectroscopic technologies are powerful methods for the determination of a drug substance chemical structure and related degradation products. Unconventional use of NMR in drug development is described in this section, namely NMR applications in the development of drugs containing cyclodextrin inclusion complexes. Cyclodextrins (CD) are cyclic oligosaccharides whose lipophilic cavity and hosting capacity make them ideal for forming inclusion complexes with lipophilic drugs. Generally, this complexation involves the inclusion of a 'guest' molecule into the cavity of a CD 'host' molecule, with no covalent bonding. Inclusion and/or molecular dispersion of a drug in cyclodextrin are made possible by the partitioning of the dissolved drug between the aqueous phase and the hydrophobic cyclodextrin cavity followed by specific molecular interactions, including hydrogen bonding and hydrophobic interactions between the drug and cyclodextrin. Proton NMR is used to confirm the host/guest inclusion by the chemical shift of the cyclodextrin internal protons, which are disturbed by the lipophilic guest. Examples of approved drug products containing cyclodextrins and two detailed case studies involving NMR applications in the development of water soluble diclofenac sodium and progesterone for parenteral use are reported in this article.

Keywords: Drug development, heparin, cyclodextrin, progesterone, diclofenac, NMR.

INTRODUCTION

In the common language, "Drug" refers to any substance which, when absorbed into the body of a living organism, alters normal bodily functions [1]. It is therefore extremely important that everything related to "Drugs" be treated with extreme caution in order to avoid any health hazards for patients. A "Drug" is the final result of extensive research, development, as well as industrial and commercial activities governed by international and local laws. It is one of the most important specific duties for local governments to take care of citizens' health with the support of international organizations such as the World Health Organization (WHO) or local health authorities, e.g. the European Medicines Agency (EMEA) in the European Union or the Food and Drug Administration (FDA) in the USA, which are involved in the definition of general directives on medicinal products. These organizations were established in the last century to coordinate both scientific knowledge and regulatory issues related to remedies against human diseases, i.e. "Drugs".

The discovery of a new drug is a process based on scientific research and drug regulatory know-how. A scientific background is essential to cover all concepts related to the chemical, pharmaceutical and therapeutic action of a drug substance, but a "Drug" must agree with human physiology and anatomy and its quality grade must meet the requirements of regulatory directives. The efficacy and tolerability of a molecule might be easy to demonstrate, but to obtain the approval of a drug product by the competent regulatory authorities might prove to be an impossible task. This concept has changed in the past twenty years and it has become very clear that an improvement of analytical tools is of paramount importance in order to meet ever increasingly restrictive regulatory requirements.

State-of-the art analytical tools allow us to measure the components of a drug substance in picograms prompting regulatory authorities to ask for a more complete impurities pattern than in the past. On the other

---

*Corresponding author: Tel: +410583601460; Fax: +410583601462; E-mail: giorgio.zoppetti@ibsa.ch

Atta-ur-Rahman / M. Iqbal Choudhary (Eds.)

hand, the same analytical progress has contributed to a parallel improvement of purification systems to obtain highly pure bulk substances to be used in pharmaceuticals. In recent years analytical methods have usually been developed in the final stage of a drug development process. This has caused reduced use of available analytical tools, with the result that some advanced techniques were ignored. In a modern view of "Drugs" quality the importance of analytical tools is growing.

The concept that the quality of a drug product should be defined during the project development phase (quality by design [2]) has recently come under discussion. ICH Q8 part 2 is an addendum document released in November 2007 by the ICH steering committee which provides further clarifications of key concepts outlined in the core guideline. The philosophy behind this addendum is that minimal approach to pharmaceutical development needs to be perfected in order to reach a quality by design approach. This concept is applicable to all drug development specific areas, such as pharmaceutical development and manufacturing process, as well as to all issues related to drug development, such as process controls, product specifications, control strategy and lifecycle management.

In the pharmaceutical development area the minimal approach recommended at present means that step-by-step empirical experiments are carried out. This simple approach will be encouraged or even become mandatory in the near future with the following recommendations: **(a)** a systematic understanding of input material attributes and process parameters to drug product "critical quality attributes" (CQA); **(b)** application of multivariate experiments to understand the product under test; **(c)** establishment of the design space and finally **(d)** adoption of Process Analytical Technology (PAT) tools to gain complete control of a drug product from the initial development stage. PAT tools are statistical methods used to acquire information from a complex population of data. The increasing attention of international authorities on the quality aspects of drug products and drug substances from an early stage of the development process requires the use of adequate instrumental tools to meet modern standards.

Nowadays, the characterization of a chemical molecule structure is performed to a minor extent by chemical methods and more by using instrumental analytical tools such as Nuclear Magnetic Resonance (NMR), infrared (IR) and UV/VIS spectroscopy, mass spectroscopy (MS), also combined with chromatographic systems (liquid HPLC or gas GC). All the available instrumental methods provide important structure information and all available techniques should be used in order to acquire a complete structure profile, especially when dealing with a complex molecule. NMR is the ultimate technique for the characterization of the fine structure of a molecule as well as for measuring intermolecular distances. In fact, while IR and UV provide information (even quantitative) on chemical groups in a molecule and sophisticated MS instruments can practically provide a single *"formula bruta"* of each entity in a multi-component solution, NMR provides a relationship between chemical groups, allowing the characterization of a complex structure.

NMR IN DRUG DEVELOPMENT

In an analytical environment, Nuclear Magnetic Resonance (NMR) is mostly used to detect proton (^1H) and carbon thirteen (^{13}C) in organic molecules. Other elements such as nitrogen or phosphorus, can be measured with special experiments. Owing to its great advantage to selectively measure individual atomic species, NMR enjoys a leadership position among analytical techniques for fine chemical structure determination. Proton and carbon spectra are complementary in final structure determination and advanced techniques can be applied to elucidate fine structures and chemical group's distances. The limitation of NMR technology is the interference of organic molecules which are mixed with the drug substance and the complex preparation process of the sample before the experiment. Proton and carbon (carbon thirteen accounts for 1.1% of the total carbon) are always present in organic molecules. This means that it could be difficult to perform a fine structural analysis of a single molecule in a complex mixture because of the highly possible overlapping of signals from other substances. Furthermore, water is very often present in drug substances and products: unless it is removed, it will interfere with the proton spectrum in an informative area from 4 to 5 ppm and if present in high amounts it can cause interferences also in the ^{13}C spectrum. The reduction of water protons as well as other "mobile protons", such as hydroxyls and amines groups is a delicate process to be performed under deuterium atmosphere, therefore it must be carried out by experienced technicians and it is time consuming. Mobile protons from chemical groups exchange freely with water in diluted water buffers, therefore the classical method to reduce those protons is an exchange with deuterium, which is not detectable in the proton or carbon NMR analysis. The sample is then diluted and exsiccated several times to reduce the residual mobile protons to a minimum. To guarantee a good result, the operation must be performed under low humidity conditions, or even better, in a deuterated water atmosphere. This operation is not necessary in carbon experiments, but it is recommended for fine structure definition during high resolution experiments to avoid carbon duplets due to the presence of both deuterium and proton in similar quantities in the same chemical group.

Owing to its peculiar qualitative structure determination and its relatively late arrival in the pharmaceutical field, NMR is currently mentioned as a characterization method in just a few official pharmacopeia monographs. It was approved in those cases where intrinsic structure heterogeneity is present. Sodium low molecular weight heparin (LMWH) [3] and hydroxypropyl-β-cyclodextrin (HPBCD) [4] are two examples. It is interesting to observe how in LMW heparin testing, NMR is used as a qualitative determination by comparing the main signals of the test sample with an internal standard of the carbon spectrum, whereas for HPBCD the test is quantitative and measures the relative proportion of the unsubstituted hydroxyl groups. Both substances are intrinsically heterogeneous, but with different degrees of complexity.

LMWH consists of a linear disaccharide repetition constituted by about 70% of a "regular region" made by N,6-disulphated –glucosamine linked 1-4 to 2-sulphated iduronic acid (Fig. **1a**) and 30% by several variations in which glucosamine can be N-acetylated or 6-O-non sulphated or 3-O-sulphated and 2-O sulphated Iduronic acid can be glucuronic acid (Fig. **1b**).

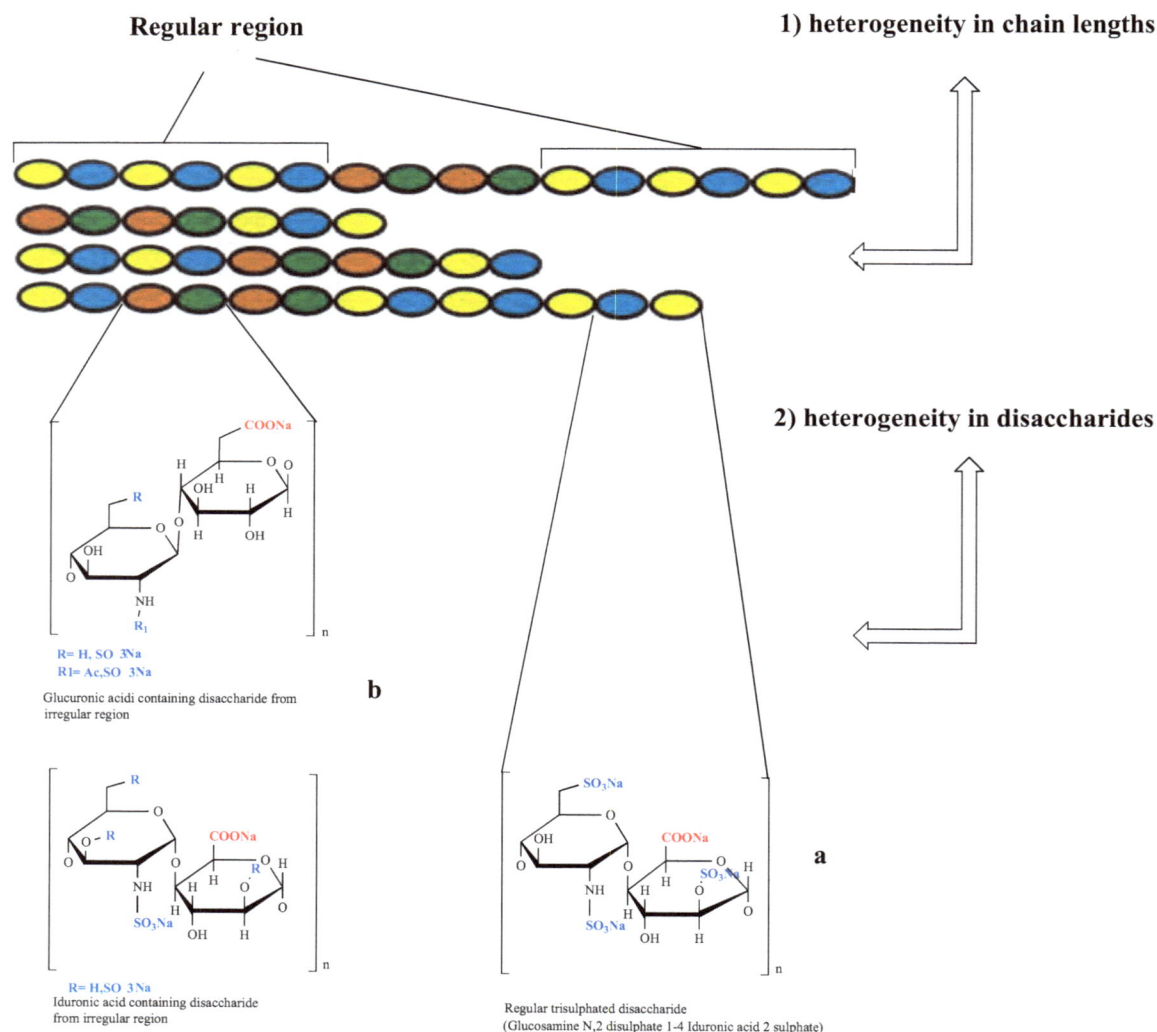

Fig. (1). Heparin structure and heterogeneity.

A lower degree of complexity is observed in HPBCD, in which the natural molecule β-cyclodextrin is a homogeneous heptasaccharide constituted by seven glucose residues linked 1-4 to form a circle without reducing units (Fig. **2a**) and the derivative is a randomly substituted hydroxypropylated product (Fig. **2b**).

reducing units (Fig. **2a**) and the derivative is a randomly substituted hydroxypropylated product (Fig. **2b**). In this case, NMR is the sole means available for measuring the hydroxypropyl groups with respect to the glycosidic protons.

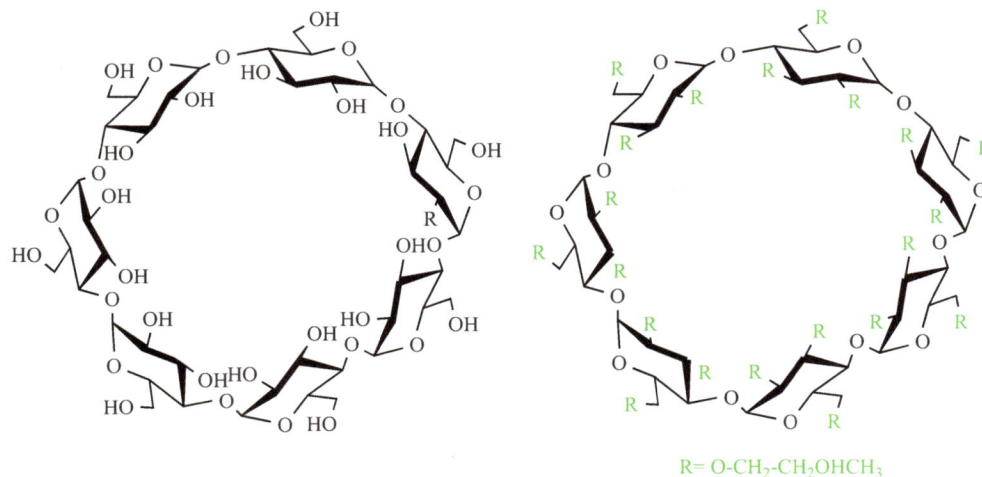

$R = O\text{-}CH_2\text{-}CH_2OHCH_3$

Fig. (2). β-cyclodextrin (**a**) and hydroxypropyl-β-cyclodextrin (**b**).

It is even more interesting to observe how NMR technology can help global drug quality, mentioning as an example the severe side effects recently experienced by a number of patients in the USA who were treated with contaminated conventional sodium heparin [5]. Without entering into specific reasons for this contamination, it was thanks to NMR that the impurity - supersulphated chondroitin sulphate - was identified [6]. In this particular case it is worth mentioning that, unlike for LMW heparin, NMR is not mandatory as a test method for drug release of conventional heparin. Supersulphated chondroitin sulphate is impossible to detect with normal assay tools, because its structure is similar to heparin (Fig. **3**), therefore only an unconventional supplement of analysis was able to discover the impurity. This is the reason why as of June 18, 2008 NMR test was introduced as a stopgap measurein the USP heparin sodium monograph [7] and revised as final version to be introduced by August 1 2009 [5]. Regardless of the differences among official pharmacopoeias, today it has become mandatory worldwide for heparin drug substances to be certified as free of supersulphated chondroitin sulphate by NMR before being used in drug product manufacture.

In modern drug development, NMR is normally included as analytical tool for testing the fine chemical structure of drug substances. The main application in drug products is the definition of the chemical structure of drug substance related impurities or degradation products. Impurities related to drug substance are generated by either the production method or the raw material, while degradation products can form during the manufacturing process of a drug product or during product shelf-life. Known impurities, as well as known degradation products related to the drug substance, are included in all official pharmacopoeia monographs of drug substances.

In most cases chromatography is used as analytical procedure and it is not mandatory that the chemical structure be confirmed during the drug development process. However, recent guidelines require also drug substance related unknown impurities/degradation products to be measured and identified in case of signals above 1% of the drug substance signal (for drug daily dosage above 1 g) [8]. The protocol to reach this stage during the drug development process usually involves a validated HPLC method to be applied to drug release specifications, in which all main signals are attributed to active, known impurities/ degradation products and excipients. New signals potentially forming during product shelf-life are detected and measured to assess compliance with the approved specifications. All new signals are attributed to unknown degradation products and each signal area is calculated as a percentage of the active substance. All signals above the identification limit will follow the procedure for the identification of a new degradation product, which is usually the same HPLC method upgraded with a mass detector (HPLC/MAS). In this way, additional information can be acquired on the same chromatographic signal and once the signal has been identified a synthesis of the molecules fitting the HPLC/MAS analysis is performed. Finally, the fine structure needs to be confirmed by NMR. HPLC is applied as a standard method on the newly synthesized molecule selected among the candidates to obtain the final confirmation in the method used for drug release.

Regular region

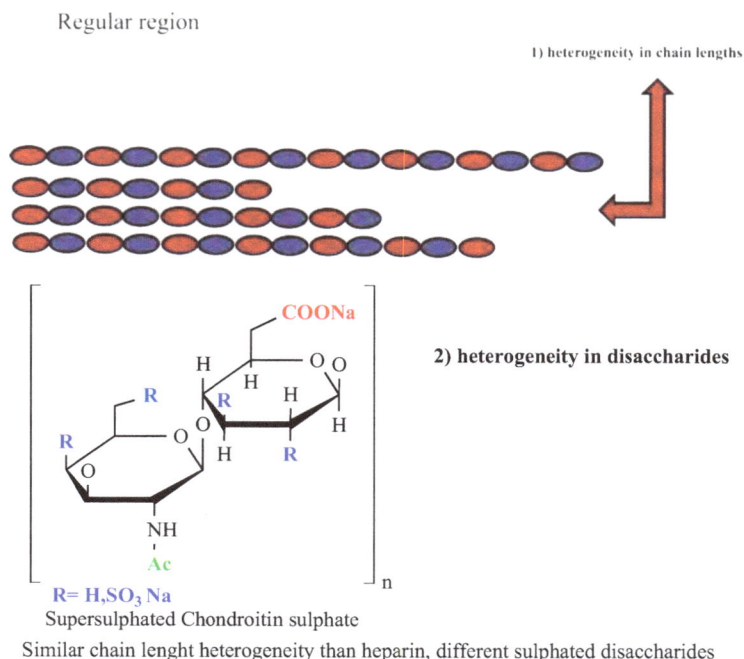

Supersulphated Chondroitin sulphate
Similar chain lenght heterogeneity than heparin, different sulphated disaccharides

Fig. (3). Supersulphated Chondroitin sulphate structure and heterogeneity.

NMR is the most frequently applied method for the qualification of drug product related impurities/ degradation products. However, NMR is only a small part of the study, since it is used only to confirm the chemical structure after a more extensive study is done by HPLC and Mass. Other applications in drug product approval processes have already been described above, while two unusual applications with a common drug development process confirm the host/guest complex formation of cyclodextrins and the study on the level of energy of water molecules in complex mixtures, such as drug products.

Host/guest complex will be described in more details later on in this chapter. Some concepts on water energy status measurable by NMR and how this technique might be useful in drug development process are presented at this point.

Water Energy Status Measured by NMR

Water energy status indicates the degree of freedom of water to exchange protons alone or with other substances. In chemicals, water behaviour with other substances depends on the physico-chemical class of a molecule. In hydrophilic substances residual water forms a network of hydrogen bonds with different energy strength, depending on the type and amount of chemical residues present in the substance. Water has the maximum degree of freedom to exchange protons with the environment when used as a solvent. In concentrated solutions this degree of freedom is reduced by molecules cohesion. A gel state is formed in special cases of molecules forming high density hydrogen bonds in an ordinate overall structure. In this state water has a minor degree of freedom. In exsiccated substances water can be present as crystallisation water, which does not exchange protons with vicinal substances and therefore has a low energy state.

Water is present in all pharmaceutical solid oral forms, and can be arbitrarily classified in three energy forms, depending on the pharmaceutical type. The minimum level is represented by crystallisation water, the intermediate energy level by a gel and the maximum level by free water, which exchanges freely with other mobile protons. Energy levels from the crystals can be detected and quantitatively measured using the Cross Polarization Magic Angle Spinning (CP-MAS) solid state NMR technique [9]. In this application ^{13}C CP/MAS high resolution solid state NMR spectra are obtained by spinning the sample at an angle of 54.7° towards the magnetic field to avoid disturbances due to the chemical shift anisotropy and the strong dipole-dipole interactions present at the solid state. The carbons relaxation time (T_1 and $T_{1\rho}$) is influenced by the physical status of the surrounding protons and hydroxyls. ^{13}C T_1 measures the surrounding status of each carbon of the molecule, providing information on each carbon, while $T_{1\rho}$ measures the relaxation time of all the nuclei allowing the detection of heterogeneous areas in the sample. If more than one relaxation time is detected, the sample is not as homogeneous as water energy status. The experiment can also quantify the relative percentage of the species. The application of CP-MAS to measure the water energy status has been described for starch [10]. The authors describe the application of T_1 and $T1_\rho$ CP-MAS techniques to measure the different energy states of water in starches from different sources. The presence of both crystalline and amorphous structures has been identified and also indication of the quaternary structure of the different starches has been reported. Similar applications using a low field NMR spectroscopy allow us to measure water energy in semisolids or suspensions by proton transverse relaxation time T_2 [11]. In this case T_2 is the direct measure of proton relaxation time of starch suspended in water. The interchange of water molecules in wheat starch suspension was demonstrated between "bound" water and "weakly bound" water. On the contrary, the exchange of "bound" water with "tightly bound" water was not detected.

This technique allows experiments to be performed directly on a pharmaceutical form, even a complex solid oral, semisolid form or suspension, if the test is carried out only to measure water in all its different energy states. Applications can be performed on solid excipients (fillers) or directly on oral forms such as tablets, powders or granulates to study the supramolecular environment rather than the chemical compatibility between the active and the excipients only. Semisolid complex pharmaceutical forms such as gelatine capsules can also benefit from this technique to study the intimate relationship of the drug substance and polymeric excipients in the final formulation.

CYCLODEXTRIN-GUEST COMPLEX IN DRUG DEVELOPMENT

Cyclodextrins are cyclic oligosaccharides formed by 1-4 linked α-glucose. Natural cyclodextrins are derived from the action of the cyclomaltodextrin glucanotransferase (E.C. 2.4.1.19; CGTase) enzyme on starch. The natural spirals formation of starches in solution leads to oligosaccharides rings, which are melted in a cyclic shape by the enzyme (Fig. **4**).

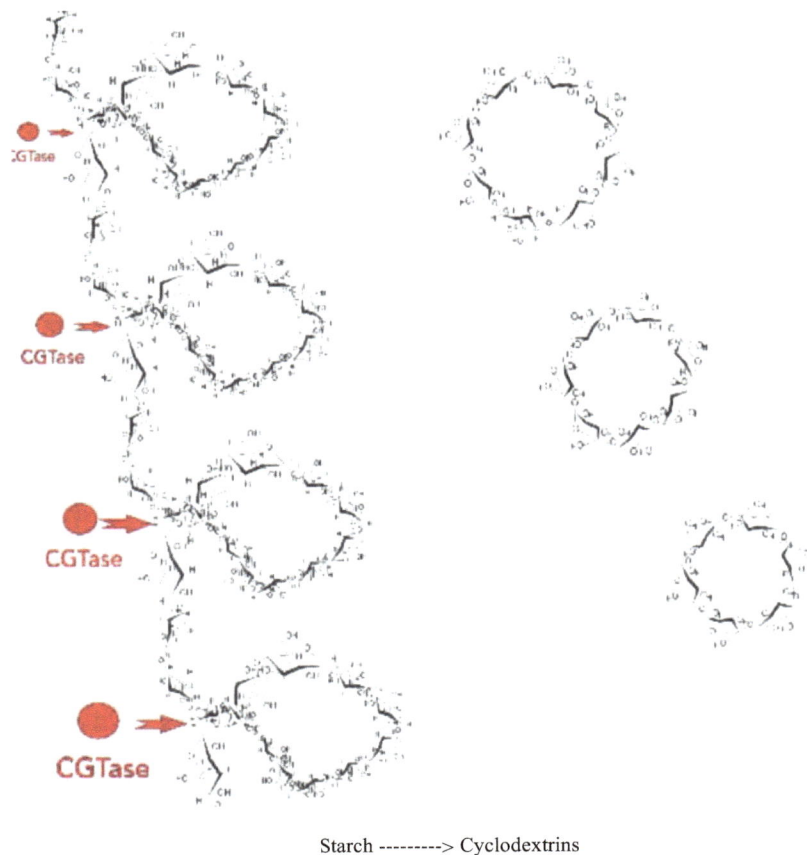

Starch --------> Cyclodextrins

Fig. (4). Action of CGTase on starch to produce cyclodextrins.

The natural cyclodextrins family consists of three industrially produced molecules (Schardinger's alpha, beta and gamma cyclodextrin) and other minor ones not developed on an industrial scale. Alpha, beta and gamma cyclodextrins are formed by six, seven and eight glucose residues respectively (Fig. **5**), leading to cavities with different diameter and different solubility in water depending on the ring flexibility.

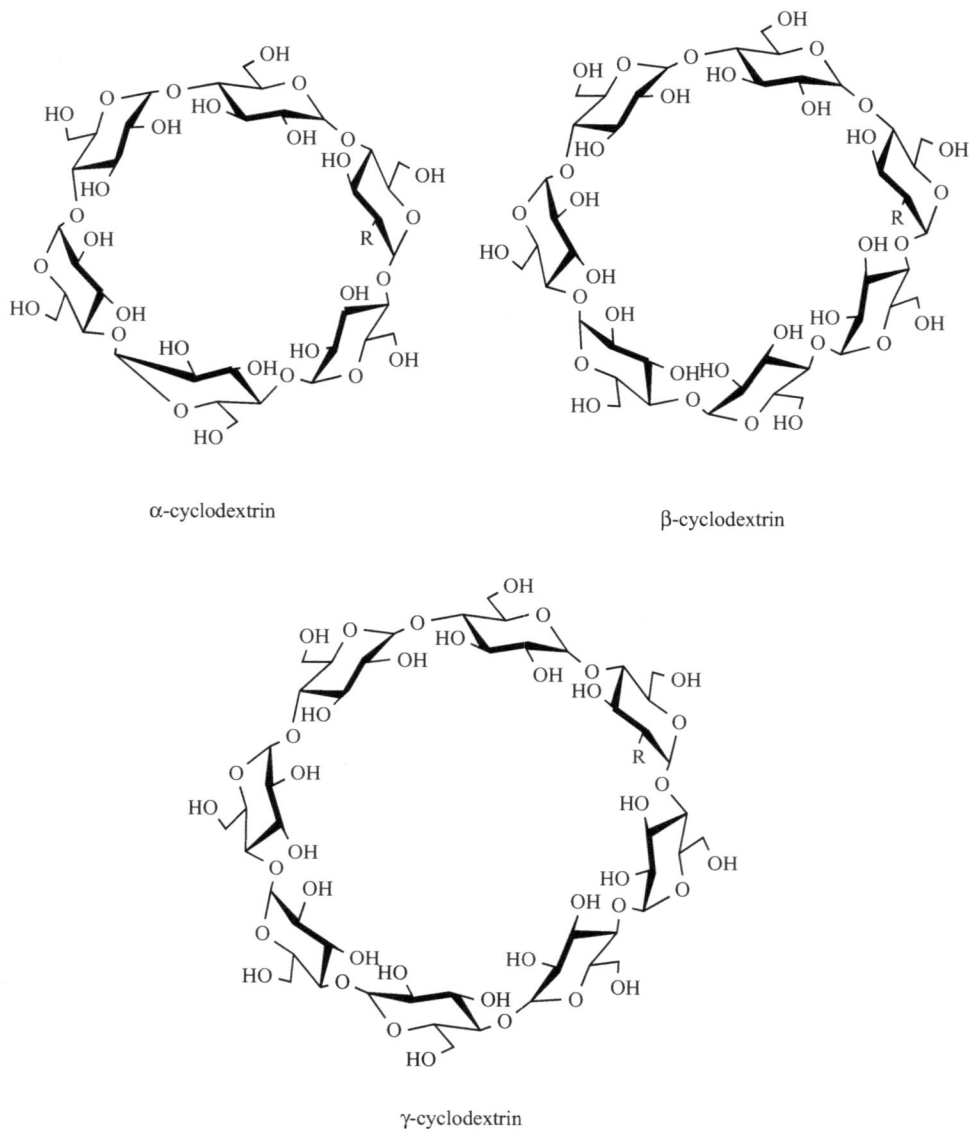

α-cyclodextrin

β-cyclodextrin

γ-cyclodextrin

Fig. (5). Natural cyclodextrin structures.

Cavities have different diameters depending on the number of glucose units (empty diameters between anomeric oxygen atoms given in Fig. (**6**) below). The side rim depth (shown below in the diagrams) is the same for all three (at about 0.8 nm).

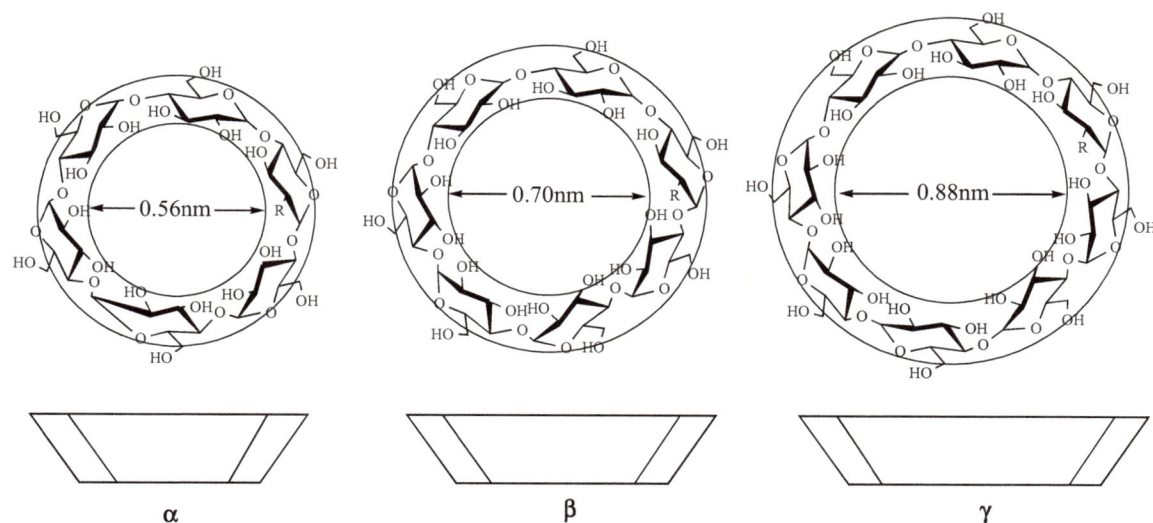

Fig. (6). Natural cyclodextrins and hosting representation.

The internal cavity is formed by 2 and 3 protons of glucose, while hydroxyl groups form the external part of the toroid, 6-OH residue in the smaller end and 2, 3 –OH on the opposite wider end (Fig. **7**).

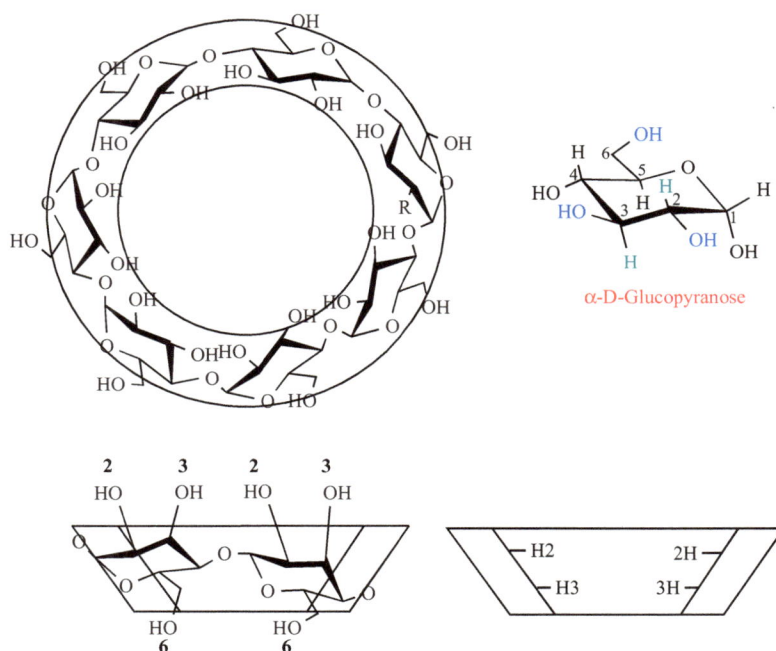

Fig. (7). Spatial representation of cyclodextrin toroid. 2 and 3 internal protons (green) form the lipophilic cavity. Hydoxypropyl groups are located externally on the wider end of the 2 and 3 hydroxyl residues and on the smaller end the 6 residues (blue).

In this peculiar configuration water molecules play structural roles, providing complete solvation to the external part of the toroid and forming a lipophilic area inside the cavity. In the crystal structure some water molecules are included in the lipophilic cavity. It has been reported that alpha cyclodextrin includes 2 water molecules while beta cyclodextrin includes 6 water molecules and gamma cyclodextrin includes 8,8 water molecules [12], conferring to cyclodextrins different solubility properties in water and a different capability to hosting molecules. The perfect guest for alpha cyclodextrin is a simple aromatic ring such as benzene or toluene. Instead, beta and gamma cyclodextrins can accommodate larger organic molecules with lipophilic properties, like naphthalene for beta or anthracene for gamma cyclodextrin [13] (Fig. **8**).

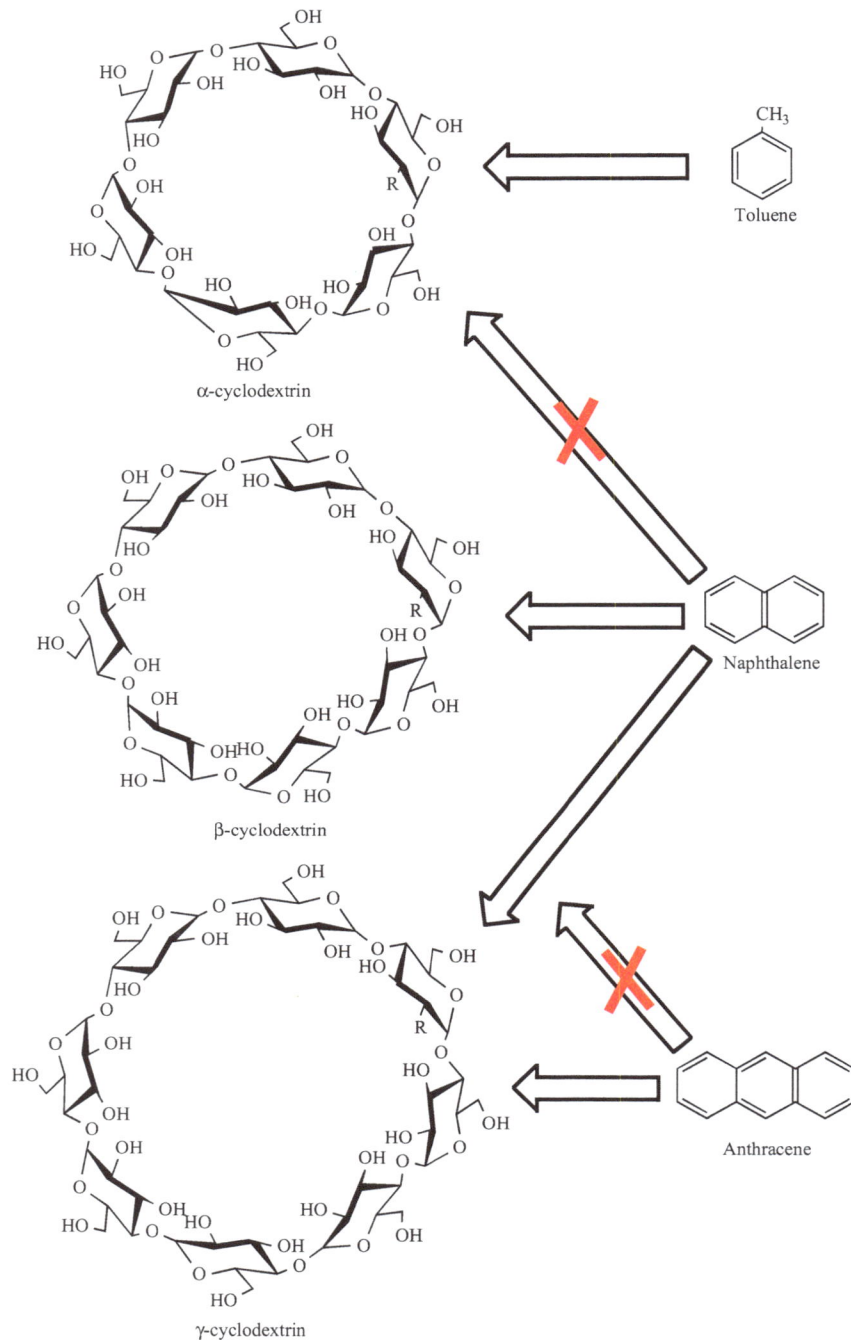

Fig. (8). Hosting capacity of natural cyclodextrins. Anthracene and Naphtalene do not fit in the smaller cyclodextrin. On the right, tridimensional representation of β-cyclodextrin (above) and benzoic acid complex (below). Tridimensional pictures have been obtained from Roquette Fréres S.A.

Cyclodextrin derivatives are the second generation of cyclodextrins used in drug development. The drive to develop synthetic derivatives was mainly due to the poor water solubility of β-cyclodextrin, which is the most widely used of the three industrial products, owing to its production yield and the flexibility to host interesting molecules. α-cyclodextrin is often too tight to host drug substances and γ-cyclodextrin is too supple to properly host a drug product. Although in most cases β-cyclodextrin would be a better choice for a stable inclusion complex with a drug substance, its solubility profile in water is very low (1-2%), precluding its use in many pharmaceutical preparations. This is due to the peculiar distance between two vicinal glucose hydroxyl group 2 and 3, which induces a tight hydrogen bonding and as a consequence a rigid conformation at temperatures up to 35 °C. At temperatures above 40°C, its solubility increases ten fold because the molecule enters a higher energy level, breaking the hydrogen bonding. The substitution of some hydroxyl groups permanently prevents the 2,3 hydrogen bonding, giving the derivate molecule a higher water solubility, saving the cavity dimension.

Among more than 180 cyclodextrin derivatives those mostly associated with solubility enhancement are: ethers such as methyl, ethyl or propyl; anionic derivatives such as sulphopropyl, sulphobutyl, carboxylate or sulphate; esters like hydroxyethyl or hydroxypropyl [14]. Due to the particular characteristics of some of the abovementioned derivatives also α and γ cyclodextrin derivatives have been investigated.

The most appropriate cyclodextrin derivatives used in drug products so far are hydroxypropyl-β-cyclodextrin (HPBCD) and sulphobutyl-β-cyclodextrin: both have been approved as solubilising excipients in drug products.

The complexation mechanism involves the displacement of the water molecules included in the cavity with the guest molecule. This is due to the dynamic exchange of hosted water molecules present in a lower energy state with the surrounding water molecules. As a consequence, van der Waals interactions and the number of hydrogen bonding decrease in the cavity and the lipophilic guest molecule can be complexed [15].

Like the simple structures reported in Fig. (**8**), complex molecules can be hosted to form inclusion complexes with different constant strengths. The ideal application is to enhance the water solubility of a poorly water soluble molecule to achieve a higher concentrated solution. It is also a rationale target to protect molecules which are sensitive to heat, pH or biological fluids from degradation by means of an inclusion complex.

Analytical tools to determine the presence of an inclusion complex involve several methods. The simplest method is the phase solubility profile according to Higuchi and Connors, which is applicable to poorly water soluble compounds [16]. The guest is measured by appropriate assay in guest super saturated water in the presence of guest precipitate, then the test is repeated after the addition of a known amount of cyclodextrin, calculated as a molar fraction of the total guest molar quantity in the reaction vessel. Cyclodextrin forms a complex with some guest precipitates, turning it into solution then the new guest concentration is measured. Cyclodextrin is added until a plateau is formed. From the flex point it is possible to calculate the host/cyclodextrin ratio (see the case studies below for examples). This technique cannot be applied to water soluble molecules; therefore other techniques must be used.

The presence of an inclusion complex can be determined quantitatively by UV/VIS spectroscopy. VIS can be used to detect the complex formation of a dye with cyclodextrin, e.g. methyl orange with α-cyclodextrin [17]. When a correct amount of α-cyclodextrin is added to a methyl orange solution the original orange colour disappears because the complexation mobilises the methyl orange π electrons. If ethanol is added, the colour reappears because the hosted molecules escape from the cavity, since the external solution is more lipophilic than in pure water. UV frequency can also be used to measure colourless substances hosted by cyclodextrins like mefenamic acid. If different amounts of β-cyclodextrin are added to a solution of mefenamic acid under appropriate buffer conditions a change in the spectrum from 240 to 360 nm is observed. In particular, a decrease of optical density correlated with the added amount of β-cyclodextrin is measured at 280nm and decreases at 340 nm, with an isosbestic point at 328 nm [18]. Other informative instrumental techniques are infrared or Raman spectroscopy, Differential Scanning Calorimetry (**DSC**) and **X-ray** spectrum. IR and Raman spectra provide information on the reactive groups of a molecule, which are disturbed by the complex formation, while DSC and X-ray techniques detect the presence of crystal structures in the mixture. In these cases a guest with a crystal structure is required. The evidence is given by the disappearance of the sharp signal from the spectra in both cases (see case study 1 for examples).

Nuclear Magnetic Resonance (NMR) provides direct evidence of the presence of a guest in the cyclodextrin cavity. In fact, all other methodologies can detect false inclusion complexes when a sort of close relationship occurs between the molecule and the external part of cyclodextrin. NMR can measure the perturbance of the internal cyclodextrin protons (3 and 5) due to the presence of a guest [19]. In Fig. (**9**) the partial proton spectrum of hydroxypropyl-β-cyclodextrin in deuterated water with the indication of protons 3 and 5 is reported.

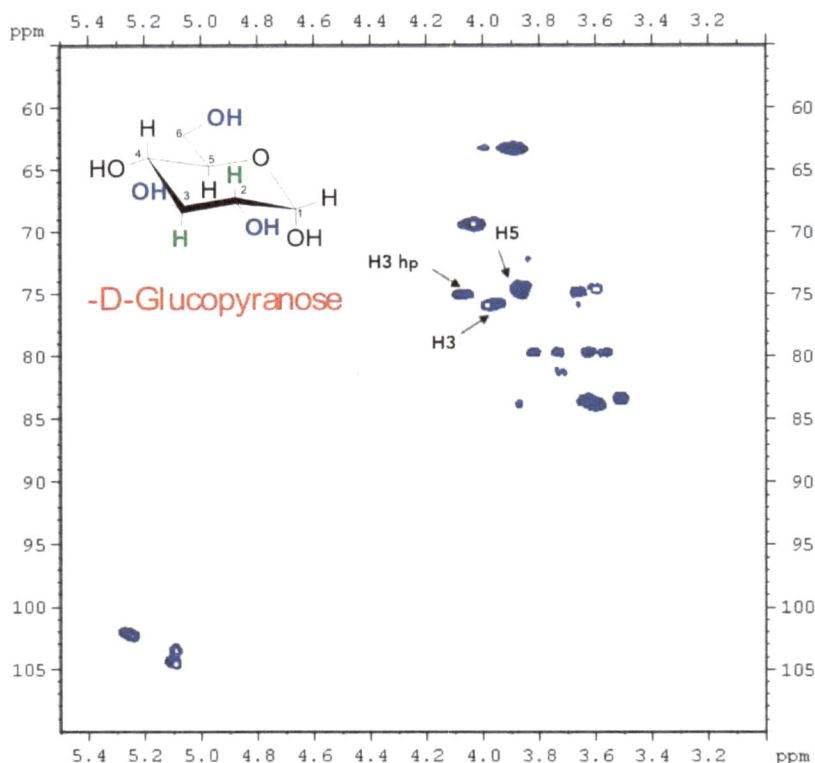

Fig. (9). HSQC spectrum of HPBCD. Arrows indicate H3 and H5 protons. Bidimensional spectrum can identify hydroxypropyl substitution on H3 hydroxyls (H3-hp).

With the availability of high field and solid state techniques, NMR has become the most useful tool to study cyclodextrin complexes. Thanks to high resolution instruments and the availability of two-dimensional techniques, the output of qualitative data has been greatly improved. Today it is possible to perform quantitative experiments on the sterical structure and kinetic parameters of cyclodextrin complexes. Therefore, not only can the direct presence of a guest molecule in the cyclodextrin cavity be detected, but, by means of more sophisticated experiments, also the complex formation constant can be calculated as alternative to the phase solubility study and each complex type can be detected when multiple inclusion complexes are present in a solution.

The use of cyclodextrins in drug development has been extensively studied to improve solubility, stability and bioavailability of many drug substances [20]. It has been reported that in at least 10 % of oral drugs, the active ingredient seems to be complexable; in half of these cases one or more physical or chemical properties can be modified by the addition of cyclodextrin [21]. Although this sounded very promising in the drug development field, only a few drug substances reached market approval. Although the first experiments in humans were carried out in the Seventies, it was not until the Nineties that natural β-cyclodextrin reached the market as drug product and food flavouring after the establishment of industrial production and the definition of non-toxic excipients. The cyclodextrin derivative hydroxypropil-β-cyclodextrin (HPBCD) was developed in the mid-Eighties as a better tolerated cyclodextrin for parenteral injection applications. A complete pharmaco-toxicological study was performed during drug development [14] to confirm better tolerability of this derivative in comparison with natural β-cyclodextrin, which was not tolerated if injected [22]. Sulphobutyl-β-cyclodextrin (SBBCD), whose safety profile is similar to HPBCD, is a recent derivative introduced in the pharmaceutical market.

After first approval of a sublingual tablet containing a Prostaglandin E_2-β-cyclodextrin complex in Japan, another tablet formulation was approved in Italy (piroxicam-β-cyclodextrin tablet). Natural β-cyclodextrin was used in both formulations to improve the bioavailability of the drug substance. More recently, the derivative HPBCD was introduced in the pharma market as an excipient of an itraconazole oral formulation to improve bioavailability and in a parenteral formulation of sodium diclofenac (see case study 2 for details). SBBCD is currently available as parenteral injection of melphalan, an anti-cancer drug (for a review of cyclodextrin in drug delivery see [20]).

CASE STUDIES

Two real examples of drug development involving the inclusion of complexes with hydroxypropyl-β-cyclodextrin are reported in detail in the last part of this section. Case study 1 deals with the development of an injectable water solution of progesterone to replace the present injectable formulation in oil. In the second case study another poorly water soluble drug substance, diclofenac sodium, was complexed with HPBCD to enhance water solubility. A stable inclusion complex is described in the first case, whereas in the second case a multiple inclusion complex is shown. However, due to the formation of a precipitate, no injectable drugs have been developed so far. The method adopted to obtain injectable solutions and the use of NMR to study the inclusion complex in the drug development phase are described in the mentioned case studies.

CASE STUDY 1 – HYDROXYPROPYL-B-CYCLODEXTRIN-PROGESTERONE COMPLEX - A WATER SOLUBLE DRUG FOR PARENTERAL INJECTIONS

In this example NMR has been used as analytical method to characterize the complex, in addition to other instrumental as well as chemical methods [23]. In particular, NMR was essential to determine the fine structure of the precipitate and the existence of a cyclodextrin complex, showing the chemical shift of some protons belonging to progesterone in the complex. Further experiments by bidimensional NMR have been performed [24] to confirm the stoichiometry as well as the binding constant of the complex. In order to provide better understanding of the progesterone HPBCD complex, besides NMR experiments, a comprehensive review of two further papers is provided here, in which other instrumentation techniques and physico-chemical methods are reported.

Progesterone (P) is a natural hormone used as a human drug to control reproductive functions, as well as in postmenopausal therapy. The currently available administration routes are: by oral route with soft gel capsules, by vaginal route with soft gel capsules or cream and by intramuscular route with oily solution or water suspension injections. The bioavailability of oral and vaginal routes is low. Furthermore, the usefulness of the oral route is limited, due to its short half-life and extensive liver degradation after absorption. Therefore, the injection route is the preferred choice, but the oily solution and the water suspension allow only intramuscular administration, causing considerable pain when injected. This has encouraged the research of a new injectable pharmaceutical form to improve patient compliance. A water soluble P would be a good solution, since it could be administered also by subcutaneous route, allowing self-medication and causing less pain during the treatment.

The use of Cyclodextrin to increase water solubility of Progesterone (P) was described by Pitha *et al.* as a complex with β-cyclodextrin and derivatives to obtain a water soluble formulation [25]. The evidence of the inclusion complex formation of progesterone with β-cyclodextrin and derivatives is also described by other authors using different physico-chemical methods [26-28].

Among the tested cyclodextrins, hydroxypropyl-β-cyclodextrin (HPBCD) is the preferred choice for a parenteral injectable formulation containing a therapeutic amount of P. The therapeutic dosage of P by injection varies from 5 to 200 mg, depending on the pathology [29]. HPBCD is highly water soluble (48% W/V), allowing the solubilisation of a considerable amount of P; considering a 1:2 guest/host complex stoichiometry, it is possible to obtain up to 50 mg/ml of P concentration, which is a considerable dosage in a therapy with progesterone.

During drug development, the HPBCD/P complex in water showed the formation of a light precipitate during stability testing at ICH conditions. A precipitate formation in progesterone complexes with various natural and modified cyclodextrins was already mentioned by Choi *et al.* [28], but its chemical structure was not elucidated. A method to obtain a HPBCD/P complex in pure water solution without insoluble particles and the physico-chemical characterization of the purified inclusion complex are described in this work.

Precipitate Characterization

The precipitate was purified from a pilot scale production and was equivalent to 0.055% of the total product quantity. The structural analysis by NMR indicates the presence of P and β-cyclodextrin, but not HPBCD [23].

The formation of an insoluble β-cyclodextrin/P complex is confirmed by Forgo *et al.* [26]. The author concluded that the progesterone molecule is deeply incorporated in the cyclodextrin cavity, therefore a strong inclusion complex with low solubility in water is formed. A pharmacopoeial grade HPBCD containing 0.8% of unsubstituted β-cyclodextrin was used in this preparation. This residue caused the formation of the precipitate in the formulation. After the elimination of the precipitate using an original procedure [30], the resulting product did not more than 0.1% of residual β-cyclodextrin, with no precipitate formation up to 24 months at ICH stability conditions. The resulting HPBCD/P complex afforded a stable injectable formulation.

Determination of Stability Constant by Phase Solubility Studies

The stoichiometry of the complex and the complex formation constant were calculated by means of a phase solubility experiment. The molecular association of progesterone and HPBCD is described in Fig. **(10)**. Three association constants are calculated: K_1 is presumed to be the constant of the initial 1:1 complex and K_2 is related to the second entering cyclodextrin molecule to form the final 1:2 complex; K_3 is the overall constant. More details are reported in the original paper [23].

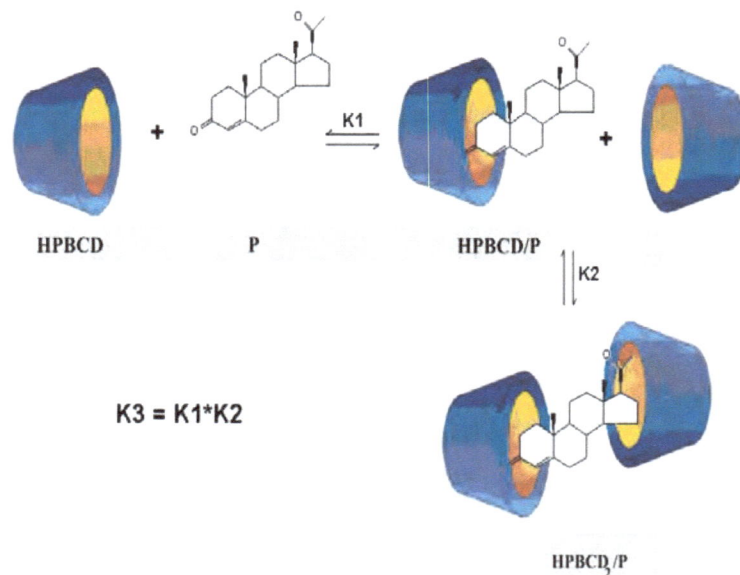

Fig. (10). Mechanism of molecular association of the complex HPBCD/P. Zoppetti *et al.* [23]. *Reprinted with permission of Springer Science B.V.*

Determination of Complex by Physico-Chemical Methods

The DSC (Fig. **11**) and X-Ray (Fig. **12**) analysis confirm the previous published data indicating the absence of crystal structures due to the complex formation by HPBCD. DSC analysis shows a sharp signal at 130°C in pure progesterone or in the physical mixture (progesterone and HPBCD powders mixed in dry condition), while only a broad signal is evident in the complex obtained by water solubilization of progesterone in the presence of HPBCD. Similar results were obtained in the X-Ray test, where sharp signals are evident only in the physical mixture.

The structural evaluation by spectroscopic analysis (Raman, FTIR and NMR) confirm the presence of an inclusion complex with a stoichiometry guest/host 1:2.

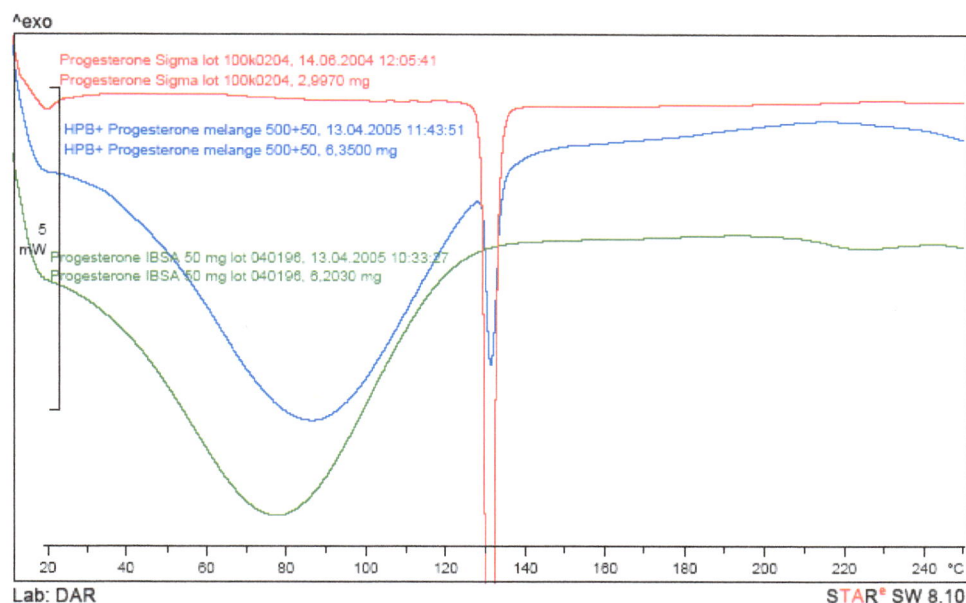

Fig. (11). Differential Scanning calorimetry (DSC) – Crystalline forms are shown only in red and blue lines representing P and the physical mixture respectively (sharp signal at 130 °C) while in the green line, representing the complex, only a broad signal is present. Zoppetti *et al.* [23]. *Reprinted with permission of Springer Science B.V.*

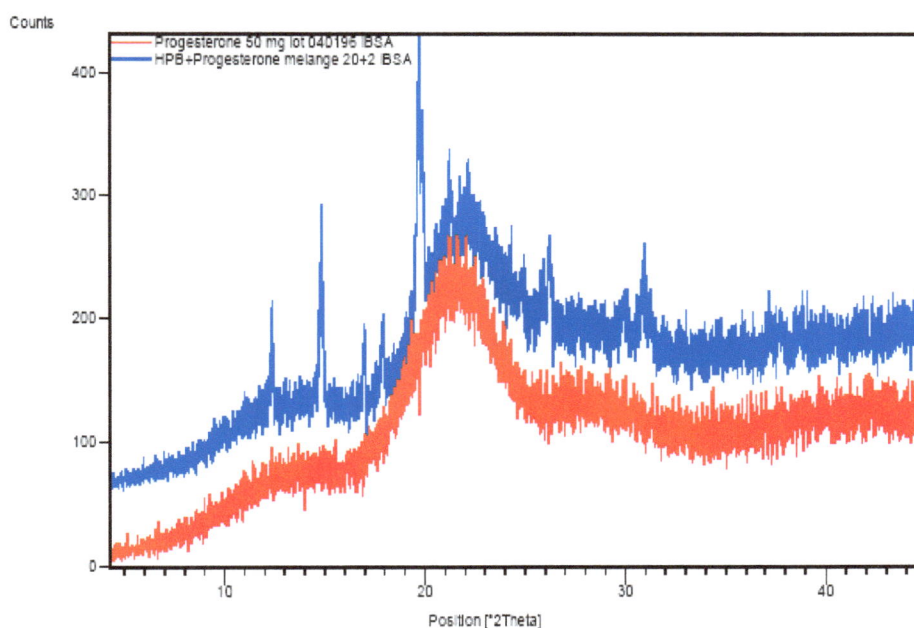

Fig. (12). X-Ray -Crystalline structures are present only in the physical mixture (blue line). Zoppetti *et al.* [23]. *Reprinted with permission of Springer Science B.V.*

In the **Raman spectra** of HPBCD/P complex (Fig. **13A**), the intensity of the corresponding pure P (Fig. **13B**), are reduced and shift towards lower cm^{-1} values (Δ=4-6 and 6-8 cm^{-1} according to the different points of the tested sample). Moreover, a broadening of these bands (stretching vibration of the C=O groups) is also observed in the Raman spectrum of the complex. The increase of the bandwidths, i.e. the decrease of the vibrational relaxation time, confirms the interaction between both C=O groups with HPBCD.

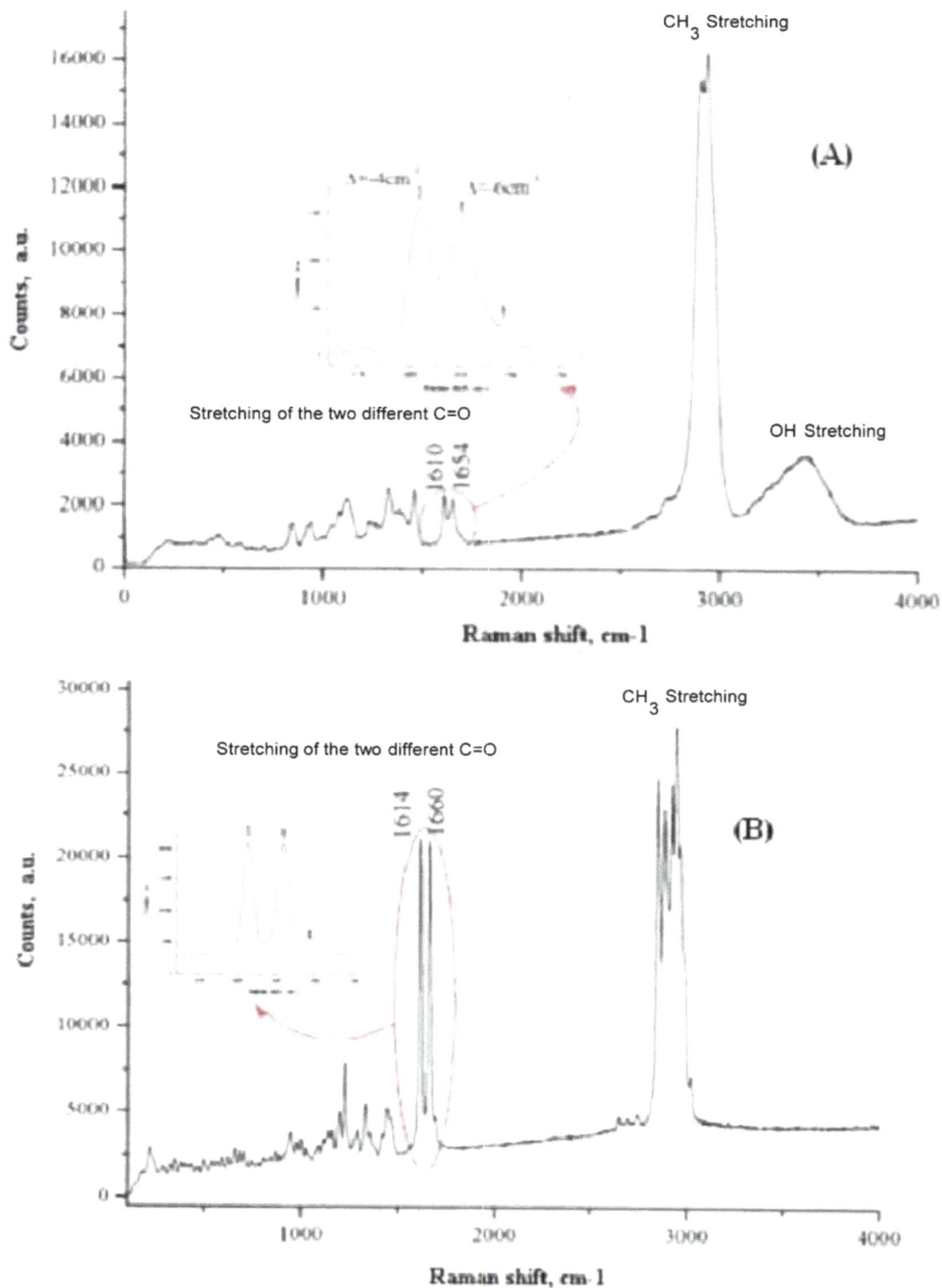

Fig. (13). Raman spectra of P/HPBCD complex (**A**) and P (**B**). Zoppetti *et al.* [23]. *Reprinted with permission of Springer Science B.V.*

DETERMINATION OF COMPLEX FORMATION, STOICHIOMETRY AND BINDING CONSTANT BY NMR

Owing to its very low water solubility, progesterone can be analysed by NMR in organic solvents.

The classical NMR experiment is performed in deuterated chloroform or dimethyl-sulphoxyde. Since in this case the experiment to define the relationship between cyclodextrin and progesterone must be carried out in water, mono dimensional proton NMR by high field instrument was performed on a saturated water solution of P(Dioshynth NL). All NMR experiments in solution were performed at 11.4 T on a Bruker Avance500 spectrometer equipped with a TXI 5 mm probe at 313 K. The HPBCD (Roquette Fr) and HPBCD/P samples [30] were characterized by homo and heteronuclear experiments (Bruker library: Correlation Spectroscopy (COSY), Total Correlation Spectroscopy (TOCSY), Heteronuclear Multiple Quantum Correlation (HSQC), and Heteronuclear Multiple Bond Correlation (HMBC) with a 7.0 mM D_2O solution. External trimethylsilyl propionate sodium salt was used as chemical shift reference. The P spectrum obtained by monodimensional technique (upper line), and the HPBCD/P complex on the lower line are reported in Fig. (**14**). The corresponding ^1H chemical shifts are reported in Table **1** [31].

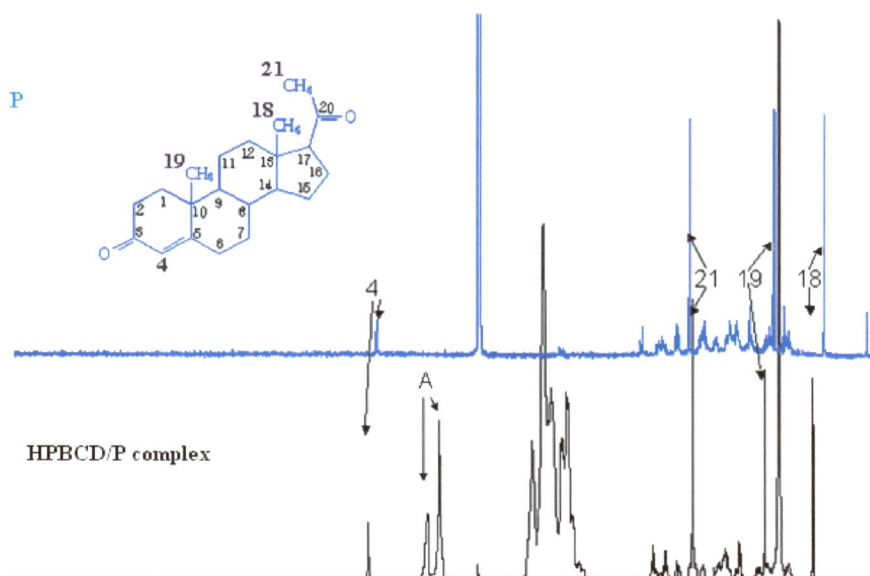

Fig. (14). Proton NMR spectra of P and HPBCD/P complex in deuterated water. Zoppetti *et al.* [23]. *Reprinted with permission of Springer Science B.V.*

Table 1. ^1H and ^{13}C Chemical Shifts (ppm) of P in a D_2O Solution and in the 2:1 HPBCD/P Complex

Atom	P		P in 2:1 complex
	1H (ppm)	13C (ppm)	1H (ppm)
1	2.17 (β) 1.75 (α)	38.52	2.10-1.98 (β) 1.78-1.61 (α)
2	2.20 (β) 2.32 (α)	36.14	2.60-2.46 (β) 2.36-2.33 (α)
4	5.90	125.05	
5		180.04	
6	2.58 (β) 2.45 (α)	35.66	2.36-2.33 (β) 2.36-2.33 (α)
7	2.04 (β) 1.14 (α)	34.80	1.93-1.87 (β) 1.11-1.02 (α)
8	1.77	38.26	
9	1.05	56.95	1.11-1.02 (α)
10		41.68	
11	1.80 (β) 1.63 (α)	24.05	1.78-1.61 (β) 1.78-1.61 (α)
12	2.16 (β) 1.60 (α)	41.41	2.10-1.98 (β) 1.78-1.61 (α)
13		47.20	

(Table 1) Contd…..

Atom	P		P in 2:1 complex
14	1.29	58.79	
15	1.88 (β) 1.39 (α)	26.84	1.25 (β) 1.78-1.61 (α)
16	2.13 (β) 1.82 (α)	26.09	1.25 (β) 1.78-1.61 (α)
17	2.61	66.36	
18	0.77	16.64	
19	1.32	20.33	
21	2.16	33.98	

The arrows indicate that the progesterone protons undergo chemical shifts due to the inclusion complex. Protons 4, 18, 19 and 21 indicate an involvement of rings A and D in the inclusion complex. Furthermore, the 1:2 guest/host molecular ratio is confirmed by the measure of the area of P proton 4 and anomeric proton of HPBCD (A). The comparison between the MSQC spectra of HPBCD and the 1:2 complex (Fig. **15**) displays a significant shift of both H3 and H5 signals of cyclodextrin, therefore confirming the formation of an inclusion complex.

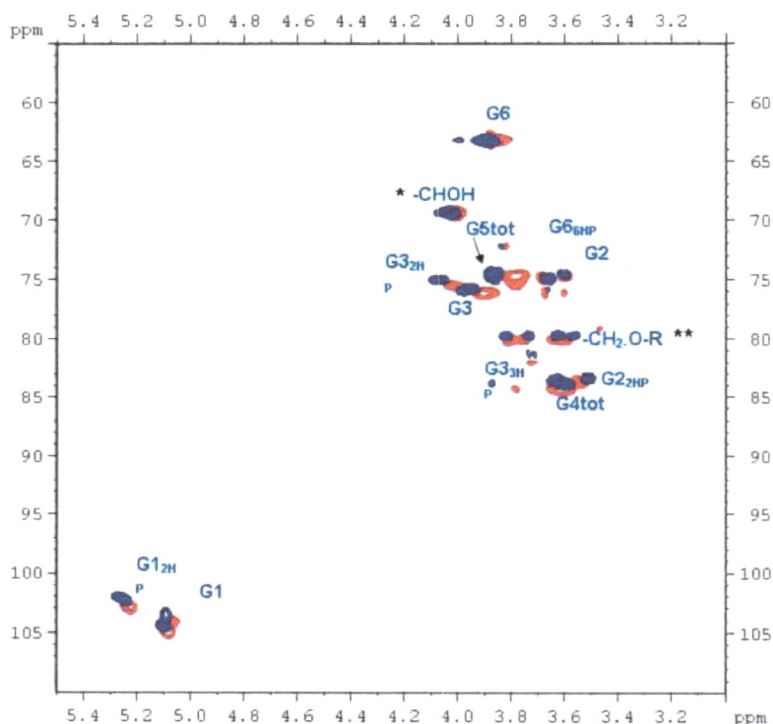

Fig. (15). Overlap of HSQC spectrum of HPBCD (blue) and its complex with P (red) (*CHOH hydroxypropyl groups; **CH2OH hydroxypropyl groups). Torri *et al.* [24]. *Reprinted with permission of Springer Science B.V.*

Additional NMR techniques have been used to further characterise the HPBCD/P complex. The formation and stoichiometry of the inclusion complexes were determined using the *continuous variation method*, through monitoring of the proton shifts of P during NMR titration. A deuterium oxide solution containing a constant concentration of P (8.9×10^{-5} mmol/ml) was titrated with increasing amounts of HPBCD (from 0 to 5.48×10^{-4}M). Due to the complexity of the HPBCD spectrum, only the ^{1}H shifts of P were measured in the HPBCD/P complex to study the inclusion process (Fig. **16**, Table **2**). Moreover, the HPBCD/P complex was also investigated by means of the solid state ^{13}C NMR spectroscopy technique. Solid state ^{13}C cross Polarization Magic Angle Spinning with dipolar decoupling (^{13}C CP-MAS) spectra were recorded with a Bruker ASX300 spectrometer operating at 75.47 MHz equipped with 4 mm CP-MAS probe. The spin rate was 8000 Hz.

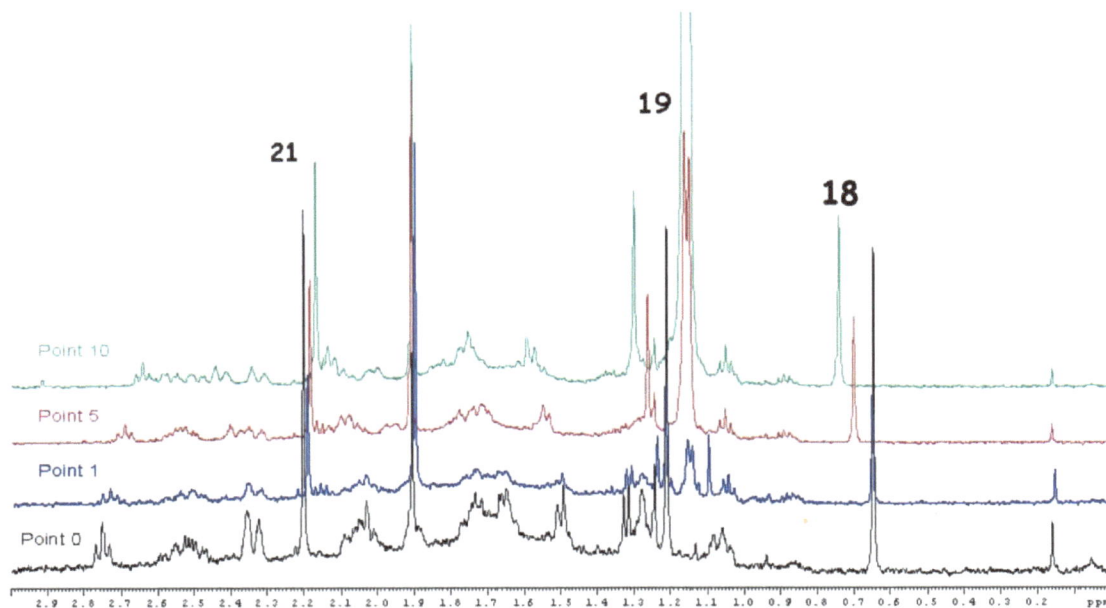

Fig. (16). Expansion of 3-0 ppm region of the spectra recorded for points 0,1,5 and 10 of the titration. Chemical shifts of progesterone protons 18, 19 and 21 are highlighted. Torri *et al.* [24]. *Reprinted with permission of Springer Science B.V.*

The 'H NMR complexation-induced shifts (CIS) of the most intense signals are reported in Table **2**.

Table 2. P Chemical Shifts in 1:2 Complex (ΔHz)

HPBCD (M)	Peak 19	Peak 18	Peak 21	Peak 4
0.000	0.000	0.000	0.000	0.000
8.55×10^{-6}	4.170	4.900	2.615	3.255
1.71×10^{-5}	6.920	8.020	3.715	5.270
3.42×10^{-5}	14.620	16.270	5.915	11.505
5.10×10^{-5}	17.185	19.935	7.745	13.885
6.84×10^{-5}	24.390	27.320	8.980	19.805
1.03×10^{-4}	26.120	29.055	9.995	20.805
1.37×10^{-4}	32.405	36.440	13.250	36.355
1.71×10^{-4}	36.070	40.105	14.530	29.290
2.14×10^{-4}	38.920	43.685	15.900	31.585
2.56×10^{-4}	43.680	48.260	16.825	35.245
3.42×10^{-4}	44.275	49.040	17.695	35.845
5.48×10^{-4}	50.055	55.375	20.350	40.340

The complexation-induced proton shifts observed by increasing the amount of HPBCD are reported in Fig. (**17**).

The complexation shifts were simulated using different approaches. First, the presence of only a 1:1 complex was considered. Then, we moved to a two step equilibrium characterised by two apparent formation constants, K_{11} and K_{12}. All proton complexation shifts of P were utilised in the non-linear least-square fitting procedure according to Connors *et al.* [32].

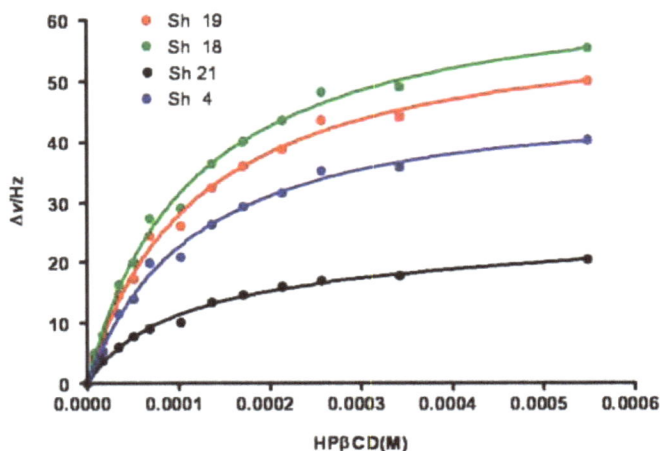

Fig. (17). Complexation shift of protons induced by HPBCD addition – non linear square fitting. Torri *et al.* [24]. *Reprinted with permission of Springer Science B.V.*

To confirm the inclusion of P into HPBCD, the ^{13}C CP-MAS solid state spectra of P, HPBCD and their complex sample were recorded. The spectrum of P shows very sharp and well defined signals; a peculiarity of a crystalline and rigid structure (Fig. **18A**). By contrast, the solid state spectrum of HPBCD (Fig. **18B**) consists of broad signals indicating an amorphous structure with very high mobility. The CP-MAS spectrum of the 2:1 complex shows the signals related to HPBCD and aliphatic carbons of P (Fig. **18C**).

Progesterone peaks are broad, reflecting loss of order structure. Moreover, the peaks attributed to carbonyl groups (carbon 3 and 20) and to the double bond (carbon 4 and 5) are not present in the spectrum recorded with the same instrumental parameters used for P. This means that, due to the higher rigidity of the included aromatic region more than for the rest of the P structure, the relaxation time of these atoms in the 1:2 complex increases. This confirms that both sides of P are included in the cavity of two HPBCD molecules.

In conclusion, several instrumental and chemical methods have been used to elucidate the HPBCD/P complex structure. Using other methods than NMR it was possible to establish that the investigated inclusion complex (HPBCD/P) has a 1:2 type guest/host complex with a medium apparent complex formation constant (K_3) of 111,473.70 m^{-1}. This constant is a product of two primary formation constants (K_1 and K_2), whose values are 3,477.96 m^{-1} and 32.05 m^{-1} respectively. The inclusion complex involving both progesterone molecular sites was suggested by Raman, DSC and X-ray techniques.

NMR spectroscopy was used as a complementary technique to define HPBCD/P complex by attribution of the most significant ^{1}H and ^{13}C signals. The complexation-induced shifts of H3 and H5 HPBCD protons observed in the MSQC experiment are those expected as a result of the interaction with those protons which are oriented towards the cyclodextrin cavity, hence clearly confirming host/guest interaction. Such a conclusion is also in agreement with the CP MAS spectra of the solid complex, where signals associated with the crystal structure of P are no longer visible. Spectra in solution confirm that two opposite sides of a single P molecule are included in two HPBCDs. The complexation-induced proton shifts of P measured upon addition of increasing amounts of HPBCD agree with the hypothesis of a multiple equilibrium involving both 1:1 and 1:2 species, as shown in Fig. (**10**).

CASE STUDY 2 – HYDROXYPROPYL-B-CYCLODEXTRIN- SODIUM DICLO-FENAC - COMPLEX – A WATER SOLUBLE DRUG FOR PARENTERAL INJECTIONS

Diclofenac sodium (Dc) is a well-known non-steroidal anti-inflammatory drug (NSAID) of the phenylacetic acid class with analgesic, antipyretic and anti-inflammatory activities.

Due to its poor water solubility, injectable forms of Dc at the currently approved dosage of 75 mg only use water/solvent mixtures. The drug product consists of 3 mL of water containing 30% of propylene glycol. This formulation causes considerable pain when injected and, because of the large volume, it can only be administered by intravenous or intramuscular route. The subcutaneous route, which is an easy route also suitable for self-administration, can be applied only to smaller injection volumes.

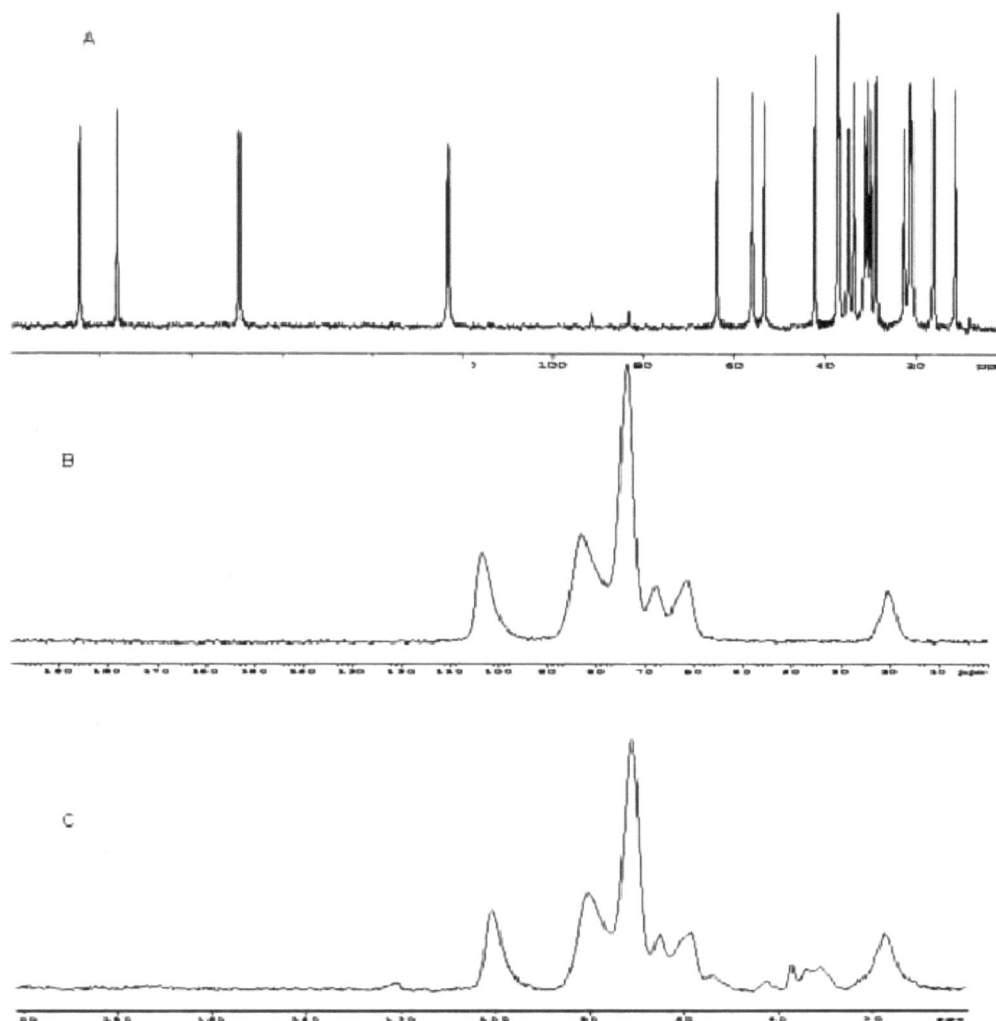

Fig. (18). 13C-CP-MAS spectra of P (**A**), HPBCD (**B**) and HPBCD/P complex (**C**). Torri *et al.* [24]. *Reprinted with permission of Springer Science B.V.*

The development of a new formulation of a more acceptable form of Dc could lead to parenteral formulations affording better patient compliance. Therefore, subcutaneous or intramuscular administration of a small volume using only water as solvent was the preferred choice. The use of complexing agents such as cyclodextrins provides a solution to enhance water solubility of Dc.

In this study a new drug product was formulated at a concentration of 75 mg/ml of Dc in pure water. This concentration was obtained with a 33% solution of hydroxypropyl-β-cyclodextrin (HPBCD) leading to a 1:1 molar complex stoichiometry. The addition of 0.018 % polyoxyethylene 20 sorbitan monolaurate (Tween® 20) in the solution inhibits the possible formation of crystals under experimental stability conditions [33]. The experiments described here confirm the data published on the complex without Tween®-20 and the negligible influence of Tween®20 in the complex formation is substantiated.

In this study NMR was used to define the presence of an inclusion complex by measuring the chemical shifts of internal cyclodextrin protons 3 and 5. Also in this case NMR was complementary to other instrumental and chemical techniques aimed to define the inclusion complex.

The evidence of the complex formation of Dc in the presence of HPBCD in water can be demonstrated simply by observing the increase of Dc concentration after addition of HPBCD. In fact, water solubility of Dc is 15 ug/ml without any co-solvent help in water. It has already been described how HPBCD enhances the water solubility of Dc by forming a 1:1 complex [34]. Considering the high solubility of HPBCD (48% W/V) and the 1:1 complex stoichiometry it is possible to obtain up to 109 mg/ml of Dc concentration. According to this author, the Dc dichlorophenyl ring is included in the HPBCD cavity. In the same paper it is also said that the phenylacetate ring can be included into one molecule of HPBCD, therefore forming a different inclusion 1:1 host-guest complex or a 2:1 host-guest complex. However, the author's conclusion is that the preferential complex is 1:1 host-guest, with the inclusion of the diclofenac dichlorophenyl ring with a complexing constant of 115.8 M^{-1}. To establish the complex structure both in water and in a solid state the author used different analytical methods such as Nuclear Magnetic Resonance (NMR), Different Scanning Calorimetry (DSC) and X ray difractometry [34]. In the DSC experiment the complex was evident by the presence of a peak in a different position from the reagent signals or by the absence of sharp peaks in the plot. In the first case the complex is considered crystalline and very strong, while in the second case the complex is not crystalline and is considered weaker, in a transition form from free to complex status. There are no sharp peaks in the DSC thermogram of HPBCD/Dc, confirming that the complex is neither crystalline nor in a transition state. As already mentioned, the formation of an inclusion complex can be confirmed by NMR using different acquisition techniques, by observing the shift of both Dc and HPBCD protons. In fact, when an inclusion complex is formed the HPBCD internal protons (the glucose H3 and H5) are disturbed and a chemical shift is observed. Also, observing which diclofenac protons (Fig. **19**) are shifted, it is possible to establish which part of the molecule is included.

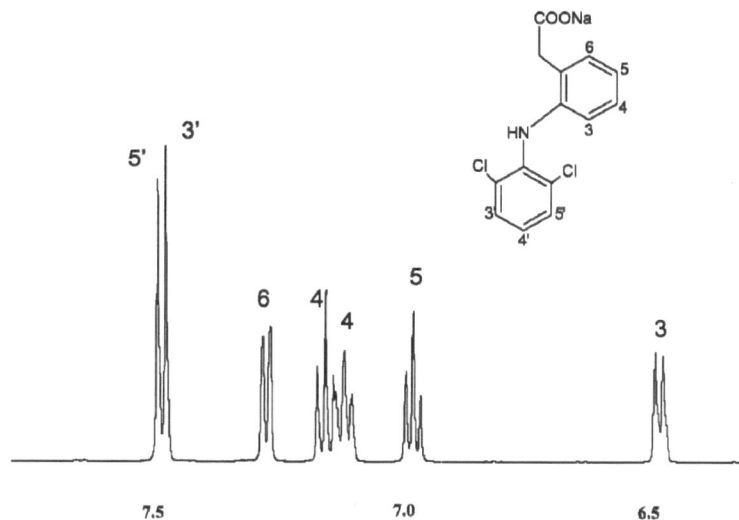

Fig. (19). Diclofenac sodium chemical structure and Proton –NMR spectrum with signal attribution (6,3-7,7 region).

According to the reported data, in the HPBCD/Dc complex, Dc protons H3' and H5' are downshifted, while H3 is upshifted (Fig. **25**), indicating the inclusion of the dichlorophenyl ring and also an interaction with the phenylacetic ring, probably due to the external hydroxypropyl residues of cyclodextrin. The inclusion of the full Dc molecule into the HPBCD cavity is unlikely, due to the high torsion angle between the two rings (Fig. **21**).

X-ray diffraction provides additional evidence that a real complex is formed by comparing the standard and complex patterns. In the presence of a complex the pattern is substantially different from the standard and the physical mixture patterns. A second paper [35] confirms the inclusion of Dc in HPBCD through the dichlorophenyl ring, but it also indicates the possible minor presence of a 1:2 host-guest complex. The author reached this conclusion because the NMR spectrum shows upshift of the H4 of diclofenac, indicating that one molecule of diclofenac is deeply complexed by one molecule of HPBCD through the dichlorophenyl ring and another diclofenac molecule is complexed by the same HPBCD from the opposite side. However, under these experimental conditions, the potential 1:2 host-guest molecule is present in a very low amount, if at all (Fig. **20**).

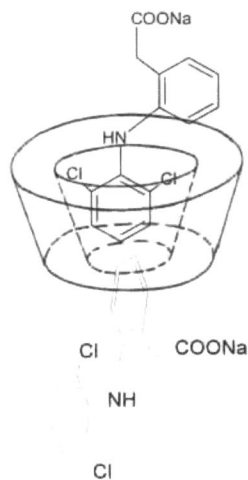

Fig. (20). HPBCD/Dc 1:2 host:guest complex model.

The HPBCD/Dc complex models suggested by Pose-Vilarnovo *et al.* [34] are reported in Fig. (**21**).

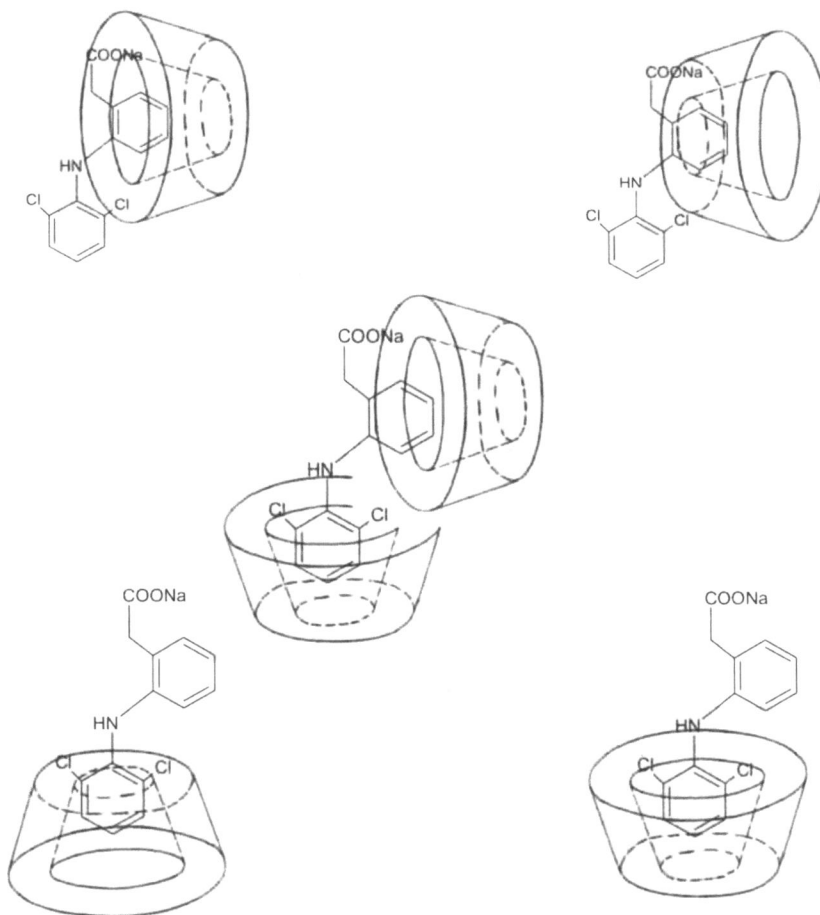

Fig. (21). HPBCD/Dc complex models.

This work confirms the inclusion complex in the original formulation (Dc-HPBCD) containing Tween[®] 20 as inhibitor of crystal formation [33] and defines the complex type and formation constant. Different physico-chemical methods were used, as well as the solubility profile according to Higuchi and Connors [16]. DSC and X-ray experiments were similar to published data [34]. NMR mono and bi-dimensional spectra were obtained with a Bruker 500 MHz instrument to definitively prove the existence of the inclusion complex.

Solubility Profile

The inclusion complex type in our product was first determined by the Higuchi and Connors method [16], in which a fixed amount of Dc was treated with a different amount of HPBCD dissolved in a fixed volume. The resulting suspensions were filtered and the amount of Dc in the solution was measured. The resulting amount (mmol) of Dc in the solution was plotted against the amount of HPBCD (mmol) and the results are reported in Fig. (**22**) below:

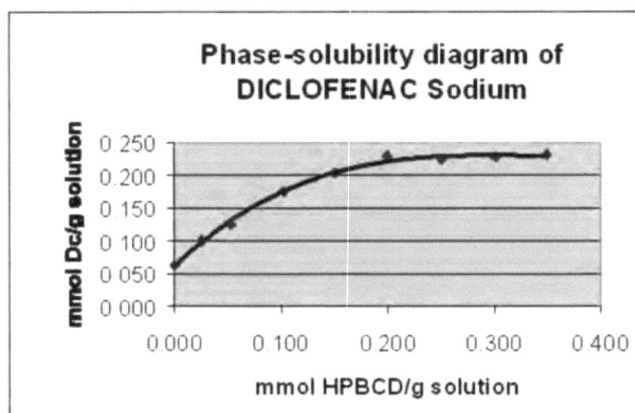

Fig. (22). Phase solubility diagram of HPBCD and Dc.

The increase of the Dc quantity by raising the amount of HPBCD indicates a type A curve according to Higuchi and Connors, therefore, when the slope is less than unity a 1:1 inclusion complex is suggested. The calculation of the complex constant ($K_{1:1}$) was made accordingly, by applying the related formula for 1:1 complexes shown below (1):

$K_{1:1}$ = slope/So(1-slope)   (1)

So refers to the solubility of Dc without HPBCD.

In our experiment the slope corresponds to 0.9843 and so to 0.0648. Applying the above formula (1) the resulting constant is obtained:

$K_{1:1}$ = 0.9843/0.0648(1-0.9843) = 967.50 m^{-1} (2)

No clear flex point is shown in the curve obtained in the previous experiment, suggesting a dynamic multiple complex with a possible change of the complex stoichiometry during the experiment itself. An unconventional solubility profile experiment was repeated with fixed HPBCD concentrations at 7 and 33% to verify this hypothesis. The experiment according to Higuchi and Connors implies the addition of different amounts of cyclodextrin at fixed volumes, thus increasing the concentration of cyclodextrin during the experiment. In the new experiment the cyclodextrin concentration was maintained and the solution volume was modified until complete Dc solubilisation. In this way it was possible to determine the influence of the cyclodextrin concentration on the solubilisation of Dc. The flex point gives a molar ratio of the HPCD/Dc complex.

The resulting amount (mmol) of diclofenac/mmol initial diclofenac in solution is plotted against the amount of HPBCD (mmol)/mmol initial diclofenac, as reported in Figs. (**23** and **24**) below:

The influence of the HPBCD concentration on the type of inclusion complex is clearly shown in this experiment. At 33 % the host/guest ratio is 1:1, while at 7 % it changes to a 1:2 complex.

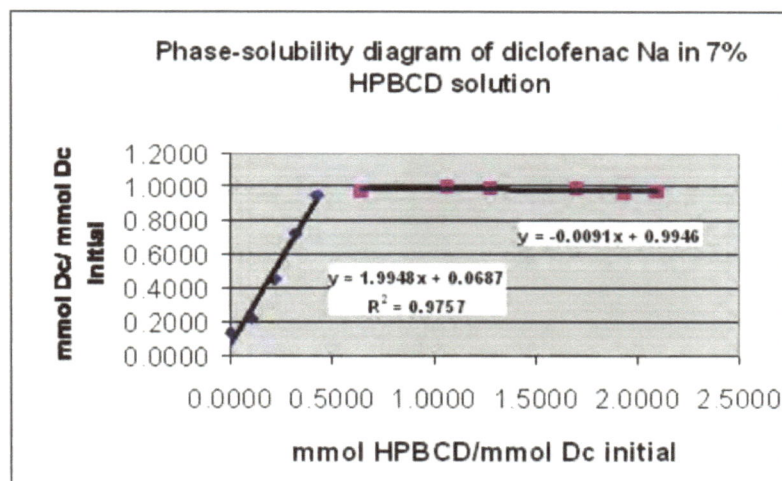

Fig. (23). Phase solubility diagram in diluted HPBCD solution.

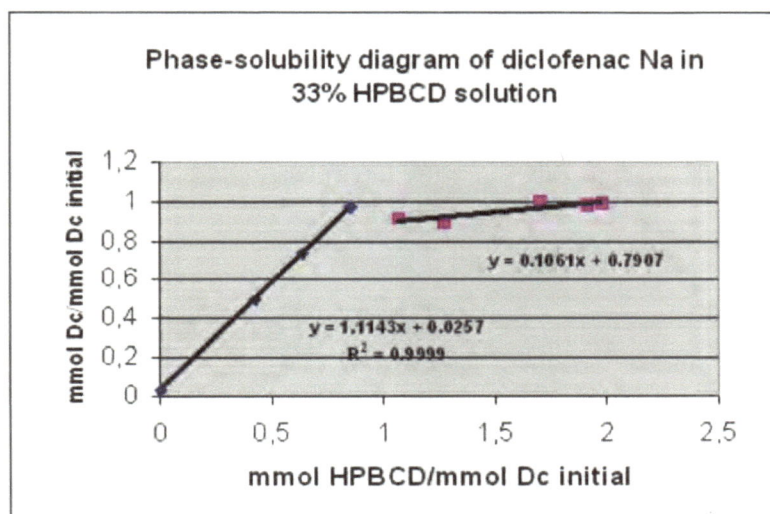

Fig. (24). Phase solubility diagram in concentrated HPBCD solution.

Tween[®] 20 at 0.018 % is added in the formulation to prevent potential crystals formation during stability [37]. To assess the influence of Tween[®] 20 on the inclusion complex type and on the constant magnitude, the same solubility profile according to Higuchi and Connors was performed in presence of Tween[®] 20. Tween[®] 20 at 0.058% of the HPBCD quantity was added at 0.058 % to reproduce the same tri-component ratio of the experiment with 33% HPBCD. The resulting plot (data not shown) indicates a similar 1:1 host/guest A-type curve and a complex constant of the same magnitude (254.79 m[-1]) as without Tween[®]20.

NMR Study

The experiment was performed with freeze-dried Dc-HPBCD reconstituted with deuterated water, measuring proton and carbon spectra by mono and bi-dimensional techniques and comparison with the standard Dc and HPBCD.

The proton spectrum from 6.3 to 7.8 ppm of Dc already shown in Fig. (**21**) above was compared with the HPBCD/Dc complex in Fig. (**25**) below. Main Dc protons are highlighted above the Dc spectrum.

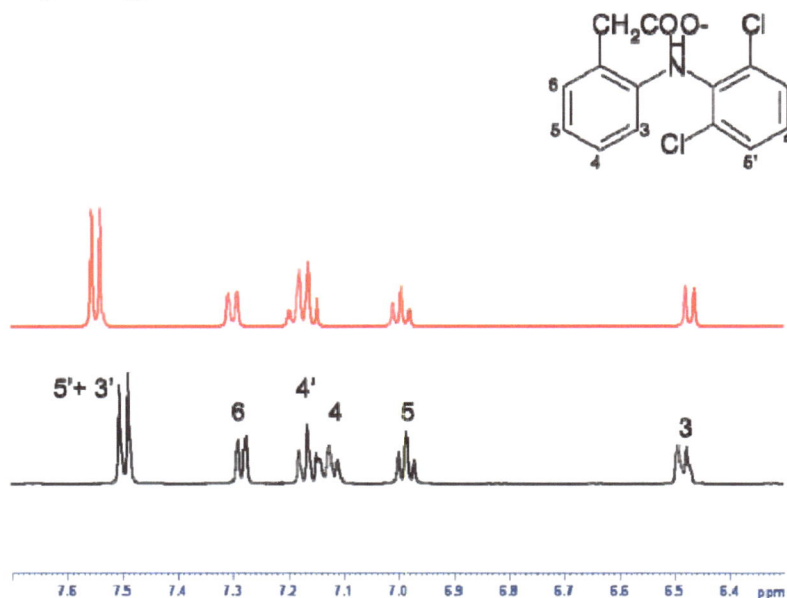

Fig. (25). Proton spectra of Dc (black) and HPBCD/Dc complex (red).

The main shift is evident for Dc protons 3', 5' and 4, indicating the influence of HPBCD on both rings with particular relevance for the dichlorophenyl residue. This confirms the preferential inclusion of the dichlorophenyl ring in the complex. This evidence is further confirmed by the etheronuclear bi-dimensional experiment (Fig. **26**).

Shifts are shown by comparing the black spots (Dc alone) with the closer red spots (HPBCD/Dc). The shifts observed in Dc indicate the influence of HPBCD.

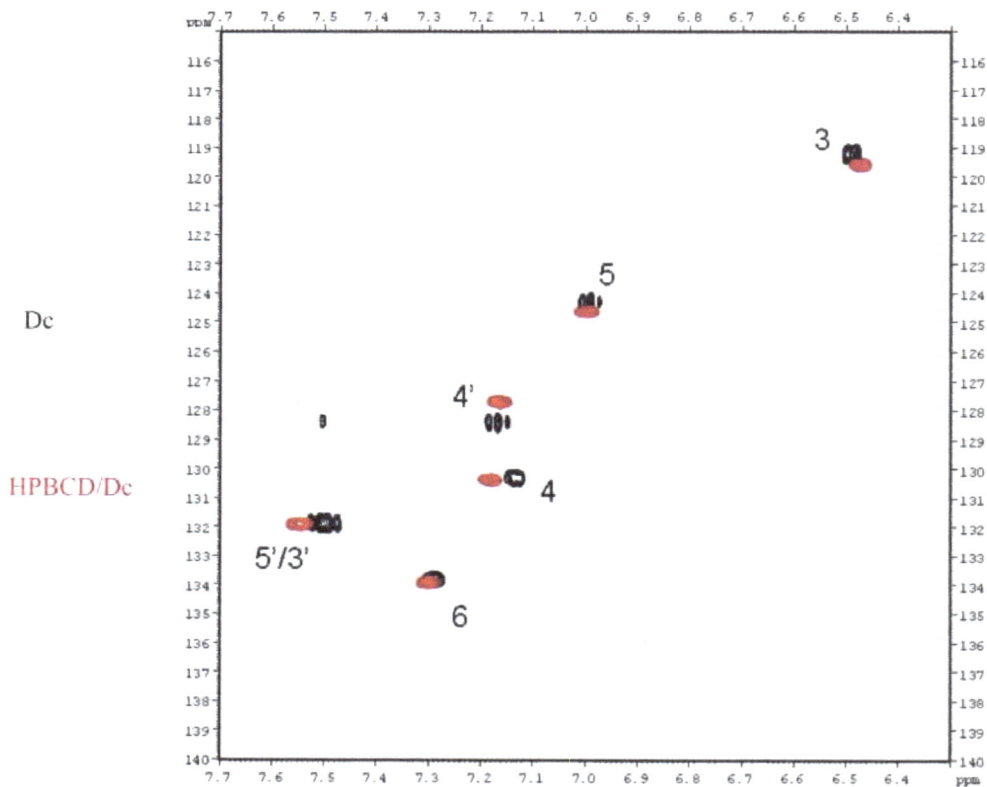

Fig. (26). HSQC (aromatic region) of Dc (black) and HPBCD/Dc complex (red).

The inclusion complex is demonstrated by the shift observed in the HPBCD internal protons 3 and 5. In fact, protons 3 and 5 belong to the lipophilic cavity, where the guest molecule is hosted. The Fig. (**27**) shows the protons 3 and 5 shifts in HPBCD/Dc (red spots) compared with HPBCD alone (blue spots).

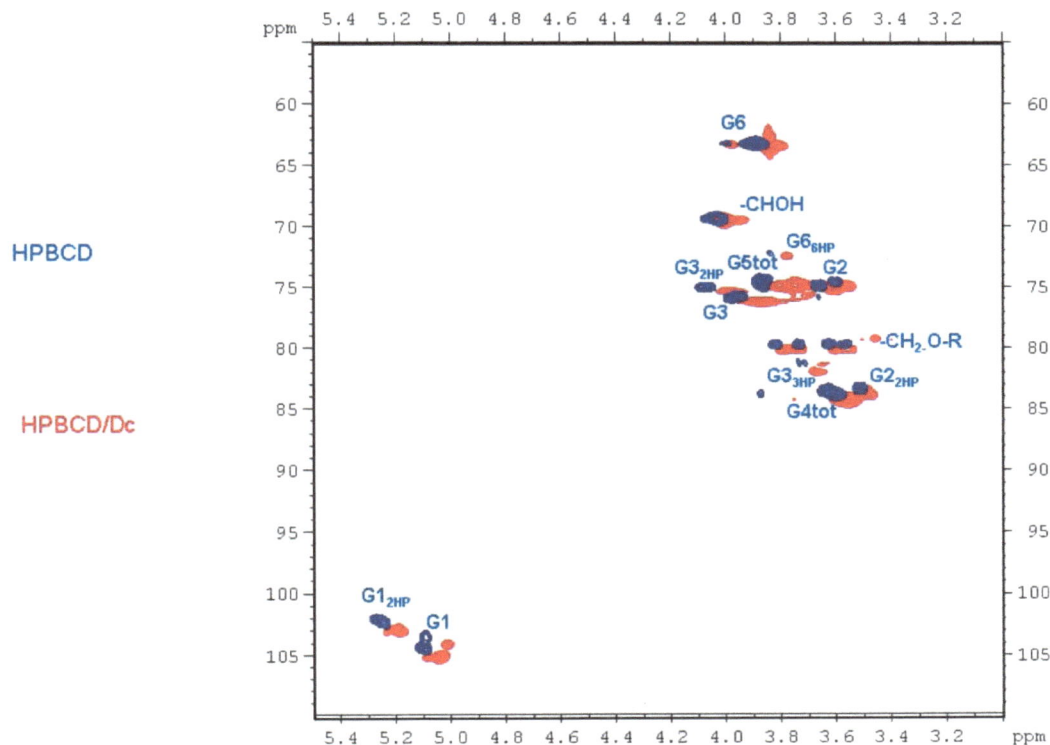

Fig. (27). HSQC (sugar region) of Dc (black) and HPBCD/Dc complex (red).

Protons 3 and 5 shifts are confirmed by the comparison of G3 and G5 blue spots with the relative closer red spot. Additional structure information is provided in this experiment on the glucose residue hydroxypropylated in 2 and 3 positions and not propylated (G3$_{2HP}$ and G3$_{3HP}$ and G3). All of these species are shifted in presence of Dc.

The inclusion complex in this study (HPBCD/Dc) has a 1:1 type host/guest complex with a weak complex formation constant of 254.79 m^{-1}. This inclusion complex is demonstrated by NMR, DSC and X-ray techniques. The complex weakness is suggested also by the phase solubility experiments, where the cyclodextrin concentration influences the complex type. In concentrated solutions cyclodextrin molecules are too close to host more than one Dc molecule. Two Dc molecules in a single cyclodextrin are shown in diluted cyclodextrin solution. The weak complex formation constant, calculated in an ideal solution such as water with a small quantity of Tween® 20, suggests a rapid decomplexation in the presence of third molecules, like in the presence of biological behavior. These physico-chemical characteristics suggest that after injection in biological tissues the complex is quickly destroyed by the competition of the biological molecules, therefore it is expected that the presence of HPBCD is not going to influence the pharmacokinetic of Dc.

CONCLUSIONS

NMR and cyclodextrin have been known for many years before entering into the drug development area. The same fate but different reasons were responsible for this huge delay. NMR was considered too much of a sophisticated technology to be implemented in a pharmacopoeia, hence in the drug development process. In addition, the standardisation of analytical procedures was very difficult and this fact did not help to convince the competent authorities experts. Cyclodextrins instead stumbled upon a false toxicology problem, which account for their belated application in drug development. The purification of α-cyclodextrin from the mixture was carried out with an insoluble complex containing toluene. Subsequently, the precipitate was heated to try and remove the guest, but this was not effective, with the result that the toxicology study was performed with contaminated material. The toxicity of cyclodextrins was given such a rushed conclusion, that for a number of years their use in drugs or food was disregarded. Nowadays, both technologies are considered very useful in drug development of new pharmaceutical forms of established drug substances. NMR is essential in the field of cyclodextrin complex to demonstrate the complex formation and represents an alternative for the calculation of the complex strength. This kind of information is required by regulatory authorities during drug product approval procedures. In fact, evidence of the complex bioavailability must be provided in approval procedures involving established drug substances. In most cases, even with high complex strengths the drug substance is captured by the tissues component to follow the normal kinetic. However, regulatory authorities can ask more data to support the fact that the complex is quickly destroyed after administration. If a cyclodextrin complex is needed to improve drug substance solubility in water aimed to reach therapeutic dosages, NMR can be used to demonstrate the actual presence of a complex. An example of tight complex and one of weak complex are described in the two case studies reported in this section. HPBCD/P complex is a high strength 1:2 complex in which the purity of HPBCD is crucial to avoid the formation of insoluble BCD/P precipitate during product shelf-life. In this particular case NMR was used also to elucidate the chemical structure of the precipitate. In the second example a complex of diclofenac sodium was shown to be a multiple complex with 1:1 stoichiometry. Multiple types of complexes have been discovered by NMR studies including the minor presence of a 1:2 host/guest complex. Today the application of cyclodextrin technology is growing in drug products applications by virtue of its favourable toxicology profile and it is expected that many new drug products developed with this technology will be approved worldwide in the near future. NMR application is growing as well in drug development allowing the improvement of drug products global quality.

REFERENCES

[1]   World Health Organization. WHO Expert Committee on Drug Dependence. Sixteenth report. (Technical report series. No. 407).Geneva: World Health Organization), 1969.
[2]   World Health Organization. WHO drug information. 2007; 00121(4): pp. 304-8.
[3]   European Pharmacopoeia – 6th ed. -Heparins, low- molecular-mass, (0828): 01/2008; pp. 2041-3.
[4]   European Pharmacopoeia – 6th ed –Hydroxypropybetadex, (1804): 01/2008; pp. 2103-5.
[5]   The Gold Sheet. USP add heparin, glycerine tests while FDA, army labs try new test methods. 2008; 43(2): 6-7.
[6]   Guerrini M, Beccati D, Zachary S, et al. Oversulphated chondroitin sulphate is a contaminantin heparin associated with adverse clinical events. Nature Biotechnol 2008; 26(6): 669-75.
[7]   The Gold Sheet. Search for heparin adulteration was complicated. 2008; 42 (12): 5-8.
[8]   ICH Harmonised Tripartite Guidelines . Impurities in new drug products (Q3B-R2) June 2, 2006.
[9]   Stejskal EO, Memory JD. High resolution NMR in the solid state. Fundamentals of CP/MAS 1994, Oxford University Press.
[10]   Torri G, Naggi A, Cosentino C, Pizzoferrato L, Aguzzi A, Capelloni M. In: Relationship between starch structure and *in vitro* digestibility: A preliminary study. Atti del 2° congresso della chimica degli alimenti. Giardini Naxos, Italy, May 24-27, 1995. Società Chimica Italiana; Gruppo Chimica degli Alimenti. 1995; pp. 1121-1127.
[11]   Le Botlan D, Rugraff Y, Martin C, Colonna P. Quantitative determination of bound water in wheat starch by time domain NMR spectroscopy. Carbohydr Res 1998; 308: 29-36.
[12]   Sabadini E, Cosgrovea T. and do Carmo Egídio, F. Solubility of cyclomaltooligosaccharides (cyclodextrins) in H_2O and D_2O: a comparative study. Carbohydr Res 2006; 341: 270-4.
[13]   Szejtli J. Cyclodextrin technology. Davies JED, Ed.; Kluvert, A.: Dordrech, The Netherland. 1988; p. 80.
[14]   Irie T, Uekama K. Pharmaceutical application of cyclodextrins. III. Toxicological issues and safety evaluation. J Parm Sci 1997; 86 (2): 147-62.
[15]   Szejtli J. Cyclodextrin technology. Davies JED, Ed.; Kluvert, A.: Dordrech, The Netherland. 1988; p. 105.
[16]   Higuchi T, Connors KA. Phase solubility techniques. Adv Anal Chem Instrum 1965; 4: 117-212.
[17]   Suzuki M, Sasaki Y. Inclusion compound of cyclodextrin and azo dye. I. Methyl orange. Chem Pharm Bull 1979; 27(3): 609-19.
[18]   Ikeda K, Uekama K, Otagiri M. Inclusion complexes of β-cyclodextrin with anti-inflammatory drugs fenamates in aqueous solution. Chem Parm Bull 1975; 23: 201-8.
[19]   Aki H, Niiya T, Iwase Y, et al. Multimodal inclusion complexes between barbiturates and 2-hydroxypropyl-beta-cyclodextrin in aqueous solution: Isothermal titration microcalorimetry, C-13 NMR spectrometry, and molecular dynamics. J Parm Sci 2001; 90: 1186-97.
[20]   Challa R, Ahuja A, Ali J, Khar R. Cyclodextrins in Drug Delivery: An Updated Review. AAPS Pharm Sci Tech 2005; 6(2): 329-57.
[21]   Stadler-Szöke A, Szejtli J. Proc. 1st. Int. Symp. Cyclodextrin. Budapest: Hungary 1981.
[22]   Frank DW, Gray JE, Weaver RN. Cyclodextrin nephrosis in the rat. Am J Pathol 1976; 83: 367-82.
[23]   Zoppetti G, Puppini N, Pizzutti M, Fini A, Giovani T, Comini S. Water soluble progesterone-hydroxypropyl-β-cyclodextrin complex for injectable formulations. J Incl Phenom Macrocycl Chem 2007; 57: 283-8.
[24]   Torri G, Bertini S, Giavana T, Guerrini M, Puppini N, Zoppetti G. Inclusion complex characterization between progesterone and hydroxypropyl- β-cyclodextrin in acqueous solution by NMR study. J Incl Phenom Macrocycl Chem 2007; 57: 317-21.
[25]   Pitha J. Pharmaceutical preparations containing cyclodextrin derivatives. US 4727064. 1988.
[26]   Forgo P, Gondos G. A study of β-cyclodextrin inclusion complexes with progesterone and hydrocortisone using Rotating Frame Overhauser Spectroscopy. Monatyshefte fur Chemie 2002; 133: 101-6.

[27]   Lin SZ, Kohyama N, Tsuruta H. Characterization of steroid/cyclodextrin inclusion compound X-ray powder diffractometry and thermal analysis. Ind Health 1996; 34(2): 143-8.

[28]   Choi HJ, Chun IK. Complexation of progesterone with cyclodextrins and design of aqueous parenteral formulations. J Korean Pharm Sci 2001; 31(3): 151-60.

[29]   Martindale. The complete drug reference. 33rd ed. Pharmaceutical Press, UK, 2002. 1489.

[30]   Zoppetti G, Puppini N, Pizzutti M. New injectable formulations containing progesterone. EU 051084945. 2005.

[31]   Shilpa S, Korde S, Katoch R. Total assignment of ^{1}H and ^{13}C NMR spectra of pregnenolone and progesterone haptens using 2D NMR spectroscopy. Magn Reson Chem 1999; 37: 594-7.

[32]   Connors KA. Binding constants. The measurement of molecular complex stability. John Wiley; New York 1987; pp. 198-201.

[33]   Zoppetti G, Puppini N, Pizzutti M. Injectable pharmaceutical compositions comprising sodium diclofenac, β-cyclodextrin and a polysorbate. EU 1609481. 2005.

[34]   Pose-Vilarnovo B, Santana-Penín L, Echezarreta-López M, Pérez-Marcos MB, Vila-Jato JL, Torres-Labandeira JJ. Interaction of doclofenac sodium with β- and hydroxypropyl-β-cyclodextrin in solution –S.T.P. Pharm Sci 1999; 9(3): 231-6.

[35]   Mucci A, Schenetti L, Vandelli MA, Ruozi B, Forni F. Evidences of the existence of 2:1 guest-host complexes between diclofenac and cyclodextrins in solutions. A ^{1}H and ^{13}C NMR study on diclofenac/β-cyclodextrin and diclofenac/2-hydroxypropyl-β-cyclodextrin system. J Chem Res 1999; 414-5.

NMR Studies in Nanoparticle Drug Delivery Systems: Paclitaxel Loaded Nanocarriers

Hoang Nam Nhat[*] and Tran Thi Hong

Department of Technical Physics and Nanotechnology, UET, Vietnam National University Hanoi, 144 Xuan Thuy, Cau Giay, Hanoi, Vietnam

Abstract: We present a concise review on current advances in the field of application of Nuclear Magnetic Resonance (NMR) in Structure-Activity Relationship (SAR) study of nanostructure drugs, particularly of paclitaxel-loaded nanocarriers. A short introduction of NMR to non-specialists is also included. The review is separated into two parts, the first discusses in broader context the concepts and potentials of drug-loaded nanocarriers as the modern tumour active targeting systems and the second focuses on the paclitaxel-loaded nanocarriers, especially on conformations and tubulin-binding forms of paclitaxel. The application of nanoscience in pharmaceutical researches explodes very fast with many anti-tumour drugs recently been approved to the market. We survey only the important and latest issues in this field.

Keywords: NMR, paclitaxel, nanocarrier, drug delivery, structure, cancer.

INTRODUCTION

In the last five years there was a robust move in biochemical researches towards the development of new drug delivery systems using nanoparticles and nanosized complexes [1-5]. The FDA approval of Abraxane [6-9], the albumin-bound paclitaxel nanoparticles for metastatic breast cancer treatment in 2005 set an important and promising landmark in the development of nanostructure drugs. Enormous efforts have been seen in the designing, preparation and characterization of the newly developed systems, and going along with this the NMR spectroscopy, as a basic analytical instrument, has heavily been exploited. The NMR spectroscopy has also undergone many significant improvements in recent years, which include building of higher field magnets, more sensitive electronics and larger databases for high throughput automatic screening. Nowadays, the NMR is utilized in a fragment-based screening of small molecules binding to a common substrate protein. The lead detection and optimization using this technique was reported by Shuker *et al.* in 1996 [10] and is referred to as the "SAR by NMR" approach. However, the structure-activity relationship (SAR) study by NMR is neither limited to nor is dominated by SAR by NMR. There is available a number of NMR studies related to the structures, internal binding and activity of nanostructure drugs where the SAR by NMR is not applicable [11-14]. This chapter provides a concise review of recent NMR advances on SAR in the area of paclitaxel-loaded therapeutic nanocarriers.

PART I. BASIC CONCEPTS OF NANOCARRIER SYSTEMS

Current Nanoscience in Pharmaceutical Research

The nanotechnology has entered rapidly in pharmaceutical research for a decade. Looking at the indexing databases like Scopus or Google Scholar, one may find how the keyword "*nano*" renders the number of biochemical publications. Before year 2000 there was only 1 among each 120 papers retrieved by matching the exact keywords "*structure activity relationship*" that has the word "*nano*" in any place of its text. The ratio is 1 per 18 for a period between 2000 and 2004 and 1 per 8 after 2004. Searching with all keywords "*NMR*", "*nano-*", and "*structure activity relationship*" revealed that almost all studies have been completed after the year 2000 and nearly 75% of the figure was from 2004-2009. The search result for "*therapeutic nanocarrier*" is featured in Fig. (**1**) where a clear trend of doubling the number of publications is observed for a period 2000-2006. The recent stagnation (2007-present) may originate in delayed data collection of indexing databases, but may also signify the diversification of researches on nanocarriers: the word "nanocarrier" can no longer serve as an effective filtering keyword for the new developments in the field.

* Corresponding author: Tel: +84-437-549-570; Fax: +84-437-547-460; E-mail: namnhat@gmail.com

Atta-ur-Rahman / M. Iqbal Choudhary (Eds.)

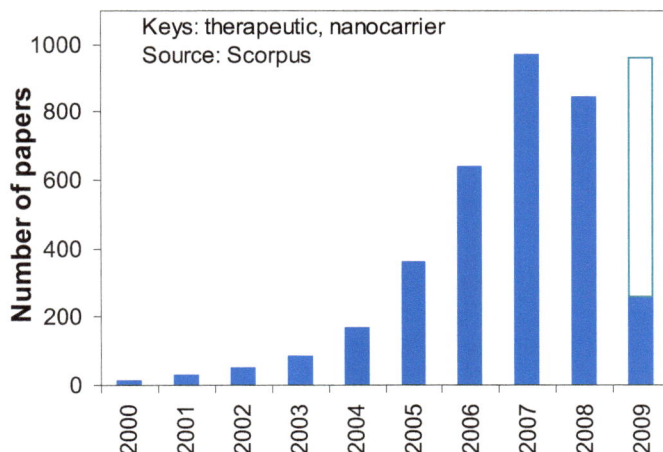

Fig. (1). Number of annual publications on the address of therapeutic nanocarriers. Data for 2009 account only for the first quarter, the whitened bar displays the expected value.

In overall, the biochemical research on nanoscale materials is believed to be still in its beginning as we may observe for a particular case of TiO_2 (or ZnO) that among 7,400 papers published in 2007-2008 on the address of TiO_2 - a common additive widely used in cosmetics and paints, there are only 9 papers dealing directly with its bioactivity. For ZnO, a constituent used in a number of medicaments for decades (or centuries?), there are only 2 papers on bioactivity among a total 8800 studies indexed.

The roar into the drug design using nanocarriers can also be seen in the industrial sectors where there is an increasing number of companies which have established the manufacturing platforms for nanostructure drugs [4].

Advances of Nanocarriers

Heavy investments in this field follow from the expected significant outcomes that the nanocarrier systems can promise (Fig. **2**). *First,* the nanocarriers can render the water-solubility of many lipophilic

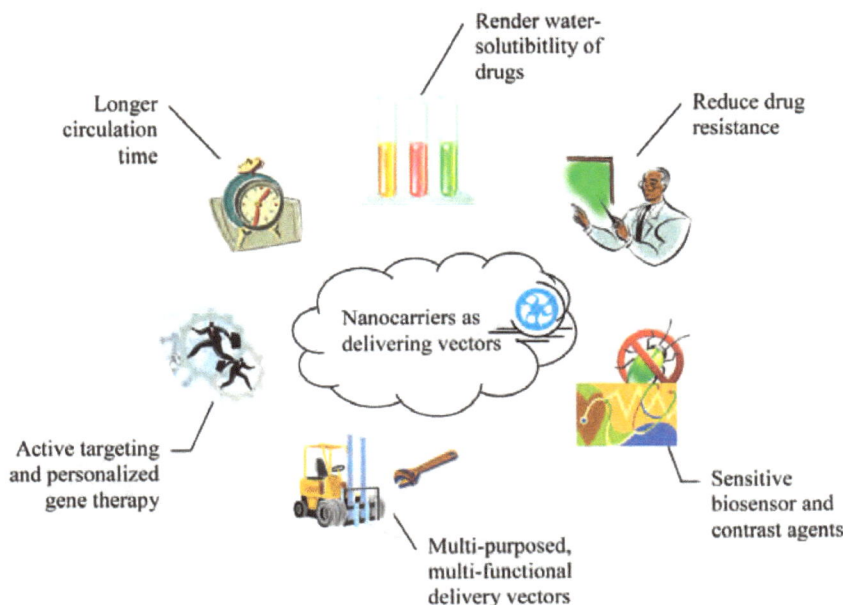

Fig. (2). Expected outcomes from the nanocarriers. A hybrid nanoparticle carrying both antigen and drug may be loaded preferably into the tumour tissues resulting in a much higher dose concentration at tumour site so that the selective killing effect on malignant tissues is achieved. This technique requires, however, a correct identification of vascular targets, particularly of the proteins, which are tumour endothelium specific markers.

drugs so making them bioavailable and reducing their toxicity. *Second*, the adjustable size of nanocarriers sufficiently increases their circulation time, therefore achieving higher concentration at target sites while reducing administration dose. Being bioavailable, all nanoscale systems should be of size of a small virus, i.e. from 10 to 100 nm. Larger systems are quickly canceled from circulation, hence are not suitable for therapeutic purposes. *Third*, the nanocarriers can be particularly coated so that they can successfully bypass the recognition by the P-glycoprotein efflux pump, one of the main resources of drug resistance. *Fourth*, the nanocarriers enable the active targeting on selected type of cells and allow the control release of therapeutic drug by mean of biochemical or physical trigger. *Fifth*, the nanocarriers can provide an alternative route to treatment of diseases by serving themselves as the nanobombs or heating sources which can be activated under applied electromagnetic field, laser pulse or other physical trigger. *Sixth*, in addition to therapeutic aspects, the metal oxide based nanocarriers are widely used as the contrast agents for CT and MRI imaging. The carbon nanotube based nanocarriers are highly sensitive to protein binding so can be used as sensors in cancer cell screening. The gold nanorodes can label one cancer cells among 10^7 normal blood cells under photoacoustic flow cytometry. *Seventh*, the nanocarriers offer an unique chance for personalized gene therapy, as the techniques for embedding DNA and antibodies were fully developed. *Eighth*, there is a strong expectation that the multifunctional nanocarriers which include drugs, antibodies, contrast agents, DNAs, receptor-recognition proteins in their core/shell structure can be successfully built up in the near future. Once such system is tested and succeeded in clinical trials, they would certainly revolutionize the practice of medicine and brought radical changes to current health-care system.

Nanocarriers in Clinical Trials

Considering the therapeutic nanocarriers from a broader context, the studies on their bioactivity account for a branch of nanoscience whose subjects are nanoparticles dispersed in liquid. This area, technically referred to as the nanofluidics, is developing very fast recently. For the nanocarriers with attached chemotherapeutic agents, a particular interest was given to the tumour targeting devices [2]. Due to the tumour specific environment, it has been demonstrated in animals that the direct targeting of tumour vascular endothelium is possible [15].

It is commonly known that the growing of a solid tumour depends substantially on development of its vasculature that is on proliferation of tumour vascular endothelia. Acquiring a surplus supply of blood, a starved, highly hypoxic tumour is characterized by the specificities, which differ from that of normal tissues, such as the high level of disorganization, arteriole and venule mixture, non-uniform capillaries, glycolysis-induced acidification etc. Hence, the tumour endothelia and surrounding stroma offer a unique chance for direct targeting by the suitable nanocarriers. The different pH at tumour site also provides an efficient trigger mechanism to control the release of therapeutic agents. So besides the search for the appropriate anti-angiogenics, which effectively inhibit the pumping of blood into the tumour vessels, the direct targeting of tumour endothelia also appeared as a possible way of cancer therapy.

Studies have been carried out to identify such genes, e.g. ROBO-4 (a roundabout-family receptor), DEL1 (a developmental endothelial locus 1), EndoPDI (endothelial protein, disulphide isomerase), TEMx family markers (x=1, 5, 7, 8), DELTA4 (a Notch-Delta signaling molecule) etc. The ROBO-4 is a gene, which is commonly expressed at the sites with active angio-genesis such as embryos or tumours. The DEL1 is a specific adhesion molecule of the embryonic endothelial cells. ROBO-4 and DEL1 are both absent in the adult human except for the tumour sites. The EndoPDI is a hypoxia-induced gene. Since the tumour endothelium is depleted of oxygen and is constantly hypoxic, the EndoPDI protects these tissues from apoptosis during the hypoxia-induced stress. The TEMx-s are the members of transmembrane domains which also showed a strong tumour-endothelial expression. The DELTA4 is a hypoxia-induced signaling molecule which is primarily located in the plasma membrane.

The methods involving in the identification of these vascular targets include the use of microarrays and bioinformatics (e.g. Ref. [16]) and serial analysis of gene expression (e.g. Ref. [17]). Excellent reviews are available for the tumour vascular targeting techniques demonstrated in the animal models (e.g. Ref. [15]).

At present, except Abraxane many other therapeutic nanocarrier systems have entered the clinical trials. These systems include the polymer-drug conjugates such as poly-L-glutamic acid (PGA) bound paclitaxel (Xyotax), PGA bound camptothecin (CT-2106) and N-(2-hydroxopropyl)-methacrylamide copolymer (HPMA) bound doxorubicin (PK1); the polymeric micelles (amphiphilic block copolymers) such as polyethylene-glycol (PEG)-poly(D,L-lactide)-paclitaxel formulation (Genexol-PM) (Fig. **3**); the liposomal formulation of anthracyclines doxorubicin (Doxil, Myocet) and daunorubicin (DaunoXome); and many other liposomes.

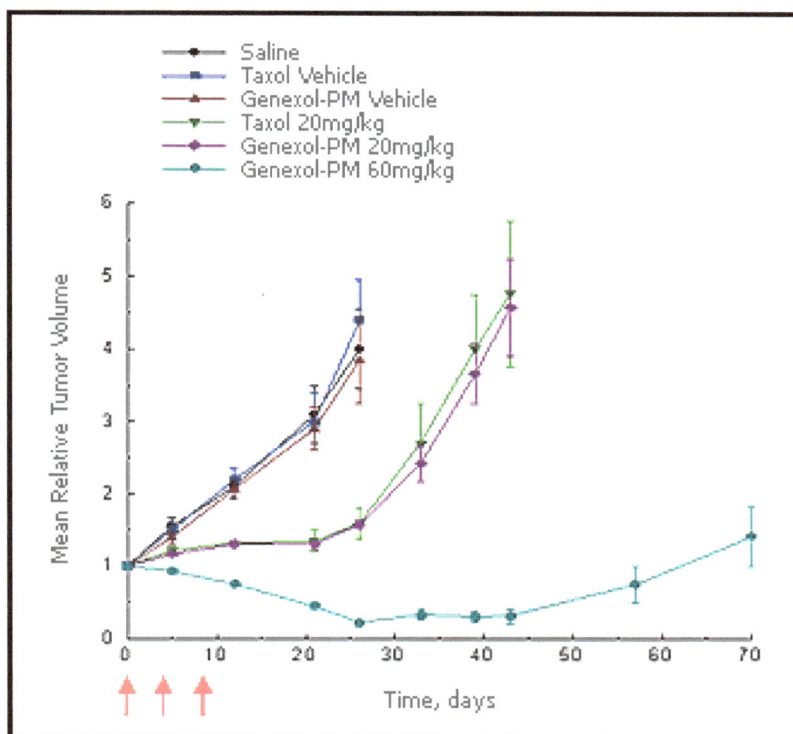

Fig. (3). Effectiveness of Genexol-PM in controlling the tumour growth in comparison with other medicals as reported by its producer, the Samyang Pharmaceuticals, South Korea (2009).

Types of Therapeutic Nanocarriers

According to their origin, the therapeutic nanocarriers may be divided into the following groups: (i) the polymeric nanoparticles, including block copolymers (micelles) and polymeric macromolecules (so called dendrimers); (ii) the lipid-based nanoparticles (liposomes); (iii) the viral nanoparticles; (iv) the carbon nanotubes; (v) the metal nanoparticles (such as gold-nanoparticle) and the metal-oxide based nanoparticles (such as iron oxide nanoparticles); and (vi) the platinum-based macromolecules and nanocomplexes.

A simplest form of the polymeric nanoparticles is a polymer-drug conjugate where the therapeutic drug is encapsulated in a nanosized spherical polymeric shell. A PGA-bound paclitaxel gives an example. Since such system does not include the pairing antigens to the tumour specific markers, it can target the tumour only in a passive way, that is, its concentration in the tumour tissues is solely depended on administration dose and circulation time. The polymeric (hydrophilic) shell has several effects. *First*, it renders the water solubility of drug; *second*, it protects the drug from opsonization (binding of antibody) by repelling plasma proteins; *third*, it limits the size of nanoparticles below 100nm, so that they can bypass the capture by fixed macrophages in liver and spleen as well as tunnel through the porous tumour endothelia. As a result, the circulation time of a polymer-bound drug is significantly improved and the selective accumulation of drug inside the tumour tissues is achieved. The release of drug from the polymeric shell is usually triggered by a degradation of the shell under reduced pH or by its cleaving under the presence of specific enzymes produced by cancer cells.

The problem of correct release of therapeutic drug inside the cancer cell is extremely important. After binding of nanocarriers to the target cells, the drugs should be released inside the cells, the incorrect release outside may cause dead to the normal tissues. This problem is commonly referred to as the *internalization* of targeted drugs and is a significant problem in drug design. It is important not only for the polymeric nanoparticles but also for the multi-functional nanocarriers with embedded drugs, contrast agents and targeting antigens. A sample mechanism of active internalization by the foliate receptor is illustrated in Ref. [18].

If the drug is lipophilic, water-insoluble or has poor water solubility (e.g. paclitaxel), then its hydrophilic nanocarriers can be synthesized by using the surfactant. A surfactant (or alias surface-active-agent) is an organic compound, which is amphiphilic, i.e. is both hydrophilic and hydrophobic. Such a molecule has two functional groups at its two ends: a hydrophobic tail is bound to drug and a hydrophilic head to hydrophilic polymer, creating a shell. The surfactant reduces the surface tension of the core oil phase and creates a structure called a micelle, which is water-soluble. An example of such system is Genexol-PM, a (PEG)-poly(D,L-lactide)-bound paclitaxel. Genexol-PM has been evaluated in Phase I clinical trial for patients with advanced refractory malignancies in 2004 [19]. Fig. (**4A**) provides illustration for the polymeric nanoparticles.

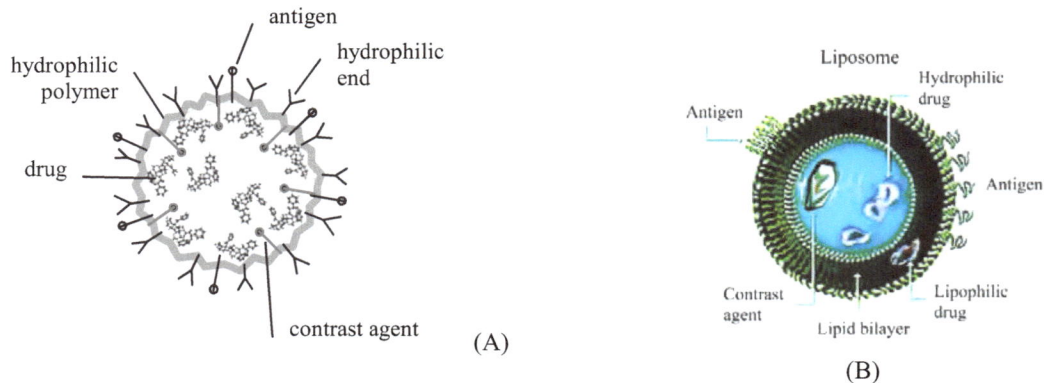

Fig. (4). Typical structure of a multi-functional polymeric nanoparticle (**A**) and a composite bilayer liposome (**B**).

A composite bilayer of amphiphilic polymers which mimics the structure of a cell membrane is called liposome. In this system, the hydrophobic tails of the two lipid (e.g. phospholipid) layers are connected to each other creating a bilayer with two hydrophilic surfaces. The inner surface encloses a small aqueous sphere (Fig. **4B**). This system can deliver both hydrophilic and lipophilic drugs: the water-soluble drug can be put inside an aqueous sphere and the lipophilic drug in spaces between the two layers. Other functional groups such as antigens, contrast agents, DNA can also be embedded. Several liposomal formulations of anthracyclines doxorubicin (Doxil, Myocet) and daunorubicin (DaunoXome) have already been approved for treatment of metastatic breast cancer.

Unlike the artificial systems for passive delivery of drug, the using of viruses for active targeting is a smart idea which rests on the nature of virus attack on target cell where the release of virus genome inside the cell is of primary consequence. During their evolution, viruses have developed a well-defined receptor-recognition surfaces which contain many spikes (about 10nm long) formed by the special proteins that mediate the entry through the cell membrane (Fig. **5A**). There is no simple one-to-one relationship between the virus protein and the cell receptor type. One receptor may often be recognized by several viruses and one virus may bind to many kinds of receptor. Some viruses, such as the human immunodeficiency virus type 1 (HIV1), may pass through a cell membrane in minutes, other (e.g. Semliki forest virus, SFV) require

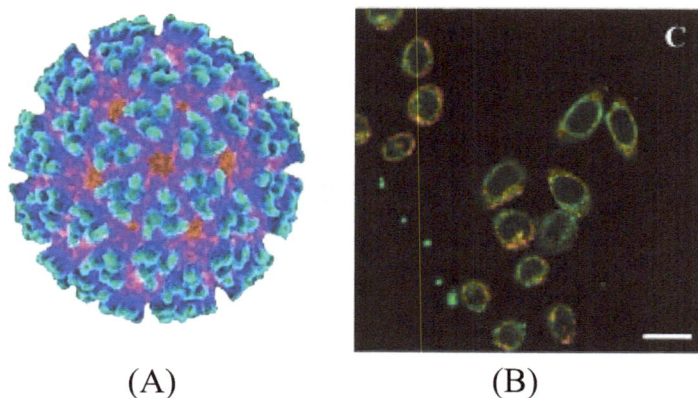

Fig. (5). Viruses as the nanocarriers promise the efficient route to targeting the malicious tissues. A Semliki forest virus (**A**) and the binding of transferrin (shown in red) to the surface of canine parvovirus (in green) (**B**) according to Ref. [21].

only a second to do so. Selecting a virus as the drug nanocarrier requires a correct identification of its receptor affinity. The canine parvovirus, for example, can bind to the transferrin receptors which are up-regulated on many kinds of tumour cells [20, 21] (Fig. **5B**). The drug and other bioactive substances can be uploaded onto the virus surface by mean of either chemical or genetic routes. At present, tests have been performed on a few types of viruses [21]. However, a drawback of this technique lies in the limited knowledge about the receptor affinity of many viruses under consideration.

The multi-functional nanocarriers may also be prepared from the core metallic nanoparticles (NP) having size from several to tens nanometres (Fig. **6**). Usually the surface of the core metallic NPs should be modified by some organic molecules which mediate the binding to drug and other functional groups, e.g. in the paclitaxel functionalized Au NPs by the 4-mercaptophenol molecule [22].

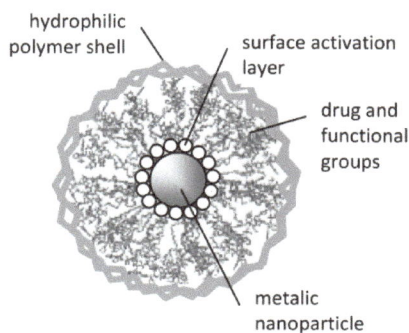

(A)

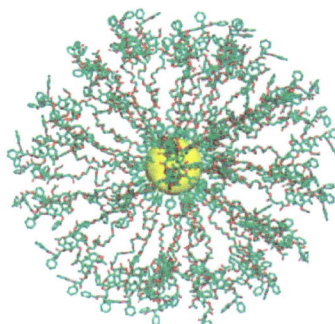

(B)

Fig. (6). Sample structure of a multi-functional metallic nanoparticle (**A**) and a real structure of Au-(HEG-paclitaxel)-bound nanoparticle (**B**) according to Ref. [22].

The possibility of using the metallic NPs such as Au, Ag and Pt as drug delivery vectors follows from their ability to covalently bind the drugs and functional groups. Such a metallic carrier is expected to exhibit many interesting quantum effects which are not cancelled out in the nanosized metallic clusters. Those effects may include the appearance of various surface plasmons (i.e. the reason why the Au NPs are of different colors according to their size), magnetism in the absence of magnetic compounds, characteristic optical excitation and thermal abnormal behaviour.

This induces that the metallic nanocarriers may serve as both delivering vectors and physicotherapeutic agents (heat sources, nanobombs), both as contrast agents and nanosensor probes. They also offer the effective trigger mechanisms based on the external physical factors such as laser-pulse or magnetic field induced release of chemotherapeutic agents. However, one of basic obstructions in the developing of these systems originates in the nanotoxicity, which remains undiscovered for many metallic systems, so the next development is probably restricted to several traditional systems such as Au, Ag, Pt and iron oxide NPs.

Basics of Nuclear Magnetic Resonance

The Nuclear Magnetic Resonance (NMR) is a resonance technique, which relies on reaction of nuclear magnetic moment on applied electromagnetic field with the energy that falls into the radio frequency (RF) band. A number of good text-books at introductory level on NMR is available, e.g. see Ref. [23] or one may read online about the basics of NMR at Ref. [24].

An intrinsic magnetic moment of a particle (such as electron, proton or neutron) is a quantity, which is derived directly from the quantum mechanical property of that particle - its spin, or intrinsic angular momentum **S** by a scaling constant. From the elementary physics, we know that if a spherical charged particle rotates around its own axis (like the earth does), the charge on its surface circulates like a small current and induces a magnetic field. Thus a rotating charged particle behaves like a tiny magnet with north and south poles. By convention a magnetic moment points from the south to North Pole as a vector. If the particle is positively charged and rotates from left to right, the north pole lies in its upper hemisphere and a south pole in lower one ($S=1/2$). The situation is inverse if the particle is negatively charged, or rotates in the opposite direction ($S=-1/2$). If the particle is a free particle (that is it does not interact with any outer electromagnetic field) then there is no way to distinguish between its two rotating (spin) states. But if we place such a particle in outer magnetic field **B**, one of the two spin states, which has magnetic moment oriented in the antiparallel direction to the south-to-north axis of **B** will have lower energy. The small energy gap between the two orientations of magnetic moment of the particle can be measured with high accuracy using the current technology. Since this gap falls within the energy of an electromagnetic wave with radio frequency (MHz) the NMR utilizes the RF signals to probe the switching between the two spin states (Fig. **7**). The NMR is specialized in detecting the changes in spin states of nuclei and the technique which measures this change for electron is called the Electron Paramagnetic Resonance (EPR). Since the magnetic moment of a nucleus like proton is about 658 times smaller than the magnetic moment of electron, the NMR measures much weaker signals in comparison with EPR.

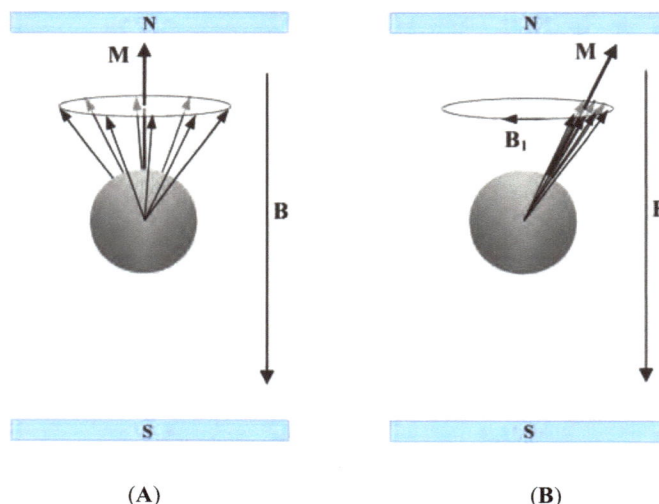

(A)   **(B)**

Fig. (7). Rotation of magnetic moment of a nucleus in outer magnetic field **B** (**A**) and the grouping of magnetic moment vectors when applying a rotating magnetic field **B₁** by switching on the RF pulse (**B**).

The energy change in NMR may be induced by two methods: either the magnitude of magnetic field **B** is varied at fixed frequency, or the frequency is tuned at fixed **B**. Suppose the latter case, the appearance of a NMR signal may be imagined as follows. At the occurrence of a RF pulse with energy equal to the energy gap between two spin states, the nucleus absorbs a pulse and excites to the higher energy state; after switching off the pulse, the magnetic moment of nucleus relaxes, it begins to oscillate around the south-to-north pole axis of **B** and after a finite relaxation time, it releases all absorbed energy, giving rise to a signal which can be measured. Of course, a signal contribution from one nucleus is extremely small so that it can not be detected, but the real materials contain a large amount of nuclei of the same kind, so the total outcome is the sum from all nuclei. Different nuclei give different resonance frequencies. The same nucleus under different chemical bonding environments may even have different resonance frequencies. To understand that, consider a nucleus of the atom as a ball putting inside the cloud shell of moving electrons.

The ^{63}Cu nucleus, for example, has 29 electrons circulating on 4 shell levels, causing a shielding effect on magnetic field for the ^{63}Cu nucleus. There is a fundamental physical law saying that there is no electromagnetic field inside a metallic ball. One may easily observe this effect by putting one's mobile phone inside a metal box. Depending on a composition of the metal cover and its thickness, the suppression of phone signal may happen.

The shielding effect on a given nucleus is reasonably different according to its chemical environment, e.g. the shielding is different for the carbon nucleus in benzene C_6H_6 and in ethane C_2H_6. This causes compound-dependent *chemical shifts* of resonance frequencies in comparison with the given standard such as TMS (tetramethylsilane, $(CH_3)_4Si$). The shift of frequency (Δv) is usually small and counted in Hz while the frequency (v) itself is of order MHz, so the relative "chemical shift" (δ) [i.e. equal to $(v_{sample}-v_{standard})/v_{standard}$] is very small quantity and must be multiplied by 10^6 to be expressed in *ppm* (i.e. parts-per-million).

The chemical shift is a sensitive imprint of the bonding environment of element, it correctly tells us in what compound the given nucleus occurs. For examples, using TMS standard the ^1H (proton) nucleus in C_6H_6 gives a sharp signal at 7.34 ppm, whereas the same nucleus in C_6H_{12} gives 1.43 ppm. Protons in $C_2H_2Cl_2$, CH_3NO_2 and C_3H_3O give 6.40, 4.33 and 2.15 ppm respectively. From these examples, one may wonder why 6 protons in C_6H_6 give rise to only one chemical shift. This is because the shielding environment is the same for all that protons (Fig. **8**).

The shielding effect of the electron cloud depends sufficiently on the valence electrons and their covalent bonding. The value 1.43 ppm from C_6H_{12} tells us that the shielding in C_6H_{12} is stronger than in C_6H_6 (7.34 ppm). The reason is simple: the π-bonding system above and below the C_6H_6 ring circulates like a current which induces a magnetic field over the ring. This local magnetic field adds to the strength of total magnetic field at the proton site so reducing the shielding effect. In C_6H_{12} there is no such π-bonding system and the shielding is of much stronger magnitude (Fig. **8**).

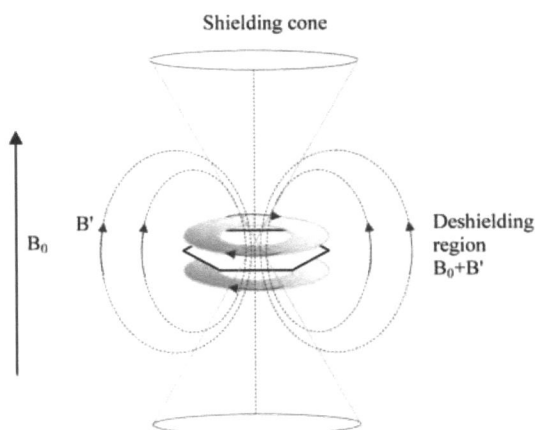

Fig. (8). Shielding and deshielding zone over the benzene ring.

In the TMS scale, the chemical shift for free proton (no shielding) is 32.775(25) ppm. This value is also given as the absolute shielding for proton in TMS (taking the shielding of free proton as zero). Some basic resonance frequencies for proton in different bonding environments are given in Table **1**. The extensive ^1H NMR data collections are now available for reference purposes [25, 26], some other may also be found in the internet [27].

Not only proton (^1H, spin 1/2, 99.99% natural abundance) that produces strong NMR signals but also the ^{13}C nuclei (spin 1/2, 1.1% natural abundance) can give rise to the well-defined NMR resonances. In fact, the ^1H and the ^{13}C NMR are most frequently used today in chemical researches. The signals from the two may even be combined together to give a two-dimensional (2D) NMR spectrum, the ^1H-^{13}C correlation spectrum. The 2D spectra may also be obtained from resonances of one nucleus type, i.e. ^1H or ^{13}C (Fig. **9**). If in both time-domains (axes) the chemical shifts are drawn, the 2D spectrum is referred to as the *shift-correlated*. If in one axis (usually y-axis) the coupling constant *J* is drawn, the resulting spectrum is called the 2D *J*-resolved correlated spectrum. In case of ^1H-^{13}C correlated spectrum, the technique is commonly referred to as the C,H-HETCOR or C,H-COSY. The abbreviations stand for the hetero-nuclear (HET) chemical shift correlation (COR) and correlation (CO) spectrum (SY).

Table 1. Summary of the ^{1}H NMR Resonances for Proton in Various Bonding Environments

Functional group	CH$_3$	CH$_2$	CH
CH$_2$R	0.8	1.3	1.6
C=C, C≡C	1.6-1.7	2.0-2.2	2.6-2.8
C$_6$H$_5$	2.3	2.6	2.9
F	4.3	4.4	4.8
Cl, Br, I	2.2 (I), 2.7 (Br), 3.0 (Cl)	3.2-3.4	4.0-4.2
OH, OR	3.3	3.4-3.5	3.7-3.8
OC$_6$H$_5$	3.8	4.0	4.3
OCOR	3.6	4.1	5.0
OCOC$_6$H$_5$, OCOCF$_3$	3.9-4.0	4.2-4.4	5.1
CHO, CONR$_2$, COOR, COR, COOH	2.0-2.2	2.2-2.4	2.4-2.6
CN, SR	2.1	2.5	3.0-3.1
NR$_2$, NH$_2$	2.2-2.5	2.4-2.7	2.8-3.0
NRC$_6$H$_5$, SOR	2.6	3.0-3.1	3.6
NR$_3^+$, NHCOR	2.9-3.0	3.1-3.3	3.6-3.7
NO$_2$	4.1	4.2	4.4

As seen in Fig. (**9**), the H(1) shows a cross-peak with H(3), H(2) with H(4), H(3) with H(1) and H(4), and H(4) with H(2) and H(3). These cross-peaks denote the nuclear spin coupling that is the magnetic interference of the near-by spins. Since all hydrogen atoms are separated by 3 bonds, e.g. H(1)-C(1)-C(3)-H(3) or H(4)-C(4)-C(3)-H(3), the spin coupling constants J between them is usually denoted as $^3J_{HH}$ to differentiate with the heteronuclear spin couplings between C and H, i.e. $^1J_{CH}$ for H(1)-C(1), C(3)-H(3), and/or $^2J_{CH}$ for C(3)-H(2).

The spin coupling forces the splitting of resonance lines, as we may observe from Fig. (**9**) that the peak no.1 is split into 3 lines of intensity ratio 1:2:1, the peak no.2 into 2 lines of equal intensity, the peak no.3 into 5 and no. 4 into 6. In general, the number of split lines of one hydrogen group is proportionate to the spin multiplicity of the near-by (no more than 3-bond away) hydrogen group. For example, in the OCH$_2$-CH$_3$ radical, the ^{1}H NMR signal of the -OCH$_2$- group is split into a quartet of intensity ratios 1:3:3:1, because the spins of three hydrogen atoms in the -CH$_3$ group may be oriented in the manner so that they create 4 distinguishable combinations with population ratio 1:3:3:1, that is to say:

- all spins down: (↓↓↓)

- two spins down: (↓↓↑)(↓↑↓)(↑↓↓)

- one spin down: (↓↑↑)(↑↓↑)(↑↑↓)

- all spins up: (↑↑↑)

Similarly, the expected splitting pattern of NMR signals for the -CH$_3$ group is a triplet 1:2:1, due to the three spin combinations of OCH$_2$- group, (↓↓):(↓↑)(↑↓):(↑↑).

Normally, one should also take into account the coupling effect between the different nuclei, e.g. between ^{13}C and ^{1}H. For ^{1}H NMR experiments, the coupling of protons to the carbon nuclei is weak but the effect is strong in ^{13}C NMR. So the H-decoupled ^{13}C spectrum is different than the undecoupled spectrum. The decoupling is achieved by a special pulse which establishes the equal population of low and high spin state for hydrogen so the coupling effect is canceled.

The change of spin state population for the protons which occurs near a given element (<5Å) usually strengthen the NMR output of that element. This effect has been discovered in 1953 by Albert Overhouser and is now referred to as the Nuclear Overhouser Effect (NOE). In NMR terminology, the NOESY (NOE

Spectroscopy) refers to the 2D method which maps the NOE correlations between protons (by the cross peaks as in COSY). The NOESY is a valuable tool in the study of large molecules.

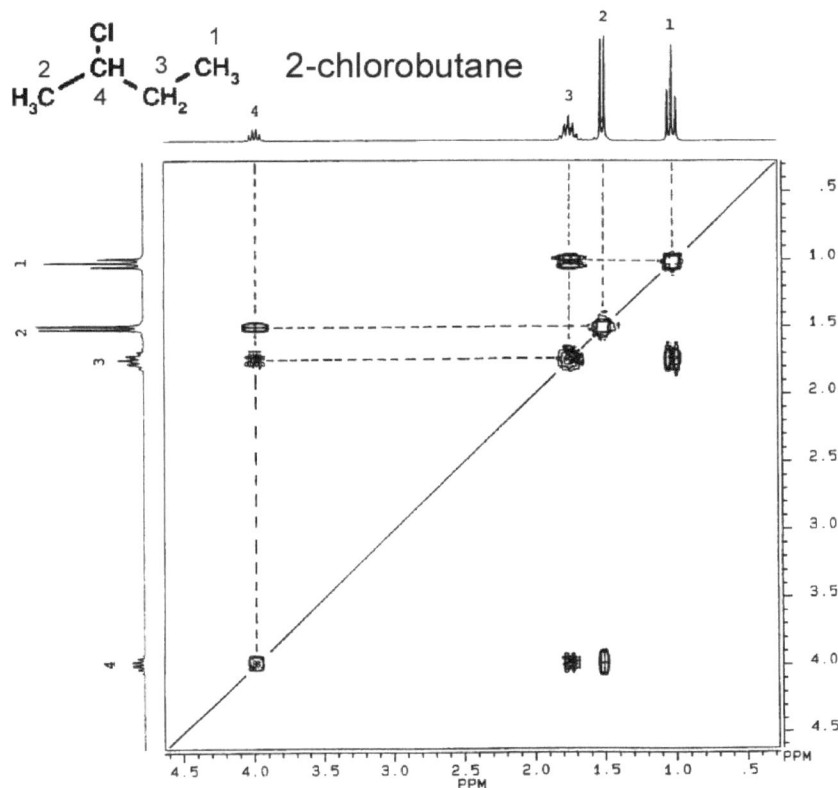

Fig. (9). A typical 2D ^{1}H-^{1}H COSY spectrum for 2-chlorobutane.

Where the "directly coupled proton" spectrum (COSY) does not provide adequate resolution, the TOCSY (Total Correlation Spectroscopy) which is capable of mapping the indirect spin coupling can be used. The TOCSY is a 2D homonuclear correlation experiment.

DEPT (Distortionless Enhancement by Polarization Transfer) is a 1D experiment which is based on the polarization transfer of magnetization from the highly populated spin state of hydrogen onto the near-by carbon. The higher gain achieved in the DEPT ^{13}C NMR experiment follows from four times higher population difference associated with hydrogen than with carbon. Furthermore, DEPT can also differentiate between the CH$_n$- groups due to the change in size and magnitude of the carbon resonances.

INEPT (Insensitive Nuclei Enhanced by Polarization transfer) is another 1D NMR experiment which is also based on the heteronuclear polarization transfer of magnetization from the nuclei with greater population differences such as ^{1}H, ^{19}F or ^{31}P onto the ^{15}N or ^{13}C near-by nuclei.

J-MOD (J-modulated spin-echo) is a spin-echo technique for identification of the CH$_n$- groups. As its name says, this technique relies on different responses of CH$_n$- groups to the different delay periods in the spin-echo.

PART II. PACLITAXEL-LOADED NANOCARRIERS

Brief History

Paclitaxel is a natural extract from the bark of the Pacific Yew tree (*Taxus brevifolia* in Latin). It has been isolated for the first time in 1967 by Wall *et al.* [28]. The discovery of paclitaxel was a result of the intensive researches taken by the National Cancer Institute (NCI) for a decade. Over 12000 natural compounds were scanned for the anti-cancer activity [29].

The paclitaxel shows sufficient anti-tumor effect in a series of trials (e.g. Phase I and Phase II clinical trails in the early and the mid 1980s respectively). It has been approved by FDA for treatment of resistant ovarian cancer in 1992 and later for treatment of breast cancer in 1994. Other paclitaxel based medical such as Abraxane [6-9] has been approved in 2005.

Nowadays, paclitaxel is widely used anti-cancer drug, not only for the treatment of breast and ovarian cancer but also for treatment of non-small cell lung cancer, colon, head cancer, Kaposi's sarcoma and other types of cancer. In 1998, its trade volume reached $1 billion and in 2007 it broke through $5 billions. The world market reported the annual growth of 30% in paclitaxel trade volume and the consumption in 2007 was about 3000kg. Only 10% of this figure was sold in USA (whose annual production capacity is believed to be less than 100kg). Paclitaxel has become one of best traded anti-cancer drugs in the history.

The production of paclitaxel from natural sources is however very limited. Although several types of *taxus* tree exist (such as *T. brevifolia, canadian, chinese, yunnanian* and *himalayan taxus, T. cuspidata, T.x. media* spp *Hicksii, T.x. dark green spreader, T. wallichiana, T. globosa, T. sumatrana, T. floridana*) which can produce various taxane analogues with comparable bioactivity (e.g. see Refs. [30, 31]), the production of drug from the natural sources cannot expand forever without the serious impacts on environment. There is approximately only 10g of paclitaxel that can be purified from 30kg crude oil extract of 1200kg of bark. That means that to produce 2g drug one need to cut down four trees. Paclitaxel can be extracted from any part of the tree which contains more than 0.005% by weight on dry basis, such as needles or barks. Therefore several total synthesis routes exist but they appear rather as expensive and complicated [32]. A semisynthethic analogue, Taxotere, which has been developed by Rhone-Poulenc Rorer, seemed to solve the supply problem since this compound can be converted to taxol in high yield from the available 10-deacetyl baccatin III [46] (Fig. **10B**).

Administration Formula

Visually, paclitaxel is a white crystalline powder which is not soluble in water (<0.6mM [33]). The melting temperature is around 216°C. Paclitaxel is soluble quite well in the organic solvents such as methanol, ethanol or DMSO (dimethylsulfoxide). The administration formula of paclitaxel is composed of 1:1 blend polyethoxylated castor oil (Cremophor EL) and ethanol. In this formula, the bioactivity of paclitaxel can be preserved for 5 years presuming that it is stored at 4°C. The only problem with this is that the Cremophor is toxic, so the premedication with corticoids (dexanethasone and antihistamine) is needed.

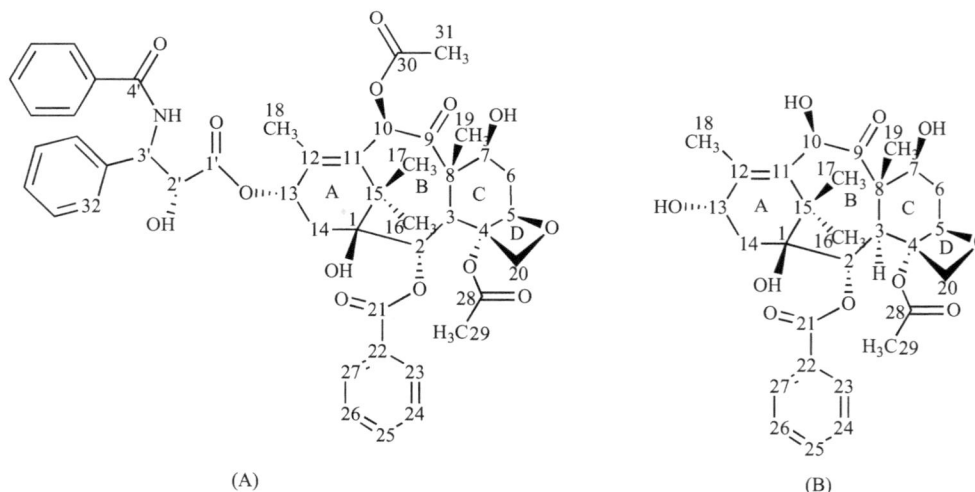

(A)   (B)

Fig. (10). Paclitaxel (**A**) and the taxane nucleus 10-deacetyl baccatin III (**B**).

Mechanism of Chemotherapeutic Action

Paclitaxel shows several chemotherapeutic mechanisms. Unlike the other anti-cancer drugs, it promotes the irreversible polymerization of tubulin [34]. The promotion of microtubule assembly by taxol prohibits the cell division process and consequently leads to the cell death. It also induces the internucleosomal DNA fragmentation associated with programmed cell death in human leukemia cells [35]. The polymerization of tubulin by paclitaxel posted a question which conformation of paclitaxel is active for the successive binding to tubulin. The determination of bioactive conformation of paclitaxel is a challenge for structural sciences and the answer is still not clear. Recent knowledge about the possible bioactive conformations mainly relies on the results of the high-field NMR measurements with *ad hoc* help of computer modeling.

Basic Geometry of Paclitaxel

Paclitaxel is a taxane diterpenoid with molecular formula $C_{47}H_{51}NO_{14}$ and molecular weight of 853 Da. The structure of paclitaxel has been identified in 1971 by Wani *et al.* [28]. It belongs to a group of 300 known natural compounds which possess the *taxane* ring (Fig. **10**). Its structure differs from the rest by having a side chain attached to the C(13) position and an *oxetane* four member ring (D-ring in Fig. **10**). The oxetane ring is quite essential for the bioactivity of paclitaxel. The existing studies showed that the removal of oxetane ring sufficiently reduced the bioactivity of paclitaxel.

It seems that the oxetane ring contributes to maintain the high rigidity of the taxane nucleus while serving itself as the hydrogen-bonding acceptor. The saddle-like taxane nucleus is highly deformed due to the *sp2* hybridization (planar triangle configuration) of C(9), C(11) and C(12) carbon atoms. All other atoms in this ring exhibit the tetrahedral *sp3* geometry. It is worth to note that this deformed geometry is not the one with the lowest energy.

Theoretical optimization of geometry with high level theory showed that there was a symmetrical conformation (with C(11) in *sp3* hybridization) that had sufficient lower ground state energy (therefore is a more stable state).

This conformation can be achieved by moving one hydrogen atom from C(18) to C(11) [36]. However, there is no experimental evidence until now for such compound. It is clear that the attached functional groups in paclitaxel have certain degrees of freedom due to the rotation axes C(2)-O, C(4)-O, C(10)-O, C(13)-O, C(2')-O, and C(4')-N. So to characterize the conformations of paclitaxel, many studies utilized the torsion angle C(3)-O-C(1')-C(2') as a characteristic parameter.

It is interesting to observe that in all crystal structures of paclitaxel which have been identified by x-ray diffraction technique until now, the free oxygen in the -COO functional groups is always oriented in a manner so that it points above the taxane ring. This is somehow difficult to explain by theory since the result of modeling for free paclitaxel molecule in absence of solvent predicted that the preferred orientation of these free oxygen radicals should be below the taxane ring [36] (Fig. **11**).

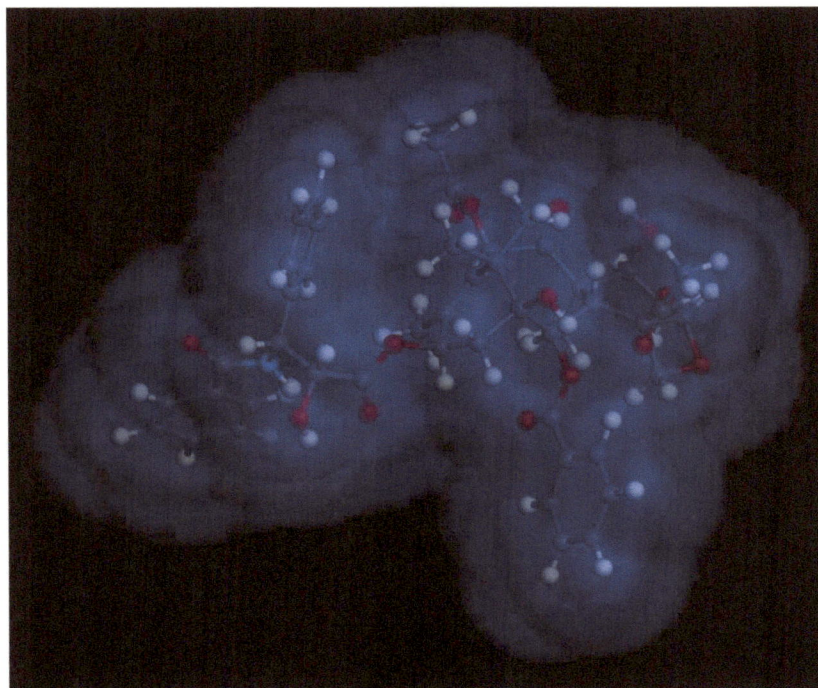

Fig. (11). An optimized molecular structure of paclitaxel drawn within its atomic volume surfaces [36].

Taxane Chemical Shift

The NMR chemical shifts (both ^1H and ^{13}C) can be calculated *ab initio* from the given taxane ring geometry [37]. The theoretical shifts are given in Table **2** and Fig. (**12B**). A specific feature of the calculated NMR shifts is that they depend only a little on the size of basis sets (e.g. 6-311G+(2dp,f) *versus* 6-311G* in Ref. [37]). This was due to the fact that electrons in the taxane ring are only weakly correlated. The *ab initio* calculation of NMR shifts for paclitaxel is considered as trivial.

Table 2. Calculated ^{13}C Chemical Shifts Relative to TMS for the Taxane Ring According to Ref. [37]. The Calculation was Carried Out Using GIAO Method at the Hartree-Fock 6-321G* Level Theory and STO-3G Optimized Geometries. The Experimental Data are from Ref. [38]

Atom No.	Calculated	Experimental
1	68.2	79.0
2	67.0	74.9
3	39.8	45.6
4	74.9	81.1
5	78.0	84.4
6	25.7	35.6
7	60.6	72.2
8	49.4	58.6
9	201.5	203.6
10	75.1	75.5
11	128.9	133.2
12	132.1	142.0
13	64.3	72.3
14	33.7	35.7
15	36.1	43.2
16	19.6	21.8
17	21.8	26.9
18	13.6	14.8

Structures of Taxane Derivatives

Since the crystallite size of pure paclitaxel is small and therefore not suitable for the single crystal x-ray crystallographic study, the structural data based on the x-ray diffraction technique are available only for the paclitaxel dissolved in solvents [39]. Several crystallographic data are available for various derivatives of paclitaxel [39-45]. As these studies showed, the paclitaxel molecule exhibits a large set of conformation geometries which depend observably on used solvents and attached functional groups. Basically, the rigid body of the taxane molecule is formed by four rings, commonly referred to as the A, B, C and D ring (Fig. **10**). Several views on the taxane nucleus are showed in Fig. (**12A**). Different structures differ each to other

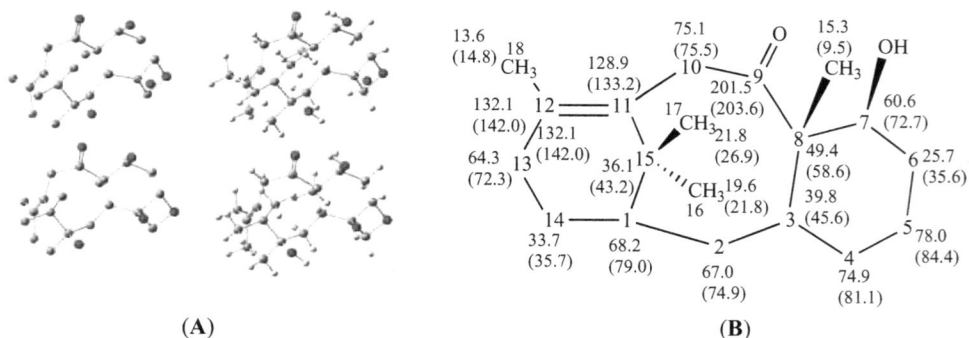

Fig. (12). Views on the taxane molecule with and without hydrogen atoms (**A**) and the taxane ^{13}C chemical shifts in brief (**B**). Experimental chemical shifts are given in the parenthesis [37].

in the orientation of the C(13)-side chain, conformation of the side chain (C(2')-C(3') rotation), orientation of C(2) and C(10) side chains. The differences may also be found between the torsion angles within the taxane rings themselves.

Among the carefully determined structures, the structure of 2-debenzoyl, 2-acetoxy paclitaxel reported by Gao *et al.* in [42] shows a typical arrangement (Fig. **13A**).

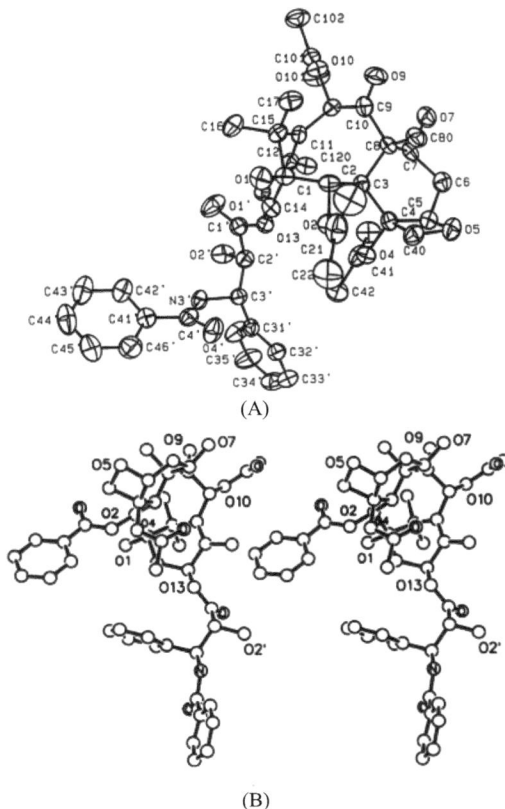

(A)

(B)

Fig. (13). The structure of 2-debenzoyl, 2-acetoxy paclitaxel as reported by Gao *et al.* in [42] (**A**) and of paclitaxel as reported by Mastropaolo *et al.* [41] (**B**).

The structural characteristics of this arrangement can be summarized as follows:

(i)   the C(13) side chain is oriented in a manner so that the O(1') oxygen points outside the taxane rings to the free space. This orientation is highly expected from the theory since it reduces the electrostatic repulsion from the rest of molecule to the side chain. If the side chain is inversely rotated by 180° along the C(13)-O(13) axis, the repulsion forces may significantly be higher;

(ii)   the same orientation (allowing the carboxyl oxygen radical to point to outer space above the taxane rings) is observed for the acetyl group attached to the C(2) position. However, in the structure of 2-debenzoyl, 2-acetoxy paclitaxel, the oxygen radical O(101) at C(10) position is pointing downside to below the taxane rings; the same is seen for the functional group at C(4) position;

(iii)   the position of C(8)-C(80)-O(7) is above the taxane rings;

(iv)   in the taxane ring, the angle C(1)-C(2)-C(3) (119.5°) is convex in the direction above taxane rings (the line connected C(15)-C(1)-C(2)-C(3) looks like a letter "h", not as "u", where C(3) lies above C(2) with respect to the taxane plane);

(v)   C(40) and O(5) lie above taxane plane.

(vi)   except C(9), C(11) and C(12), which exhibit the *sp2* hybridization (triangular conformation), all other carbon atoms in the taxane ring occur in the *sp3* hybridization (tetrahedral conformation). The *sp3* geometry allows a flexible arrangement of attached functional groups and hydrogen.

The structure of paclitaxel reported by Mastropaolo *et al.* [41] (Fig. **13B**) possesses almost the same characteristic features. The different hybridization of the carbon atoms in the taxane nucleus (ring B) was observed for the paclitaxel analogue $C_{19}H_{20}BrNO_6.0.5C_3H_8O$ [43] (Fig. **14B**). The atoms in *sp2* are C(8), C(9), C(11), C(15) instead of C(9), C(11) and C(12) as for the paclitaxel. The occurrence of C(12) in *sp3* induces higher regularity of the A-ring in the analogue than in the paclitaxel. The quantum chemistry calculation also showed that the paclitaxel conformation with higher regularity of the A-ring exhibits a lower ground state energy.

Gao an Parker [44] reported a structure conformation of paclitaxel (10-deacetyl-7-epitaxol) whose molecular arrangement is similar to 2-debenzoyl, 2-acetoxy paclitaxel and can be obtained from Fig. (**13A**) by exchanging the two positions of N(3') and H(3') (not showed in Fig. (**13A**)) which are attached to the tetrahedral *sp3* C(3') carbon Fig.(**14A**). This structure is referred to as the "hydrophobic collapse" (i.e. two hydrophobic groups are so close each to other to exclude an intervening solvent molecule) which is believed to propagate the polymerization of tubulin.

Another structural modification of paclitaxel derivatives, a taxamine A and taxamine B (Fig. **15**) was reported by Morita *et al.* [45]. These taxamines were isolated from the *Taxus cuspidata* and were shown to exhibit the bioactivity upon P-glycoprotein. Both taxamines possess the cinnamoyl ligand at C(5) position whereas the C(13) is terminated by a double bond to an oxygen and there is no four-member oxetane ring (D). Therefore, the taxamines contain no C(13) side chain and only the following rings: A (6-member), B (8 member) and C (6-member). The geometry of the taxane rings here is very similar to the one in 10-deacetyl-7-epitaxol (Fig. **14A**), including the orientation of the acetyl ligands. The oxygen radical in the carboxyl group of cinnamoyl ligand (at C(5) position) is also pointing outside the taxane rings as C_6H_6COO- ligand in the 10-deacetyl-7-epitaxol.

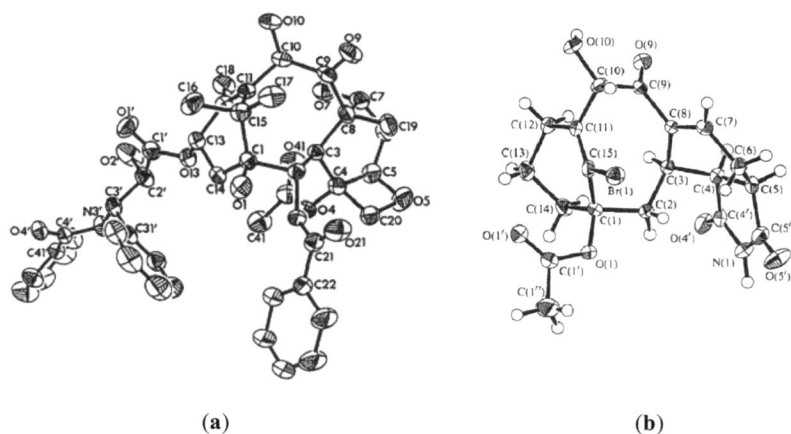

(a)   (b)

Fig. (14). The structure of 10-deacetyl-7-epitaxol as reported by Gao and Parker in [44] (**a**) and the structure of a C(10)-oxygenated ABC-ring analogue of paclitaxel [43] (**b**).

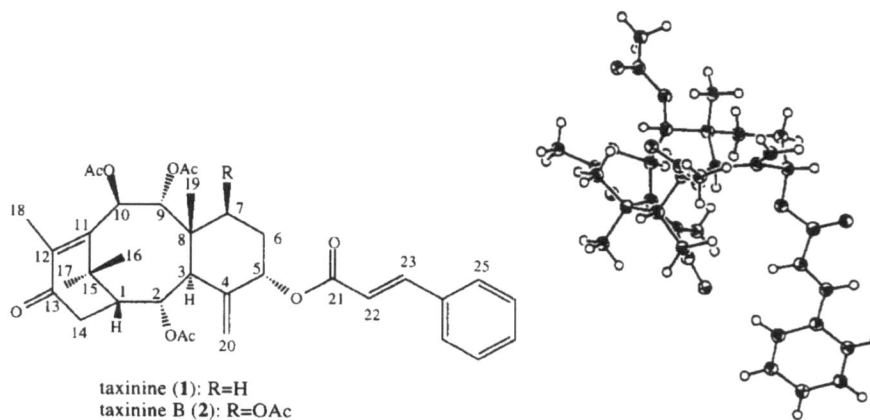

taxinine (**1**): R=H
taxinine B (**2**): R=OAc

Fig. (15). The numbering system for the taxamine molecule and the structure of taxamine A as given in Ref. [45].

The authors in Ref. [45] also reported the characterization of geometry conformation by the NMR spectroscopy using ROESY technique. The ROESY (alias NOESY in rotating frame) spectra show the correlation of nuclei through space (distance less than 5Å), so it allows to deduce the distance between the correlated nuclei.

ROESY technique utilizes almost the same pulse sequence as of TOCSY. The ROE correlation is featured in Fig. (16) and as seen here, there is a correlation between (i) H-2, H-9, Me-16 and Me-19; (ii) H-3 and H-7a; Me-19 and H-20; (iii) Me-18 and H-22. All these observed correlations agree quite well with the crystal structure as obtained from the x-ray diffraction technique (Fig. 15). The authors in Ref. [45] also demonstrated that the obtained geometries largely coincide with the structures with lowest ground state energy as optimized by the computer simulation technique.

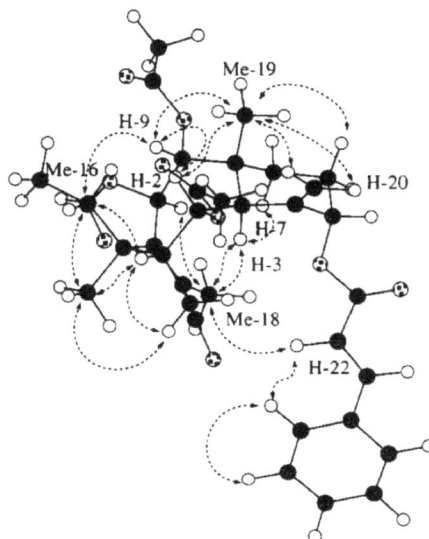

Fig. (16). ROE correlations as observed in the taxamine A according to Ref. [45].

Different geometries have been observed for the paclitaxel crystallized from different solvents [39]. The common solvents for paclitaxel include water, ethanol, dimethylsulfoxide (DMSO), N,N'-dimethylform-amide (DMF), N,N'-dimethylacetamide (DMAC), N-methyl-2-pyrrolidone (NMPO), 1,3-dimethyl-2-imidazolidinone (DMEU), 1,3-dimethyl-3,4,5,6-tetrahydro-2-(1H)-pyrimidinone (DMPU), acetonitrile and the mixtures thereof.

Fig. (17) illustrates the paclitaxel form that was crystallized from DMF. This form (denoted form E in Ref. [39]) contains 8 paclitaxel, 4 DMF and 8 H_2O molecules in one unit cell (orthorhombic space group $P2_12_12_1$). The ^1H NMR spectroscopy has identified that the ratio of DMF to paclitaxel was around 0.5:1, consistent with x-ray structural result.

What is obvious from Fig. (17) is that two paclitaxel molecules in the dimer exhibit two different C(13) side chain orientations. Whereas the side chain in molecule A is locked in a direction as illustrated in Fig. (11), the same group appears in the molecule B in an arrangement that is featured in Fig. (13A) and Fig. (14).

It is apparent that the conformation A can be obtained from the conformation B by rotating the side chain by 180° around C(13)-O(13) axis and consequently exchanging the two O(13) and H(13) positions. The orientation of the acetyl group at C(10) position in molecule A is also different than in molecule B.

The conformation A has not been reported in any study before Ref. [39] and contributes to the state with lower energy as of conformation B ("hydrophobic collapse" structure). We have showed that such rotating of the C(13) side chain in A reduced the ground state energy by about 2eV [36]. In some studies, the conformation like this is referred to as the "hydrophilic" structure.

Although the crystal structures of paclitaxel derivatives determined by the x-ray diffraction technique on single crystals are of great values for the SAR studies, the x-ray technique itself has a limitation in that the position of the hydrogen cannot be accurately determined. The forming of hydrogen bonding for paclitaxel is very important in the SAR studies. Usually, the position of hydrogen atoms can be retrieved from the structural optimization of paclitaxel using the *ab initio* technique. To date, there were reported only the geometry optimizations for the paclitaxel single molecule, not for the paclitaxel in crystal packing (that is optimization with periodic boundary condition - PBC). Since the paclitaxel is quite large system for the *ab initio* quantum chemistry calculation, the accurate theoretical results were not available for many important cases.

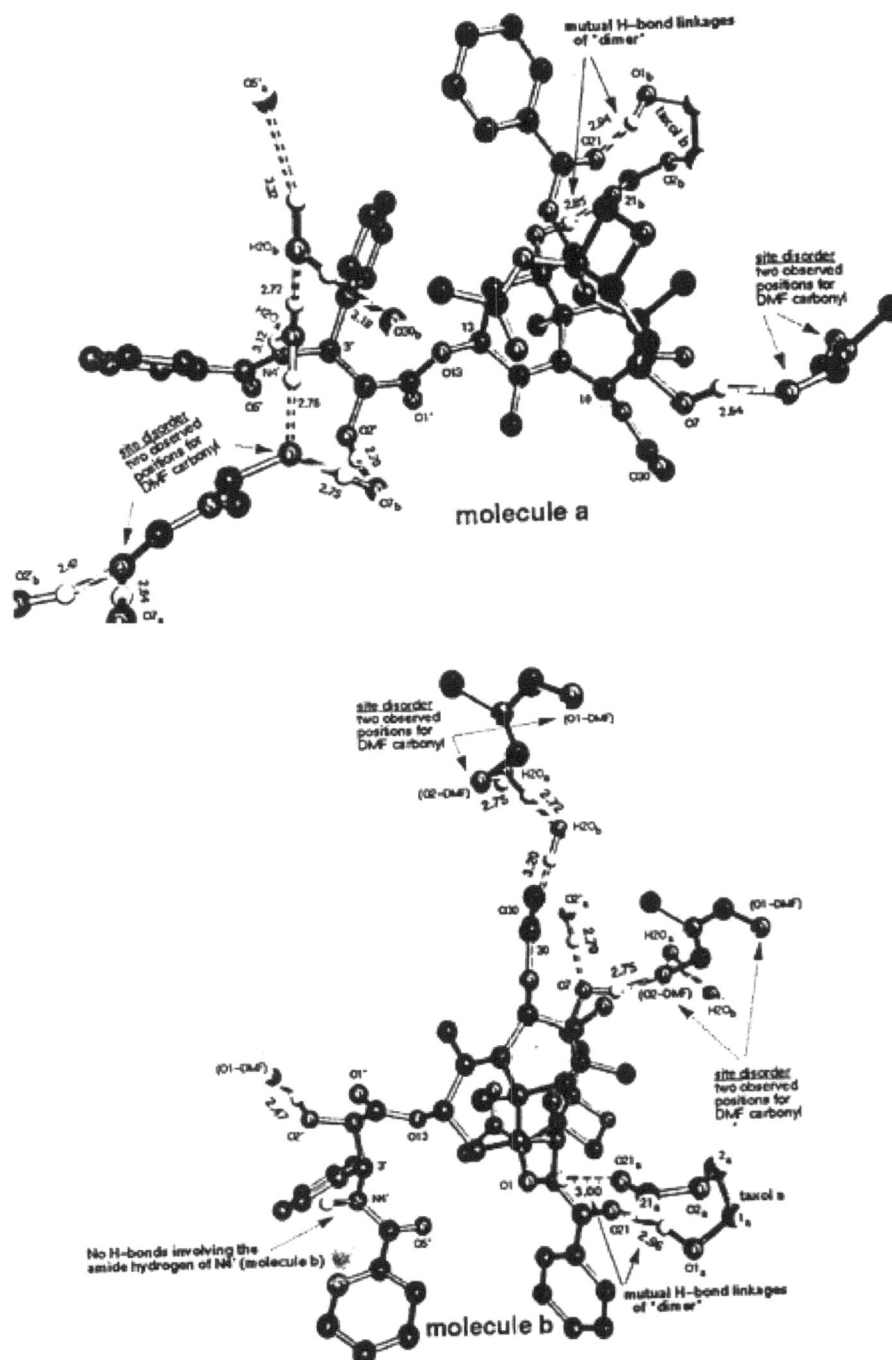

Fig. (17). Two conformations of paclitaxel as crystallized from DMF showing two different orientations of C(13) side chain (Ref. [39]).

The next limitation from the discussed structural conformations that were obtained from crystalline paclitaxel is that they do not reflect the bioactive form of paclitaxel. It is almost sure that none of the crystalline forms constitutes the tubulin-binding paclitaxel. So the structural studies of binding paclitaxel need to be carried out. Indeed, several such studies exist and they offer the valuable insights into the SAR for paclitaxel.

SAR for Paclitaxel

The Oxetane Ring

The available studies showed that the opening of this 4-member ring (D-ring) sufficiently reduced the bioactivity of paclitaxel. In the solid paclitaxel, the oxygen O(5) of this ring points upwards to the "northern hemisphere" i.e. the angles C(19)-C(5)-O(5) and C(3)-C(4)-C(20) are greater than 90° and less than 180° (Fig. **14A**). In some analogues these angles may be greater than 180° so that the O(5) lies downwards in the "southern hemisphere" of the taxane planes. There are studies which reported the lowering of the paclitaxel bioactivity by a factor of 10 in the tubulin-assembly assay when O(5) was replaced by S (sulfetane) or N (azetidine) [46, 47]. It is commonly accepted that the hydrogen-bond acceptor ability of the oxetane ring is important for the bioactivity of paclitaxel.

Taxol SAR Summary

Fig. (**18**) summaries the known SAR for paclitaxel according to Kingston [46]. Other reviews on SAR for taxol are also available but this one given by Kingston seems to be most complete.

From Fig. (**18**), the -OH groups at C(1), C(7), C(9) and the -AcO group at C(10) position are less significant in reducing the bioactivity therefore can be used as the binding acceptors in the paclitaxel-loaded nanostructures [48]. The other positions, such as the C(2) acyloxy group, C(4) acetyl group, oxetane oxygen and C(13) side chain are essential for the bioactivity of paclitaxel.

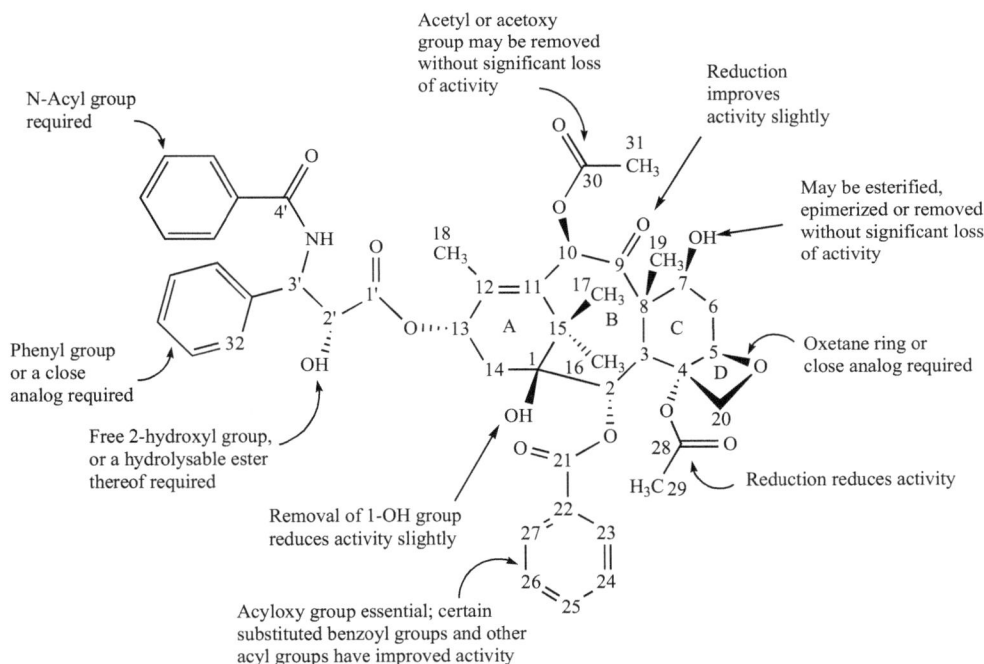

Fig. (18). SAR for Taxol in summary (*Reprinted from Ref. [46]*).

The conformation of micro-tubule bound paclitaxel was studied by the REDOR NMR spectroscopy together with the fluorescence spectroscopy [49]. REDOR (or Rotational Echo Double Resonance) is a high resolution solid state NMR technique which employs the magic angle spinning and cross polarization to detect the distance between two labeled nuclei, such as ^{13}C - ^{15}N, ^{13}C-^{19}F etc. The fluorescent derivative of paclitaxel, 3'-N-debenzoyl-3'-N-(*m*-aminobenzoyl) paclitaxel (N-AB-PT), was studied. The authors in [49] claimed that the adding of the amino group resulted in no difference in the microtubule-promoting activity of the compound. The distance between the fluorophore N-AB-PT and the colchicine binding site on tubulin polymers was determined by the time-resolved fluorescence resonance to be 29 ± 2Å.

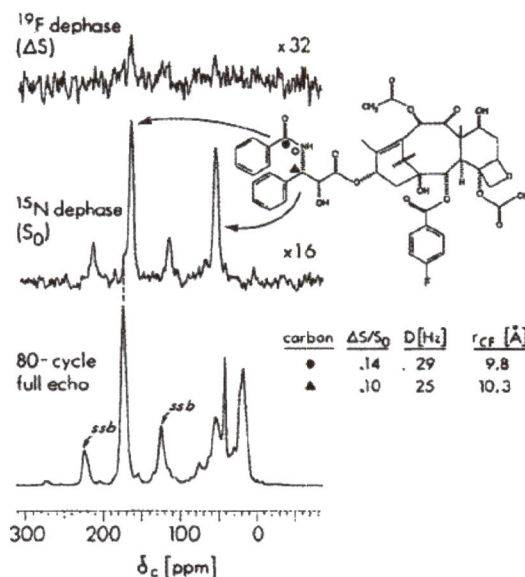

Fig. (19). The 125MHz $^{13}C\{^{15}N$ or $^{19}F\}$ double REDOR spectra of a micro-tubule bound complex 2-FB-PT (shown in the inset) according to [49]. ΔS an S_0 are the $^{13}C\{^{19}F\}$ and $^{13}C\{^{15}N\}$ REDOR differences correspondingly. The determined distances are given in the inset table.

For the REDOR measurement, the paclitaxel was labeled with ^{19}F at the para position of the C(2) benzoyl substituent and with ^{13}C and ^{15}N in the side chain. The ^{13}C (3'-amide carbonyl group) -^{19}F distance was determined to be 9.8±0.5Å, and the distance between the same fluorine atom and the 3'-methine carbon was 10.3±0.5 Å. These distances corresponded very well to the hydrophobic collapse structure labeled Taxol B in Ref. [41] (see also Fig. (**13A**), Fig. (**14A**) and the molecule B in Fig. (**17**)). The REDOR experiment was carried out in the sampling rate of 1s and in total 8 million scans through 3 month period.

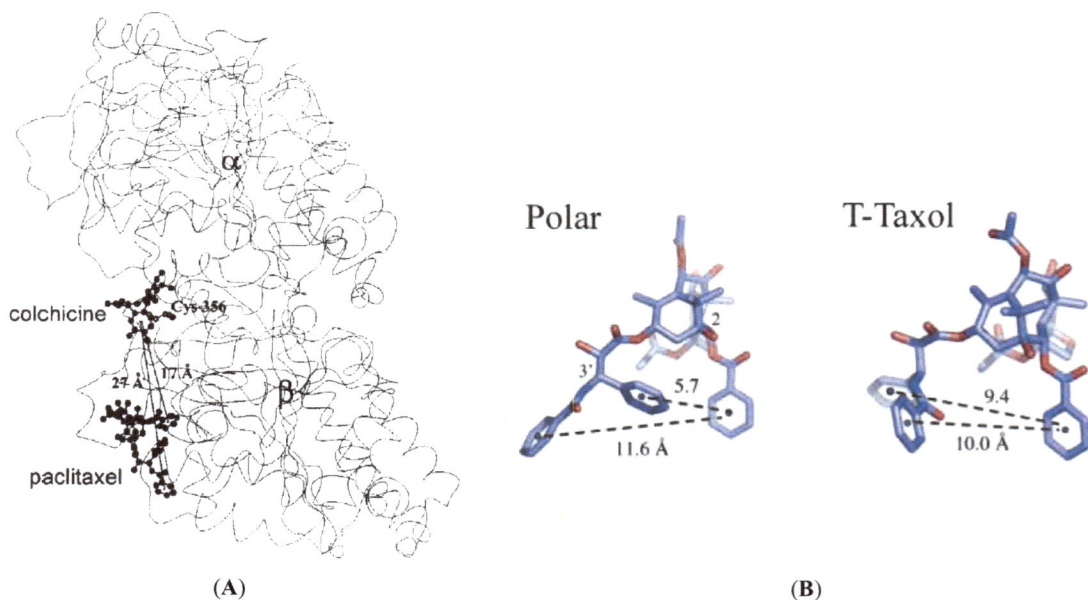

(A)

(B)

Fig. (20). The proposed relative position of paclitaxel when binding to the tubulin according to [49] (**A**). The Cys 356 molecule of the β-tubulin is highlighted. Two conformations of paclitaxel, the polar form and the T-Taxol [50] (**B**).

The resulting spectra are shown in Fig. (**19**). For the $^{13}C\{^{19}F\}$ curve (top curve in Fig. (**19**)), the SNR (signal-to-noise ratio) is only 3:1 at 175 ppm and 2:1 at 50 ppm. The positions of peaks, however, are exact as of the ones from the stronger $^{13}C\{^{15}N\}$ REDOR difference spectrum (middle curve in Fig. (**19**)). Since the uncertainty in the interatomic distance translates in the same way as the peak intensities (determined from the peak areas) which have the SNR about 5:1, the corresponding error level in determined distance is only 6%.

With the assistance of molecular dynamic simulation, Li *et al.* [49] were able to propose a schematic position of paclitaxel binding to tubulin as given in Fig. (**20A**). To date, these results are still significant as there was not much similar advance in this field.

In 2006, Snyder's group [50] has reviewed the calculation of REDOR error in [49] and showed that the corrected error level should rather be 6.7%, which contributes to the error in the interatomic distances of ±0.7 Å instead of ±0.5 Å as reported in [49]. Based on this higher error level, they argued that their molecular dynamic calculation of the T-Taxol conformation geometries (Fig. **20B**) felt within the REDOR error limit. Therefore the T-Taxol conformation cannot be ruled out from the bioactive forms of paclitaxel. Although a careful analysis of errors has been presented by this group, including those of recalculated REDOR dephasing curves, most of quantum chemistry results were performed on the basis of the molecular dynamic calculation, which is only an approximate level theory and is not accurate enough for quantitative analysis of paclitaxel binding mechanism.

An analysis of ^{13}C solid-state NMR chemical shifts for 10-deacetyl baccatin III was reported by Harper *et al.* [51] and later by Heider *et al.* [52]. The geometry for 10-deacetyl baccatin III is shown in Fig. (**10B**). Harper *et al.* studied the chemical shift differences caused by two conformations of paclitaxel: a pure compound and a crystallized paclitaxel with dimethyl sulfoxide (DMSO) using a technique called FIREMAT NMR [53]. The FIREMAT is a high-resolution two-dimensional (2D) magic-angle-turning (MAT) experiment, which measures chemical shift tensor principal values (CSPV) in powdered solids. The method's sensitivity results from combining the 5π magic-angle-turning pulse sequence and the TIGER protocol for processing 2D data. The obtained FIREMAT spectrum for the paclitaxel dissolved in DMSO is shown in Fig. (**21**).

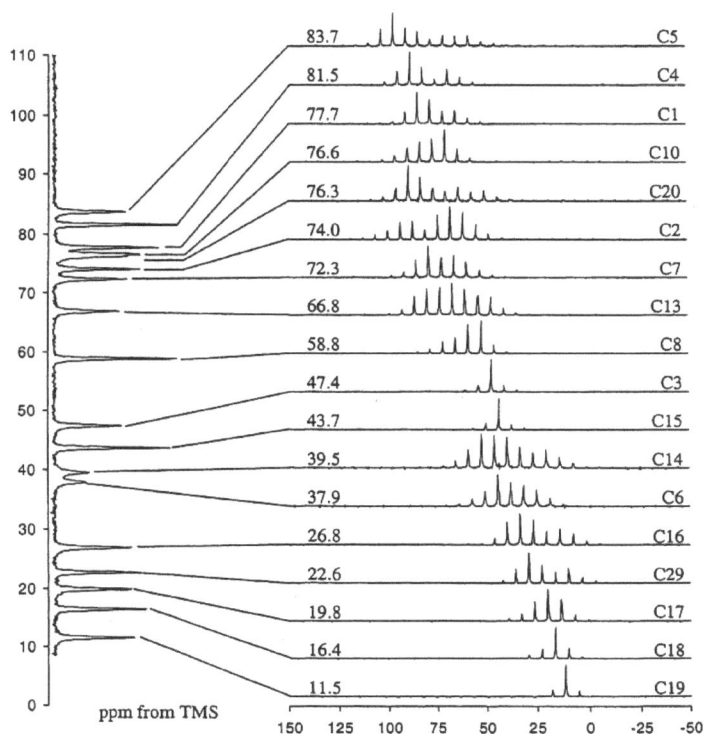

Fig. (21). The FIREMAT spectrum as obtained for the paclitaxel dissolved in DMSO according to Ref. [51]. The assigned tensor patterns are shown in the right side.

By comparing the differences in CSPV of two polymorphs with the ones computed by *ab initio* technique, the authors in [51] stated that the structural variations occur primarily in the ring substituents and the cyclohexenyl ring, whereas the taxane nucleus appear fairly rigid.

The techniques of the solid-state NMR were further exploited by Heider *et al.* [52] to determine the conformations of anhydrous paclitaxel that was crystallized in the $P2_12_12_1$ space group with two molecules in the asymmetric unit of unit cell (Z'=2). Complete molecular geometries have been predicted by combining the CSPV with the results of *ab initio* calculation. In their experiments, the FIREMAT data were collected using a spinning speed of 950 Hz and a TPPM phase angle of $\pm18°$. The HETCOR analysis was performed on a 600 MHz spectrometer using a pulse sequence of van Rossum *et al.* [54]. A total of 28 [1]H-^{13}C correlations, schematically given in Fig. (**22A**), was obtained to provide shift assignments. The typical HETCOR resolution is shown in Fig. (**22B**) for the C(18) and C(19) positions where one may clearly reveal the effect of coupling of the two paclitaxel molecules in the asymmetric unit.

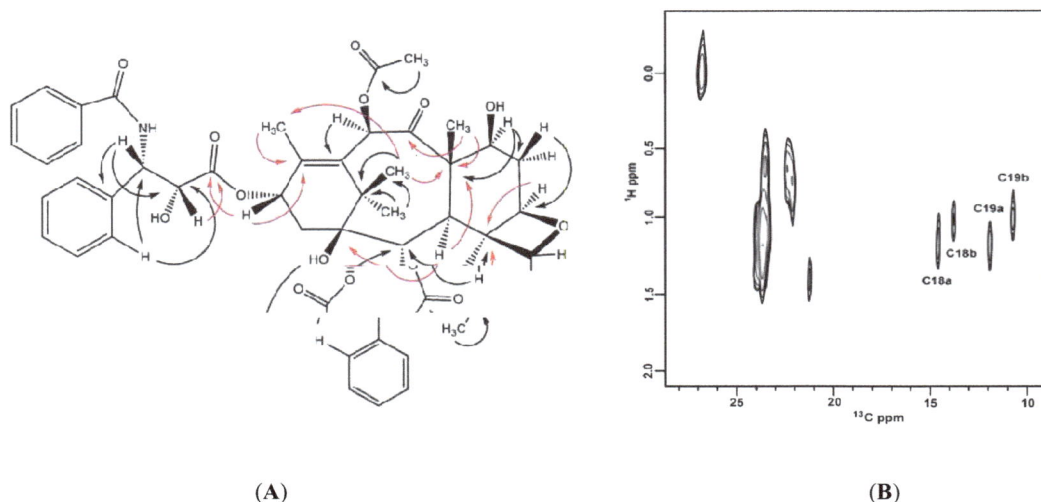

(**A**)   (**B**)

Fig. (22). The [1]H-^{13}C correlations as obtained from HETCOR experiment [51] (black arrows denote the weak couplings) (**A**) and the illustration of [1]H-^{13}C HETCOR resolution for C(18) and C(19) - two molecules in asymmetric unit are clearly observed (**B**).

On the computational basis, the shielding tensors were computed at the B3PW91/D95* level theory (B3PW91 denotes a Beck's style 3 parameters hybrid functional with Perdew–Wang correlation functionals; D95* means the atomic orbital basis set) using the Gaussian 03 code [55]. Before the calculation of NMR shielding, the corresponding structure of paclitaxel has been optimized by semiempirical model AM1. The translation between shieldings and chemical shifts was carried out separately for the *sp2* and *sp3* carbons using the linear relationship (shift = shielding - *k*, where *k* = 186.86 for *sp2* and 185.88 for *sp3*). The ^{15}N chemical shift was calculated as -(shielding+135.8).

It is worth to note that despite a careful treatment of theoretical CSPV, the work of Heider *et al.* [52] lacks a consistency by fitting the experimental tensors to the computed CSPV values (from 650 model structures) since the geometries matching the CSPVs may not belong to the ones with well-defined ground state. The establishment of such ground state geometry is critical for every computational issue.

Taking into account that the T-Taxol like conformations, that satisfy the REDOR constraints on geometry [49], may be the ones active for tubulin binding, several groups were joined together for the synthesis of a number of C(4) to C(3') bridged paclitaxel analogues [56]. The presence of bridges fixes the T-Taxol geometry within the REDOR limit, therefore may introduce the paclitaxel compatible bioactive forms. The major parts of the synthesized analogues showed a sufficient reduction in bioactivity in a series of assays, as the obtained IC_{50} values were often hundred or thousand times greater than that of paclitaxel. However, a few exceptions appeared more active, e.g. the authors in Ref. [56] have obtained a compound, which was 1200-fold more active than paclitaxel in the test against the paclitaxel resistant cell lines 1A9-PTX10. For the structural analysis, the 2D ROESY NMR technique was used with the 1D ^{1}H assignment by COSY.

The debate continues on whether or not the T-Taxol or other REDOR constraint taxol conformations constitute the tubulin-binding form by a recent contribution from Snyder group [57]. The authors presented a high-level theory study of ground state energy (B3LYP functional with large basis sets such as 6-311+G) with molecular dynamic simulation to support the T-Taxol as the bound conformation of paclitaxel on β-tubulin in microtubules. No prediction on NMR chemical shifts of the bound complex was reported. In the absence of the high-resolution structural data, the decisive answer which conformation is the true bound form seems not be soon achieved.

Paclitaxel-Loaded Nanocarriers

A number of paclitaxel loaded nanocarriers has been prepared in the last decade. Some of them have already entered the market such as Genexol-PM (a paclitaxel loaded polymeric micelle from Samyang Pharmaceuticals, South Korea) or Abraxane (an albumin bound paclitaxel from Abraxis, USA). Many other have just been subjected for the clinical trials or are under test. We will focus on several systems that were recently presented.

The ^{31}P-NMR spectroscopy was involved in the study of the interaction of paclitaxel with the carrying liposomal bilayer [58]. The ^{31}P-NMR spectroscopy provides important information on conformation change of phospholipid headgroup of the lipid bilayers due to the change in concentration of dissolved paclitaxel. As the paclitaxel dissolves in the lipid bilayers, it has fluidizing effect in the upper region of the bilayer whereas the hydrophobic core is slightly rigidized. Paclitaxel showed a tendency to accumulate at the interface and to partition out of bilayer at the higher concentration than 3%.

Fig. (23) shows the narrowing of peak width of the ^{31}P-NMR spectra as obtained for three different paclitaxel concentrations. The spectra suggest that (i) the large vesicles (400nm in diameter) undergo motional averaging due to tumbling motion; (ii) the presence of smaller vesicles due to the occurrence of small peak at 0 ppm; (iii) the fluidization of bilayers due to the slight shift towards isotropic signal of peak positions; (iv) a concentration-dependent effects of the molecule on PC:PG lipid bilayers due to the small but observable increase in the low field shoulder (arrow in Fig. (23)); (v) the hydrogen bonding to the P=O end of phosphate could be ruled out since there was no distinct associated P-peak observed in the spectra.

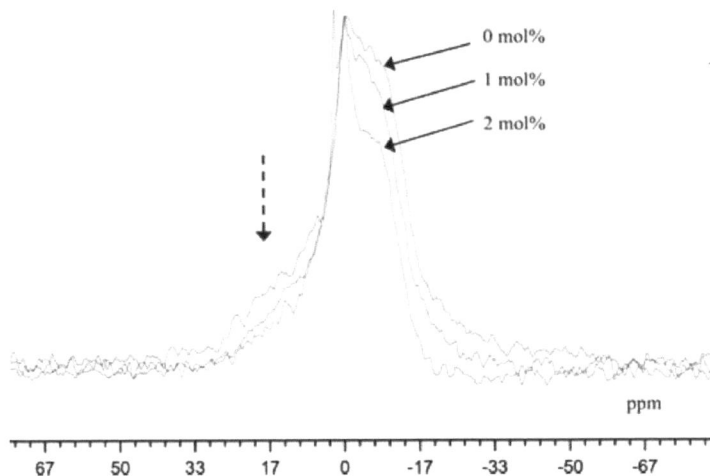

Fig. (23). The ^{13}P-NMR spectra for various paclitaxel concentration in the lipid bilayer [58].

The ^{1}H-NMR spectroscopy has been exploited as characterization tool in a series of studies concerning with paclitaxel loaded in various nanostructures [59-71].

Kim *et al.* [59] has prepared the diblock copolymer core-shell structure nanospheres composed of the biodegradable hydrophilic MePEG [methoxy poly(ethylene glycol)] and hydrophobic PCL [poly(ε-caprolactone)] by the ring-opening polymerization of ε-caprolactone in the presence of MePEG without catalyst. Several works concerning with the copolymeric nanocarriers are known from the Kim's group [72-77]. The taxol-loaded nanospheres were obtained by adding of taxol in continuous stirring at room temperature to the solution of MePEG/PCL and dimethylformamide (DMF). High-resolution ^{1}H NMR spectra, shown in Fig. (**24**), were detected in deuterated water (D_2O) (a) and deuterated chloroform ($CDCl_3$) (b, c). Since PCL and taxol are not soluble in water, the ^1H spectrum (a) shows only the hydrophilic part (MePEG) of the nanosphere. Different features were obtained from a mixture in $CDCl_3$, for which both taxol and PCL are completely dissolved (b, c). Therefore the given spectra demonstrate the core-shell structure of the prepared nanospheres.

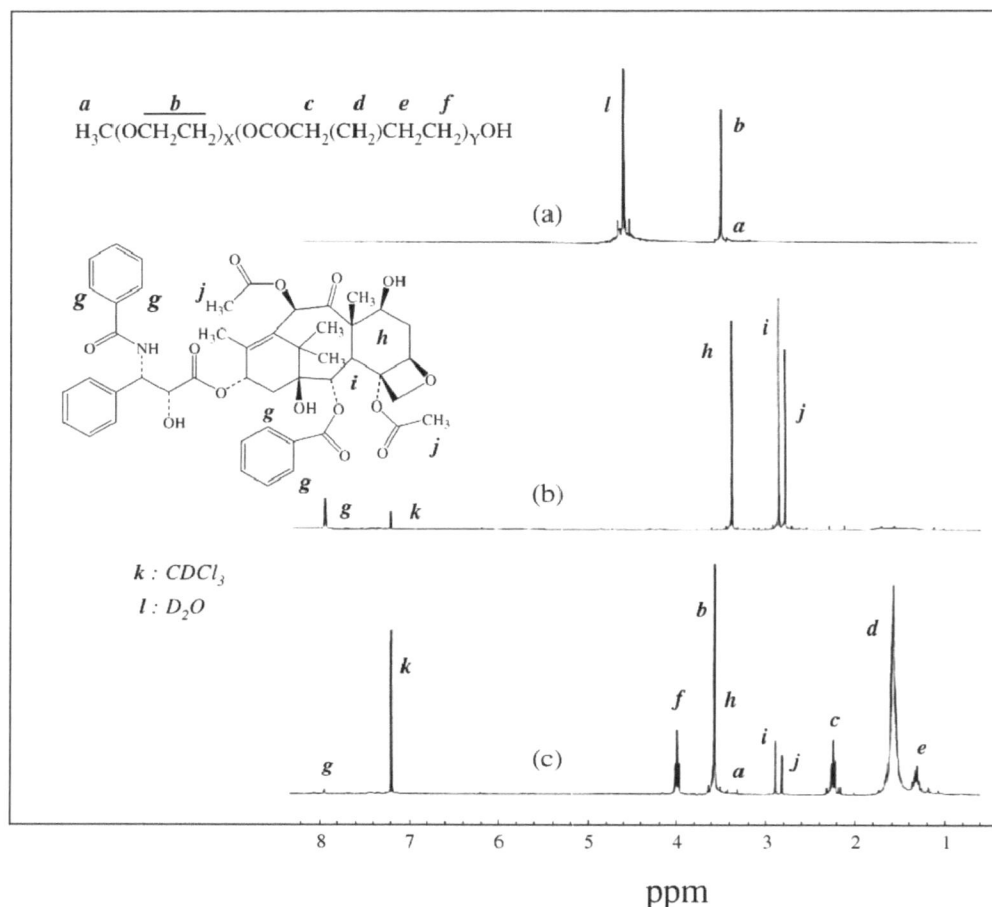

Fig. (24). ^1H-NMR spectra with assignments for the taxol-loaded MePEG/PCL nanoparticles in D_2O (**a**), free taxol unloaded in $CDCl_3$ (**b**) and taxol-loaded MePEG/PCL in $CDCl_3$ (**c**) according to Ref. [58].

Sharma *et al.* [71] reported a study of the paclitaxel/cyclodextrins complex. Cyclo-dextrins (CyDs) are the cyclic oligosaccharides consisting of linked glucopyranose rings which are frequently used to improve the water-solubility of drugs. ^1H NMR spectra were recorded in D_2O for a concentration of 10^{-5} M paclitaxel and 10^{-3} M HPβCyDs (modified CyDs). The authors in [71] asserted that the distinctive changes have been observed in the aromatic and other proton resonances of paclitaxel upon transfer of drug to a hydrophobic environment. However, they also stated that there was a lack of a well defined resonance for paclitaxel and this might be caused, as they suggested, by a rapid turnover of the paclitaxel:HPβCyD complex compared to the ∝s NMR temporal resolution.

The core-stabilized spherical copolymeric micelles of size ~30nm were prepared by Kim J.-H. *et al.* [60]. The micelle formulation was poly(ethylene glycol-b-lactide) which contains a methoxy group at the poly(ethylene glycol) (PEG) chain end and a polymerizable methacryloyl group at the poly(lactic acid) (PLA) end (MeO-PEG/PLA-methacryloyl). The taxol was entrapped into the micelles by the oil/water emulsion method. ^1H NMR spectra which were recorded before and after the polymerization of MeO-PEG/PLA-Methyacryloyl clearly showed the evidence for the presence of the polymerization iniciator group. However, no trace for paclitaxel was seen.

The next advance was marked by the 2nm paclitaxel-functionalized gold nanoparticles prepared by Zubarev's group in 2007 [22] (Fig. **6B**). This carefully presented work was accompanied by a series of supporting information by both ^{1}H and ^{13}C NMR data for paclitaxel-gold nanoparticle (AuNP) conjugates. With the help of NMR data, the authors clearly proved that the conjugates were created at the C(7)-OH group of paclitaxel (which is known not to affect directly the bioactivity of paclitaxel) (Fig. **25**). The synthetic steps are as follows. First, the AuNPs were modified by the 4-mercaptophenol to form the phenol (OH) terminated surfaces and the hexaethylene glycol linkers were attached to the C(7)-OH position of paclitaxel. Then the two blocks were coupled together in a mild esterification, yielding a product which contains approximately 67% loaded paclitaxel, i.e. 70 paclitaxel molecules per 1 nanoparticle. The presence of paclitaxel was clearly resolved in both ^{1}H and ^{13}C NMR spectra of final conjugates. This work brilliantly demonstrates the efficiency of NMR spectroscopy in characterization of drug-loaded nanocarriers.

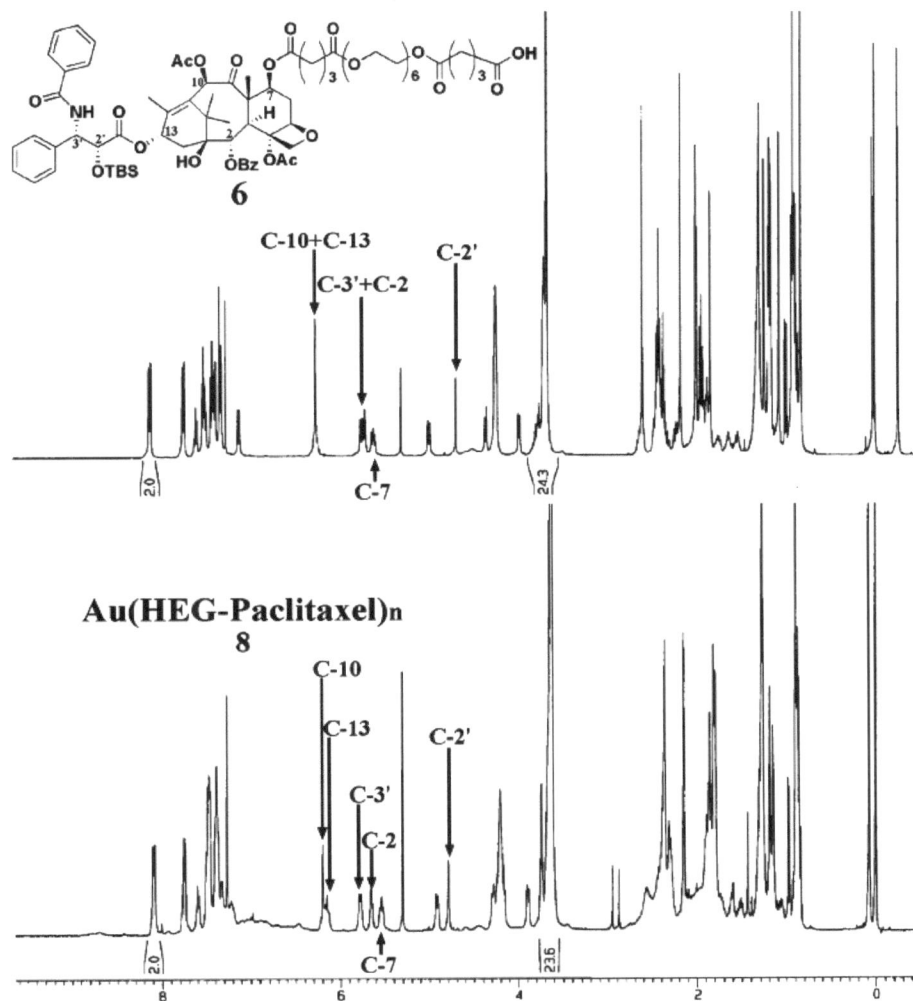

Fig. (25). ^{1}H-NMR spectra of a modified paclitaxel (**6**) and paclitaxel-AuNP conjugate (**8**) according to Ref. [22].

Zubarev *et al.* [22] also deposited the excellent ^{1}H NMR data for paclitaxel (**1**) and ^{13}C NMR data for one paclitaxel derivative (**2**). The detailed assignments are shown in Fig. (**26**). The AuNPs have extensively been reviewed, for recent summaries the readers may refer to Refs. [78, 79].

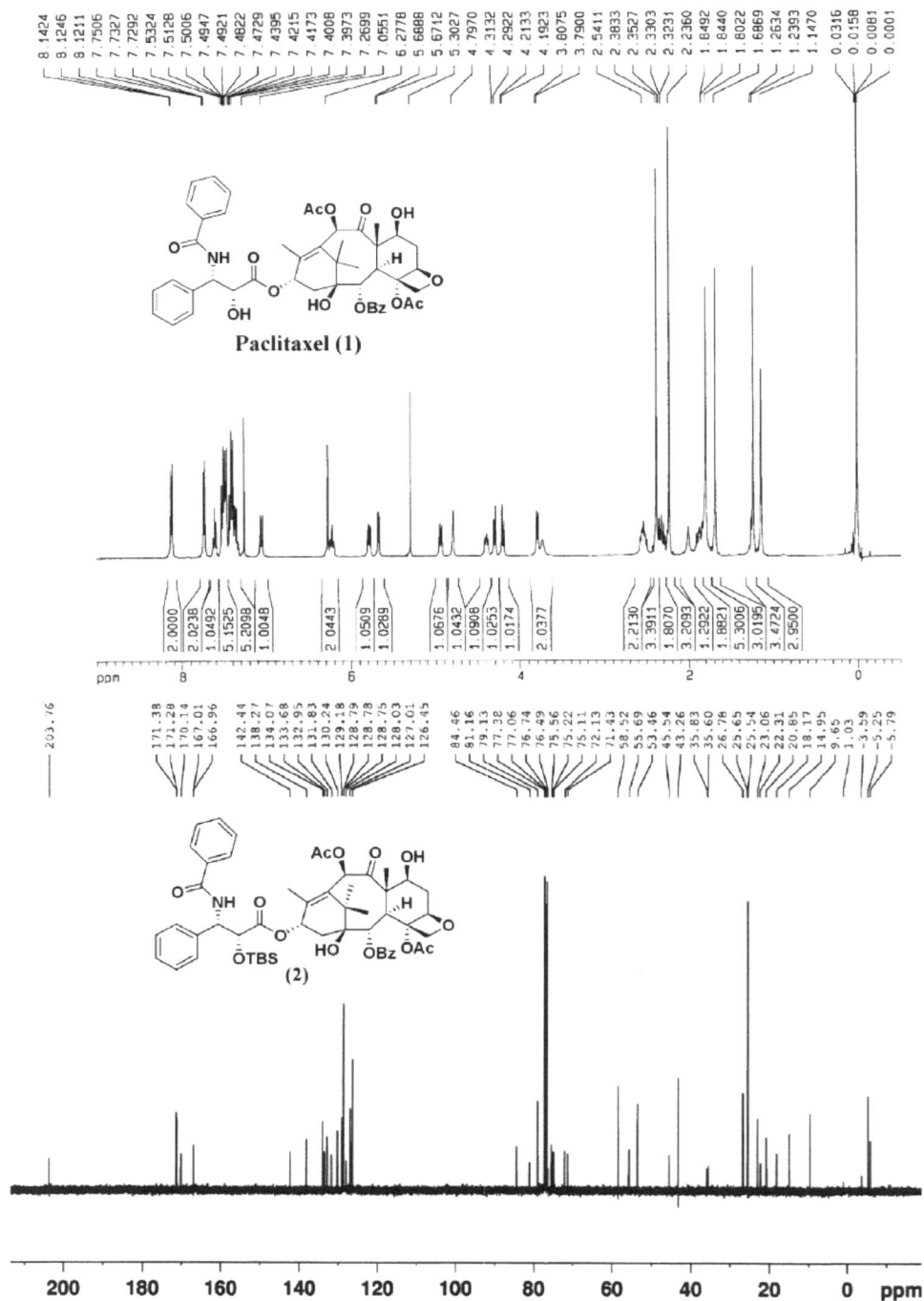

Fig. (26). ^{1}H and ^{13}C NMR spectra of paclitaxel (**1**) and of a modified paclitaxel (**2**) according to Ref. [22].

Another interesting work which involved the 1D/2D solution based ^{1}H and ^{13}C NMR and solid-state based CP/MAS (crossing polarization/magic-angle-spinning) NMR techniques for analysis of the paclitaxel complex with poly(styrene-isobutylene-styrene) (SIBS) is presented by Xie *et al.* in Ref. [61].

The obtained NMR data showed no change in chemical shifts between the neat and incorporated paclitaxel, which suggested the limited non-bonding character between SIBS and paclitaxel. The analysis of the spin-lattice relaxation for carbon and proton in rotating frame confirmed that the mobility of paclitaxel increased in the paclitaxel-SIBS conjugate. In the solution NMR, the used 2D techniques were 2D ^{1}H-^{1}H COSY-DQF, 2D [1]H-^{13}C HMQC and 2D ^{1}H-^{13}C HMBC. The [13]C T_1 (spin-lattice relaxation time) was measured at 125MHz with delays ranging from 0.01s to 40s. For the solid-state NMR, the CP/MAS spectra were recorded at 75MHz and the T_1 was also measured for both carbon and proton.

Fig. (**27**) shows the ^{13}C NMR spectra for paclitaxel in $CDCl_3$ solution (A) and the solid-state CP/MAS for the same at 298K and magic angle spin rate 10kHz (B).

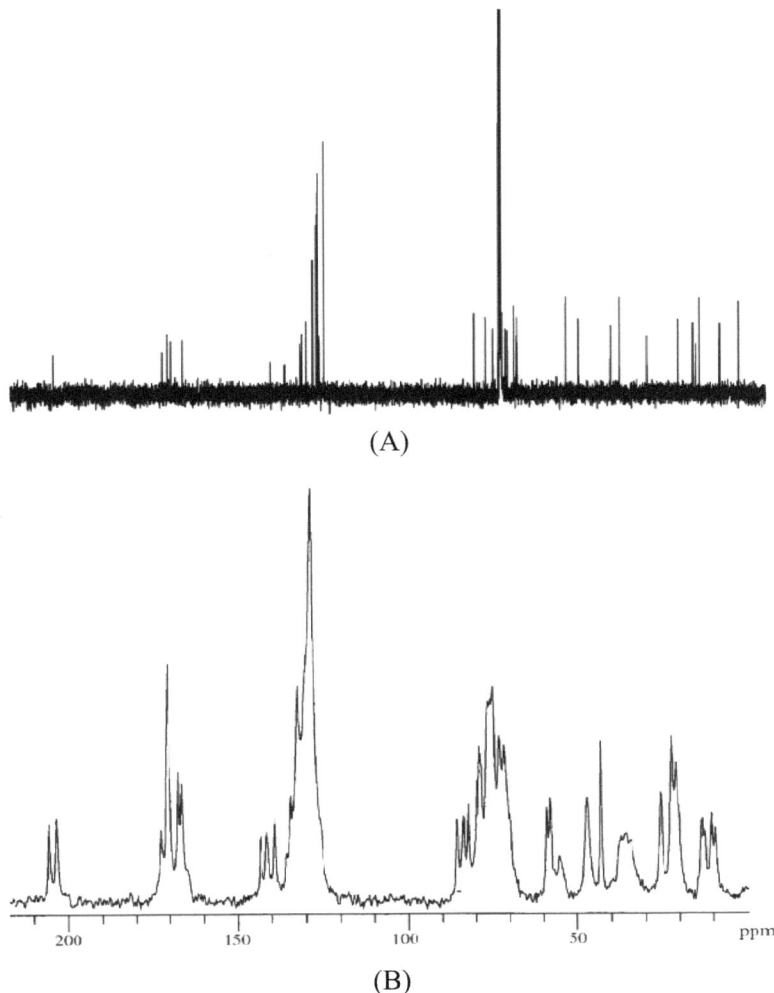

(A)

(B)

Fig. (27). ^{13}C NMR spectrum for paclitaxel in $CDCl_3$ (**A**) and the solid-state ^{13}C CP/MAS NMR spectrum for paclitaxel at 298K [61].

Table (**3**) extracts some ^{13}C solid-state CP/MAS NMR data for paclitaxel as reported in Ref. [61]. For the neat paclitaxel dissolved in $CDCl_3$ at various concentrations, the authors in [61] showed that there were the upfield shifts for three OH groups whereas all other features remained unchanged. The upfield shifts for the hydroxyl groups at higher concentration may be explained by the forming of the hydrogen bonding between the paclitaxel molecules. The same effect may be expected for the paclitaxel-SIBS conjugates. However, the downfield shifts were observed for this case. This led the authors in [61] to the conclusion that the SIBS might interrupt the hydrogen bonding system of paclitaxel molecules, or inversely, the paclitaxel interrupted the SIBS polymer chain packing through π–π interaction of aromatic rings. Therefore, the interaction between SIBS and paclitaxel should be considered as non-bonding.

The analysis of T_1 relaxation time showed the decrease from the free state to the SIBS-bound state for paclitaxel. This revealed the higher mobility of paclitaxel, which might associate with the change from the crystalline form of the pure paclitaxel to the amorphous state in the paclitaxel-SIBS complex.

Table 3. Summary of ^{13}C NMR Solution-Based (CDCl$_3$) and Solid-State Based CP/MAS Data for Paclitaxel [61]

Carbons	Solution NMR	^{13}C Solid-state CP/MAS NMR		
	Paclitaxel	Paclitaxel		
	^{13}C (ppm)	^{13}C (ppm)	T_I (s)	$T_{I\rho H}$ (ms)
C-9	203.57	205.7	35.6	14.4
		203.5	35.6	14.4
C-1'	172.63	172.8	79	14.5
C-30	171.18	170.8	79	14.5
C-21	167.14			
C-5'	166.87	166.9	42	13.9
		165.2	42	13.9
C-12	141.88	143.8	87.3	11.4
		141.9	87.3	11.8
C-38	137.94	139.5	60.5	11.5
		136.1	55	
C-25	133.64	134.8	38.8	
C-11	133.07	133.5	34.8	
C-41	131.88	132.5	51.5	14.1
C-24, 26	128.66	128.6	39.7	14.6
C-5	84.33	85.8	88.7	15.3
		84	70.2	15.3
C-4	81.04	82.7	79.2	15.2
		80.2	76.3	
C-1	78.88	79.1	48.9	14.9
C-20	76.42	76.8	57.4	
C-10	75.51	75.9	50.6	15.4
		75.2	49.7	15.4
C-2'	73.16	73.6	55.1	15.3
C-13	72.15	72.1	39.4	15.5
C-7	72.06	69.2	40.3	
C-8	58.47	59.3	41.5	13.8
		58.3	35.6	138
C-3'	55.06	55.8	43.7	16.1
C-3	45.61	47.6	84	15.9
C-15	43.09	43.5	11.5	138
C-6	35.62	37.6	82.5	18.8
C-14	35.58	36.2	38.4	18.7
		35	36.4	18.7
C-17	26.76	26	11.7	13.8
C-29	22.53	22.7	11.6	13.8
C-16	21.75	21.2	7.54	13.8
C-18	14.75	13.7	15.2	12.4
		12.9	20.9	12.7
C-19	9.52			10.9

The differences between the solution NMR and solid-state NMR data (Fig. **27**) attributed for the usual crystalline packing of two paclitaxel molecules per unit cell. For example, the C(9) resonance (203.57 ppm) (Fig. **27B**) is split into two peaks, indicating the presence of the conformation differences. Many other interesting features about the dynamics of molecules might also be retrieved from measured spin-lattice relaxation time T_1 for both carbon and proton, or by comparison of line broadenings for paclitaxel and SIBS in the CP/MAS spectra.

The paclitaxel bound biotinylated PLA-PEG nanoparticles of average size ~100nm were reported by Pulkkinen *et al.* [62]. The loading efficiency achieved over 90%. The authors claimed that the *in vitro* test against brain tumour cells (BT4C) by three-step avidin-biotin technology using transferrin as the targeting ligand expressed an increased anti-tumour activity of paclitaxel in comparison with the commercial paclitaxel Taxol®. The ^{1}H spectroscopy was used to characterize the synthesized PLA and PEG-biotin structures.

Bilensoy *et al.* [63] developed a non-surfactant biodegradable cyclodextrin (6-O-CAPRO-β-CD) amphiphilic nanoparticles of size 150-200 nm loaded with paclitaxel. The ^{1}H spectroscopy of the paclitaxel-cyclodextrin complex was taken at 400MHz. The slightly shifts to lower values (of order 0.01-0.05 ppm) were observed for the internal cyclodextrin's protons H(1)-H(6) after complexation with paclitaxel. The increase in paclitaxel-to-cyclodextrin molar ratio from 1:1 to 1:2 showed only a minor effect in varying the proton H(3) and H(5) shifts from 0.02 to 0.03 ppm. However, with the assistance of additional characterization by FT-IR, the authors in [63] argued for a possible capture of paclitaxel into the cyclodextrin cavity.

The ^{1}H spectroscopy was also used as characterization tool in the PEG-docetaxel conjugate [64] and paclitaxel/MPEG-PLA conjugate [65]. The high loading efficiency (drug-to-polymer ratio of 4:3) was achieved in [64]. The authors reported a 1.8 fold higher concentration of the prepared conjugate in the intravenous administration when compared with the original Taxotere®. The obtained ^{1}H spectra showed a clear evidence for the incorporation of docetaxel into the polymer network.

FINAL REMARK

This review does not contain many interesting works on SAR of paclitaxel, especially from the group of Ojima [66]. Since we were concentrated on the application of NMR technique, the focus was paid mainly on the studies involved more less with the NMR as characterization tool. The readers who are interested in the works of Ojima's and other groups may refer to the recent review by Ojima [66].

CONCLUSION

The search for new anticancer drugs based on paclitaxel loaded nanocarriers advances very fast recently. Heavy investments worldwide in this direction led to the intensification of researches for which the NMR as characterization tool continues to be a major analytical technique for obtaining the quantitative information about the structure-activity relationships. There are several reasons that this trend will continue in the future: (i) large magnets (up to 21 Tesla) are now commercially available for accessible price (i.e. ~1 mil. US$); (ii) large databases are also available for fast screening of large amount of data; (iii) the development of solid state NMR and other NMR-related techniques continues to explode recently as NMR promises to provide a physical platform for a new class of devices called the quantum devices (such as quantum computers); (iv) together with the development of new drug delivering systems, the NMR based devices can also be involved as imaging devices or directly as therapeutical devices. With the help of modern NMR advances, it is reasonable to believe that the dilemma of which form of paclitaxel is bioactive will soon be solved. Keeping in mind that the trade volume of paclitaxel increases approximately 20-30% annually we may expect that the multifunctional paclitaxel loaded nanocarriers will be available in the next coming years. In this amazing development, with intense SAR by NMR researches, the fundamental break-through may be soon achieved.

ACKNOWLEDGEMENT

The author gratefully acknowledges the financial support from the National Foundation for Science and Technology Development (NAFOSTED) of Vietnam, research project on Nanofluids and Application (2009).

REFERENCES

[1] Jain KK. Recent advances in nanooncology. Technol Cancer Res Treat 2008; 7(1): 1.

[2] Kwangjae C, Wang X, Nie S, Chen Z, Shin DM. Therapeutic nanoparticles for drug delivery in cancer. Clin Cancer Res 2008; 14(5): 1310.

[3] Sinha R, Kim GJ, Nie S, Shin DM. Nanotechnology in cancer therapeutics: bioconjugated nanoparticles for drug delivery. Mol Cancer Ther 2006; 5(8): 1909.

[4] Emerich DF, Thanos CG. The pinpoint promise of nanoparticle-based drug delivery and molecular diagnosis. Biomol Eng 2006; 23: 171.

[5] Rawat M, Singh D, Saraf S, Saraf S. Nanocarriers: promising vehicle for bioactive drugs. Biol Pharm Bull 2006, 29: 1790.

[6] Perez E. Presentation of Abraxane survival data by American Pharmaceutical Partners. 22nd Annual Miami Breast Cancer Conference, Miami, Florida, 2005.

[7] Gradishar WJ, Tjulandin S, Davidson N, et al. Phase III trial of nanoparticle albumin-bound paclitaxel compared with polyethylated castor oil-based paclitaxel in women with breast cancer. J Clin Oncol 2005; 23: 7794.

[8] Green MR, Manikhas GM, Orlov S, et al. Abraxane, a novel Cremophor-free, albumin-bound particle form of paclitaxel for the treatment of advanced non-small-cell lung cancer. Ann Oncol 2006; 17: 1263.

[9] Nyman DW, Campbell KJ, Hersh E, et al. Phase I and pharmacokinetics trial of ABI-007, a novel nanoparticle formulation of paclitaxel in patients with advanced nonhematologic malignancies. J Clin Oncol 2005; 23: 7785.

[10] Shuker SB, Hajduk PJ, Meadows RP, Fesik SW. Science 1996; 274: 1531.

[11] Heller M, Kessler H. NMR spectroscopy in drug design. Pure Appl Chem 2001; 73(9): 1429.

[12] Du F, Zhou H, Chen L, Zhang B, Yan B. Structure elucidation of nanoparticle-bound organic molecules by ^{1}H NMR. Trends Anal Chem 2009; 28(1): 88.

[13] Vinje J, Sletten E. NMR spectroscopy of anticancer platinum drugs. Anti-cancer Agents Med Chem 2007; 7: 35.

[14] Dalvit C. Ligand- and substrate-based ^{19}F NMR screening: Principles and applications to drug discovery. Prog Nuc Magn Reson Spectrosc 2007; 51: 243.

[15] Neri D, Bicknell R. Tumour vascular targetting. Nature 2005; 5: 436.

[16] Huminiecki L, Bicknell R. In silico cloning of novel endothelial-specific genes. Genome Res 2000; 10: 1796.

[17] St Croix B, Rago C, Velculescu V, et al. Genes expressed in human tumor endothelium. Science 2000; 289: 1197.

[18] Leamon CP, Reddy JA. Folate-targeted chemotherapy. Adv Deliv Rev 2004; 56: 1127.

[19] Kim TY, Kim DW, Chung JY, et al. Phase I and pharmacokinetic study of Genexol-PM, a cremophor-free, polymeric micelle-formulated paclitaxel, in patients with advanced malignancies. Clin Cancer Res 2004; 10: 3708.

[20] Dimitrov DS. Virus entry: molecular mechanism and biomedical applications. Nat Rev Microbiol 2004; 2: 109.

[21] Singh P, Destito G, Schneemann A, Manchester M. Canine parvovirus-like particles, a novel nanomaterial for tumor targeting. J Nanobiotechnology 2006; 4: 2 (available at: http://www.jnanobiotechnology. com/content/4/1/2).

[22] Gibson JD, Khanal BP, Zubarev ER. Paclitaxel-functionalized gold nanoparticles. J Am Chem Soc 2007; 129: 11653-11661.

[23] Macomber RS. A complete introduction to modern NMR spectroscopy. John Willey & Sons Inc., NY, 1998.

[24] Hornak JP. The basics of NMR. Department of Chemistry, Rochester Institute of Technology, Rochester, NY 14623-5603. (available at: http://www.cis.rit.edu/htbooks/nmr/bnmr.htm).

[25] Silverstein RM, Bassler GC, Morill TC. Spectrometric Identification of Organic Compounds, John Wiley & Sons, NY. 1991.

[26] Pouchert CJ. The Aldrich Library of NMR Spectra. The Aldrich Chemical Company, Milwaukee, 1983.

[27] NMR and ESR Data. Digital Engineering Library. McGraw-Hill 2004. (Available at: http://www. digitalengineeringlibrary.com)

[28] Wani MC, Taylor HL, Wall ME, Coggon P, McPhail AT. Plant antitumor agents. VI. The isolation and structure of taxol, a novel antileukemic and antitumor agent from Taxus brevifolia. J Am Chem Soc 1971; 93: 2325-2327.

[29] Appendino G. Taxol (R) (paclitaxel): historical and economical aspects. Fitoterapia 1993; 45: 5-27.

[30] Zhang J, Sauriol F, Mamer O, et al. New taxane analogues from the needles of taxus canadensis. J Nat Prod 2001, 64, 450-455.

[31] Ojima I, Lin S, Chakravarty S, et al. Syntheses and structure-activity relationships of novel nor-seco taxoids. J Org Chem 1998; 63: 1637-1645.

[32] Guo X, Paquette LA. Concise means for accessing an advanced precursor to 1-deoxypaclitaxel. J Org Chem 2005; 70: 315-320.

[33] Swindell CS, Krauss NE. Biological active taxol analogues with deleted A-ring side chain substituent and variable. J Med Chem 1991; 34: 1176.

[34] Schiff PG, Fant J, Horwitz SB. Promotion of microtubule assembly in vitro by taxol. Nature 1979; 277: 665-667.

[35] Bhalla K, Ibrado AM, Tourkina E, Tang C, Mahoney ME, Huang Y. Taxol induces the internucleosomal DNA fragmentation associated with programmed cell death in human leukemia cells. Leukemia 1979; 7: 563-568.

[36] Hoang NN. Conformation optimization of paclitaxel. To be published. Hoang NN. Unpublished data. 2009.

[37] Cheeseman JR, Trucks GW, Keith TA, Frisch MJ. A comparison of models for calculating nuclear magnetic resonance shielding tensors. J Chem Phys 1996; 104(14): 5497.

[38] Kingston DGI. The chemistry of Taxol. Pharmaceut Ther 1991; 52: 1.

[39] Benigni DA, Gougoutas JZ, DiMarco JD. Paclitaxel solvates. US Patent No. 2003/0144344 A1, Jul. 31, 2003.

[40] Peterson JR, Do HD, Rogers RD. X-Ray structure and crystal lattice interactions of the Taxol side chain methyl ester. Pharmaceutical Res 1991; 8(7): 908-912.

[41] Mastropaolo D, Camerman A, Luo Y, Brayer GD, Camerman N. Crystal and molecular structure of paclitaxel (taxol). Proc Natl Acad Sci USA 1995; 92: 6920-6924.

[42] Gao Q, Wei JM, Chen SH. Crystal structure of 2-Debenzoyl, 2-Acetoxy Paclitaxel (Taxol): Conformation of the Paclitaxel side chain. Pharmaceut Res 1995; 12(3): 337-341.

[43] Banwell MG, Hockless DCR, Peters SC. A C(10)-Oxygenated ABC-ring analogue of paclitaxel (Taxol^{TM}). Acta Cryst 1996; C52: 1832-1834.

[44] Gao Q, Parker WL. The "hydrophobic collapse" conformation of paclitaxel (Taxol^{TM}) has been observed in a non-aqueous environment: crystal structure of 10-deacetyl-7-epitaxol. Tetrahedron 1996; 52(7): 2291-2300.

[45]   Morita H, Wei L, Gonda A, *et al.* Crystal and solution state conformation of two taxoids, taxinine and taximine B. Tetrahedron 1997; 53(13): 4621-4626.

[46]   Kingston DGI. Recent advances in the chemistry of taxol. J Nat Prod 2000; 63: 726-734.

[47]   Wang M, Cornett B, Nettles J, Liotta DC, Snyder JP. The Oxetane Ring in Taxol. J Org Chem 2000; 65: 1059-1068.

[48]   Gibson JD, Khanal BP, Zubarev ER. Paclitaxel-functionalized Gold nanoparticles. J Am Chem Soc 2007; 129: 11653-11661.

[49]   Li Y, Poliks B, Cegelski L, *et al.* Conformation of microtubule-bound paclitaxel determined by fluorescence spectroscopy and REDOR NMR. Biochemistry 2000; 39: 281-291.

[50]   Alcaraz A, Mehta AK, Johnson SA, Snyder JP. The T-Taxol conformation. J Med Chem 2006; 49: 2478-2488.

[51]   Harper JK, Facelli JC, Barish DH, McGeorge G, Mulgrew AE, Grant DM. 13C NMR investigation of solid-state polymorphism in 10-deacetyl baccatin III. J Am Chem Soc 2002; 124: 10589-10595.

[52]   Heider EM, Harper JK, Grant DM. Structural characterization of an anhydrous polymorph of paclitaxel by solid-state NMR. Phys Chem Chem Phys 2007; 9: 6083-6097.

[53]   Alderman DW, McGeorge G, Hu JZ, Pugmire RJ, Grant DM. A sensitive, high resolution magic angle turning experiment for measuring chemical shift tensor principal values. Mol Phys 1998; 95(6): 1113-1126.

[54]   Van Rossum BJ, Forster H, de Groot JM. High-field and high-speed CP-MAS ^{13}C NMR heteronuclear dipolar-correlation spectroscopy of solids with frequency-switched Lee–Goldburg homonuclear decoupling. J Magn Reson 1997; 124: 516-519.

[55]   Frisch MJ, Trucks GW, Schlegel HB, *et al.* Gaussian 03, Revision B.03, Gaussian, Inc., Pittsburgh PA, 2003.

[56]   Ganesh T, Yang C, Norris A, *et al.* Evaluation of the tubulin-bound paclitaxel conformation: synthesis, biology, and SAR studies of C-4 to C-3' bridged paclitaxel analogues. J Med Chem 2007; 50: 713-725.

[57]   Yang Y, Alcaraz AA, Snyder JP. The tubulin-bound conformation of paclitaxel: T-taxol *vs.* "PTX-NY". J Nat Prod 2009; 72: 422-429.

[58]   Dhalikula AB, Panchagnula R. Fluorescence anisotropy, FT-IR spectroscopy and 31-P NMR studies on the interaction of paclitaxel with lipid bilayers. Lipids 2008; 43: 569-579.

[59]   Kim SY, Lee YM. Taxol-loaded block copolymer nanospheres composed of methoxy poly(ethylene glycol) and poly(ε-caprolactone) as novel anticancer drug carriers. Biomaterials 2001; 22: 1697-1704.

[60]   Kim J-H, Emoto K, Lijima M, *et al.* Core-stabilized polymeric micelle as potential drug carrier: increased solubilization of taxol. Polym Adv Technol 1999; 10: 647-654.

[61]   Chen J-Z, Ranade S, Xie X-Q. NMR characterization of paclitaxel/poly(styrene-isobutylene-styrene) formulations. Int J Pharm 2005; 305: 129-144.

[62]   Pulkkinen M, Pikkarainen J, Wirth T, *et al.* Three-step tumor targeting of paclitaxel using biotinylated PLA-PEG nanoparticles and avidin-biotin technology: formulation development and *in vitro* anticancer activity. Eur J Pharm Biopharm 2008; 70: 66-74.

[63]   Bilensoy E, Gurkaynak O, Ertan M, Sen M, Hincal AA. Development of nonsurfactant cyclodextrin nanoparticles loaded with anticancer drug paclitaxel. J Pharm Sci 2008; 97(4): 1519-1529.

[64]   Liu J, Zahedi P, Zeng F, Allen C. Nano-sized assemblies of a PEG-Docetaxel congujate as a formulation strategy for docetaxel. J Pharm Sci 2008; 97(8): 3274-3290.

[65]   Zhang X, Li Y, Chen X, *et al.* Synthesis and characterization of the paclitaxel/MPEG-PLA block copolymer conjugate. Biomaterials 2005; 26: 2121-2128.

[66]   Ojima I. Guided molecular missiles for tumor-targeting chemotherapy-case studies using the second generation taxoids as warheads. Account of Chem Res 2008; 41: 108-119.

[67]   Sharma A, Straubinger RM. Novel taxol formulation: preparation and characterization of taxol-containing liposomes. Pharm Res 1994; 11(6): 889-896.

[68]   Sahoo SK, Ma W, Labhasetwar V. Efficiency of transferrin-conjugated paclitaxel-loaded nanoparticles in a murine model of prostate cancer. Int J Cancer 2004; 112: 335-340.

[69]   Sharma A, Mayhew E, Bolcsak L, *et al.* Activity of paclitaxel liposome formulations against human ovarian tumor xenografts. Int J Cancer 1997; 71: 103-107.

[70]   Hwu J-R, Lin Y-S, Josephrajan T, *et al.* Targeted paclitaxel by conjugation to iron oxide and gold nanoparticles. J Am Chem Soc 2009; 131: 66-68.

[71]   Sharma SU, Balasubramanian SV, Straubinger RM. Pharmaceutical and physical properties of paclitaxel (Taxol) complexes with cyclodextrines. J Pharm Sci 1995; 84(10): 1223-1230.

[72]   Shin IG, Kim SY, Lee YM, Cho CS, Sung YK. Methoxy poly (ethylene Glycol)/caprolactone amphiphilic block copolymeric micelle containing indomethacin: I. preparation and characterization. J Contr Rel 1998; 51: 1-11.

[73]   Kim SY, Shin IG, Lee YM, Cho CS, Sung UK. Methoxy poly(ethylene glycol)/caprolactone amphiphilic block copolymeric micelle containing indomethacin: II. Micelle formation and drug release behaviours. J Contr Rel 1998; 51: 13-22.

[74]   Kim SY, Lee YM. Preparation and characterization of biodegradable nanoshperes composed of methoxy poly(ethylene glycol) and D,L-lactide block copolymer as novel drug carriers. J Contr Rel 1998; 56: 197-208.

[75]   Kim SY, Shin IG, Lee YM. Amphiphilic diblock copolymeric nanospheres composed of methoxy poly(ethylene glycol) and glycolide: properties, cytotoxicity and drug release behavior. Biomater 1999; 20: 1033-42.

[76]   Ha JC, Kim SY, Lee YM. Poly(ethylene oxide)-poly(propylene oxide)-poly(ethylene oxide) (Pluronic)/poly(caprolactone) (PCL) amphiphilic block copolymeric nanospheres: I preparation and characterization. J Contr Rel 1999; 62: 381-392.

[77]   Kim SY, Ha JC, Lee YM. Poly(ethylene oxide)-poly(propylene oxide)-poly(propylene oxide) (Pluronic)/poly(caprolactone) (PCL) amphiphilic block copolymeric nanospheres: II Thermo-responsive drug release behaviors. J Contr Rel 2000; 65: 345-358.

[78]   Milacic V, Dou PQ. The tumor proteasome as a novel target for gold(III) complexes: implications for breast cancer therapy. Coordination Chem Rev, 2009; 253: 1649-1660.

[79]   Chen PC, Mwakwari SC, Oyelere AK. Gold nanoparticles: from nanomedicine to nanosensing. Nanotechnol Sci Appl 2008; 1: 45-66.

Author Index